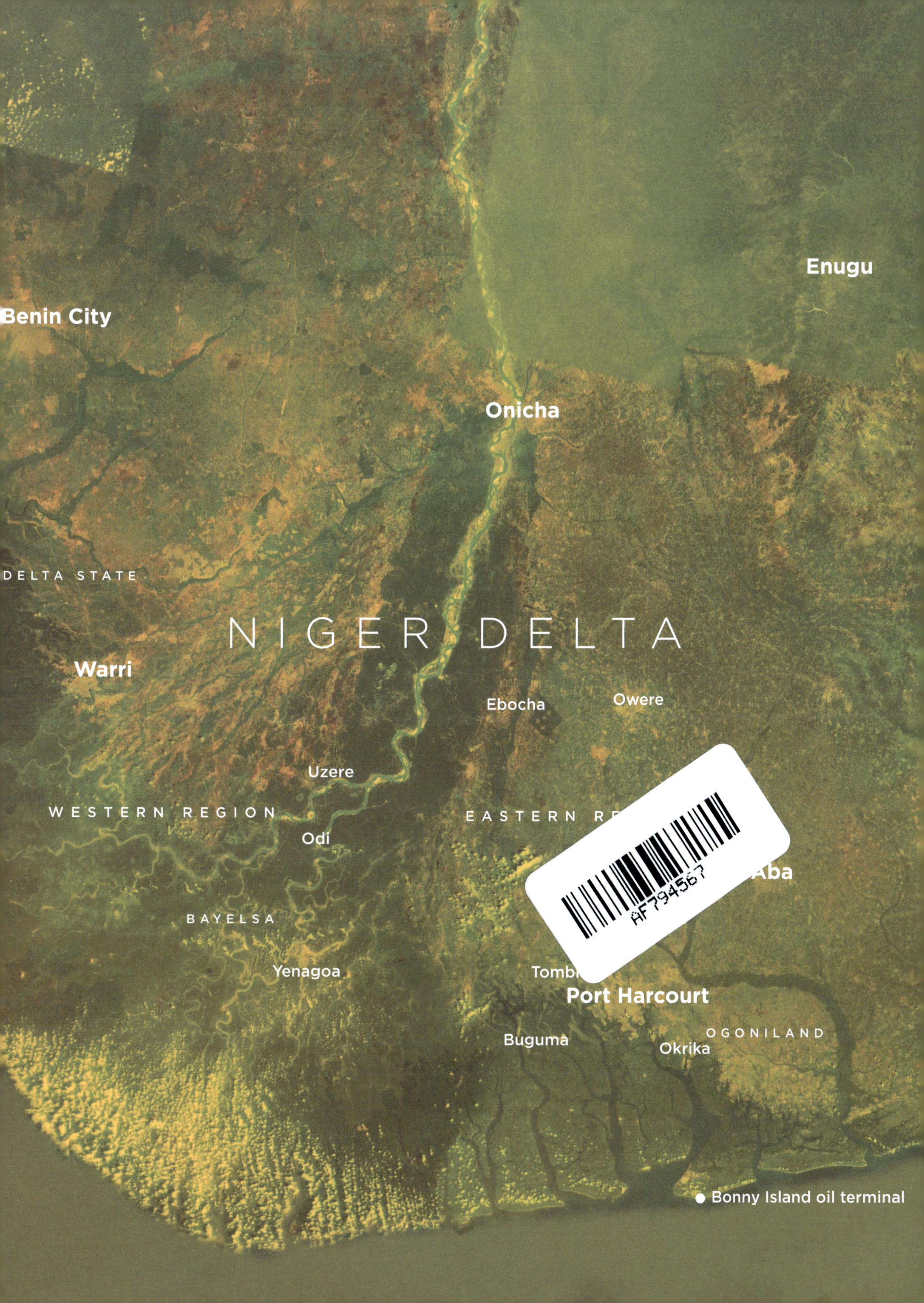
Enugu
Benin City
Onicha
DELTA STATE
NIGER DELTA
Warri
Ebocha
Owere
Uzere
WESTERN REGION
EASTERN
Odi
Aba
BAYELSA
Yenagoa
Port Harcourt
Buguma
OGONILAND
Okrika
Bonny Island oil terminal

"I do not want any blood spilt, not of Ogoni man not of any strangers among us. We are going to demand our right peacefully non-violently and we shall win."

KEN SARO-WIWA, 1993.

Author, television producer & environmental activist, hanged by the government in 1995

DELTA NIGERIA
THE RAPE OF PARADISE

TROLLEY

NNEKA EGBUNA - NIGER DELTA

From the album NO LONGER AT EASE (2007)

dem come fish our water empty.
dem come chop our oil plenty,
come take resources away.
come take all our riches
dem come use us as dem store, for their raw material to make dem rich
oh brother, brother, bros WETTIN?
sister Sister sis WETTIN?

WETTIN we dey do,e don teh, teh teh
WETTIN we dey do, when dem thief thief,
WHERE we dey when dem own us, see what they sow ,na so dem go reap.

WETTIN, we do them,
COME SEE, WETTIN dem do us.

Now tell me wettin we go do, for we are lost in the sins,in the flesh of the world,
lord forgive save these souls from dying.
broda, oga, suffri ohhhhhh, white man, VIP, suffri ohhhhh.
e don teh, when u dey do me like this!
As for right now, i no go take am no more, i go open my mouth, i go tell you wettin i dey feel.

WETTIN we dey do dem,e don teh, teh teh
WETTIN we dey do, when dem thief thief!
WHERE we dey when dem own us, see what they sow ,na so dem go reap.

WETTIN, we do them,
COME SEE, WETTIN dem do us.

as dem carry our spirit go, go far ,wettin we do them,
as dem carry everything we own, wettin we do them.
as we dey, we just dey how we dey, see wettin dem do us.

NIGER DELTA
SEEN THROUGH GEORGE OSODI'S LENS

BY FRANCIS UGIOMOH

George Osodi's photographs in this book focus on the dire and derelict conditions in the Niger Delta region of Nigeria, and perplexingly depict the human neglect and numbing ecological disasters there. Consisting of more than 200 photographs, this body of work was shown at Documenta 12, Kassel, Germany in 2007, as well as at the Bienal Del Fin del Mundo, Argentina, and Forum Stadtpark, Graz, Austria in 2009. His images engage a mounting global concern over the ruin, neglect and abandonment of these vital human communities, and the likely impact of this on social and political harmony. It also provides support to the cause of Kenule Saro-Wiwa, who since the first presence of the international oil companies in the Delta in the 1950s, called to draw attention to the ills perpetrated by them in the region; while the Nigerian government then appeared unconcerned. This unconcerned attitude by the government has changed little since then.

The region derives its name from the tributaries of the Niger River which forms an estuary in what used to be known as the Bight of Biafra on the Atlantic Ocean, now the Gulf of Guinea. The River Niger is the longest river in West Africa and it traverses almost the entire length of the sub-region as it empties into the Atlantic Ocean. The Niger Delta lies at the heart of economic prosperity in Nigeria, both before and after the amalgamation of Nigeria as a British protectorate, historical sources and statistics show that the region has welded great commercial and economic influence. Its location in Nigeria's tropical region bequeaths to it an enviable biodiversity. It boasts one of the most biologically unique terrestrial freshwater, marine ecology, and variety of habitats, that ranks as the third largest body of wetland in the world. This complex biomass since 1956 offered Nigeria an alternative to dependence on agricultural produce to fund development initiatives of government, when the first oil well to be explored in Nigeria was cited in the town of Oloibiri. Osodi's photographs of the Niger Delta region of Nigeria provide visual documents that mirror the outcome of oil exploration on the environment and people since then. The photographs embody narratives of diverse contexts that modulate our concept of the nation as idea and its reality as an agglomeration of homelands (Leopold Sedar Senghor, 1965). The ruins, neglect and abandonment in the region, which the photographs exemplify, are a direct affront on the delicate cohesion the Nigerian nation has sustained since the amalgamation of its disparate homelands or nationalities in 1914 up till the present.

However, my specific objective is to call attention to the contexts these photographs address and the upheavals associated with such backgrounds.

Bearing the above concerns in mind, my initial theoretical position is to note recent interests in the aesthetics of everyday life as well as eco-aesthetics (Bohme 1997, Lingner 1997, McFague 1993 & Paetzold 1997), which administer Osodi's work primarily. However, my specific objective is to call attention to the contexts these photographs address and the upheavals associated with such backgrounds. The reality is that they query the idea of the Nigerian nation supposedly conceived within the concept of universal suffrage and participatory democracy. Hence, in this essay I examine the historical conditions that have funded the circumstances which Osodi's photographs privilege. I engage in this task by searching out the indexical signs the photographs harbour. Thus this essay alternates between historical overviews and specific photographs and the emotions they are bound to inspire. The supposition to be pursued thus is; values that are beneficial to the Niger Delta region can only be found in restoring harmony and balance to the environment and its ecosystem; where the attainment of such harmony will in turn reshape the aesthetics of living in the region; as well as bolster the appreciation of the legacies that feed collective conscience towards a less turbulent region and the Nigerian nation. Osodi's work and the conditions it addresses are briefly reviewed here as environmental neglect and degradation, neglect of human capital and the emergence of deviant behaviours that they now sponsor. 'The Ogoni Boy', one of Osodi's photographs, is a scene where a young boy looks askance at whirling smoke soaring once more into the sky in the precincts of his home, probably with the question "When shall this unfriendly environmental practice cease?" Obviously, even in the absence of environmental laws against such acts in Nigeria, the multi-national oil corporations know better that the scene and thematic reference this photograph addresses is detrimental to humans and the environment. It was bringing such environmentally unfriendly acts in Ogoni to a world stage at the world summit on the environment in Rio de Janeiro in Brazil in 1993, that lead to the murder of Kenule Saro-Wiwa in 1995 by the military regime of General Sani Abacha (1993-1998).

'Bath in Port Harcourt' captures the effects of environmental pollution, often casued by acid rain from petroleum product wastes such as fumes. The photograph, as composed, presents a bather with a bucket used previously for stocking up paint, which now holds bathwater as paint residue is still caked on the bucket's inner wall. The squatting figure in the picture is pouring the bathwater over his body from the bucket. The environment the bathing figure is set in is mainly grey. Such intensely grey colouration often results from acid rain, for normally an uncared-for cemented block wall would appear greenish due to the growth of the green algae in a humid region like Port Harcourt. Inevitably, the bathing bucket, with its ochre-coloured paint residue, becomes the receptacle for polluted bathing water in a squalid setting. The photograph therefore metaphorically foregrounds several characteristics of a highly polluted environment.

What the region needs beyond the rhetoric of development, is how to restore harmony to the environment, contain poverty and insecurity and conflict arising from psychopathy in an oil-rich region.

Until recently, approaches by the government to address instances of agitation in the Niger Delta, have often been severely repressed. The murder of Saro-Wiwa and the sack of Odi town by the administration of President Olusegun Obasonjo in 1999, all in the Niger Delta, are typical examples along with up till recently, having to grant amnesty to supposed militants and self-styled Movement for the Emancipation of the Niger Delta (MEND) and Movement for the Survival of the Ogoni People (MOSOP). What the region needs beyond the rhetoric of development, as the government's apparent concern in recent times portrays, is how to restore harmony to the environment, contain poverty, and insecurity and conflict arising from psychopathy in an oil-rich region. Indeed Nigeria has a long history of ethnic conflicts and instability, and the reason behind them, while frivolous, was distant from the criminal neglect and insensitivity to the brazen expropriation of resources in the Niger Delta region, which now must be stopped.Osodi's photographs remain a visual testament of this criminal neglect. On the other hand, recent heightened agitations for autonomy that is ethnic-based and emanating from the Niger Delta, as enshrined in the objectives of MEND and MOSOP, are directed at the need to rectify the continued reality of abject poverty and criminal neglect that is now associated with the region and its ethnic minorities (cf. the various Bill of Rights posted on the web by the concerned ethnic minorities). The persistence of under-development in the region equates to nothing else but a programmed marginalization of the region and its minority ethnic groups. A palpable unfairness that currently underpins the style of governance projected in the Niger Delta remains one of the primary sources of tension in Nigeria. It now creates the impression of inter-ethnic disharmonies that was hardly thought of in the past, while mounting pressure on the ideal of heterogeneous make-up of the Nigerian nation as well as the concept of the nation as an idea.

But we may be asking if heterogeneity is itself the problem when indeed it is the character of the nation in many instances. Nigeria's multi-ethnic configuration is evidently not unique (cf. Senghor, 1965: 68 & Anderson, 1983). Elias Jose Palti (2001: 326-329), identifies multiculturalism as one of the component ideas of the nation, contrary to the homogeneous origin of its idea during the enlightenment in France. Yet, as Senghor (1967) defines the nation, it is designed to group "together the homelands in order to transcend them. Homelands arise naturally and are expressions of a particular place and environment, the state is a deliberately willed construction, or rather reconstruction… the nation depends on them, or rather it depends on their virtues, their real character and so on the strength of the emotion, which surrounds them." But in many instances and in Africa especially, Edward Said (293-316), notes that such wilful surrender of homelands as strategy to fight back colonialism led to the illusion that colonised states in Africa were homogeneous. How can the contraption that lays claim to homogeneity in a definitely heterogeneous constitution be an appropriate instrument for the anti-colonial confrontation in Africa, he queries. The adoption of this mixed composition of ethnicities that make up Nigeria as a monolithic block has been to the undoing of the minorities in Nigeria especially those of the Niger Delta. The Willinks Commission set up by the colonial administration in 1958 called attention to the fears of the minority ethnic groups in an independent Nigeria, highlighting the fear of domination and its consequences in the Niger Delta (Aaron and George, 2010). The notes of this commission currently validate the persistent agitations the nation is embroiled in from the region.

Osodi's work documented in the photographs here is driven by the sole objective to inspire a positive political action that is rooted in a collective national will to redress the effects of environmental and human neglect in the

Niger Delta region. The programmes the work initiates are as follows; the first is how to get the government to change its strategy with regard to the region, in such a way that allows for rethinking past ills and to create an identity that the region identifies with; secondly, the above initiative calls for the restoration of harmony to the environment, which will allow for an appreciation of good life that in turn can bolster enhanced life that impacts positively on the environment. Where these points are addressed Nigeria then would have moved beyond "placebo effects" that dominate the current approach at redressing the age-long ills visited on the region. From another perspective, the failure by government to secure an identity the people of the region can recognize as holding out a place for them, continues to problematize idea of the nation, owing to varying tensions associated with Nigeria's heterogeneous composition. This tension continues to be fanned by the injustice of governance in the Niger Delta region. Pope John Paul II (1998) tacitly recalls that "from the justice of each comes peace for all." Concretely the Pontiff furthers this thought thus; "Justice goes hand in hand with peace and is permanently and actively linked to peace. Justice and peace seek the good of one and all, and for this reason they demand order and truth. When one is threatened, both falter; when justice is offended, peace is also placed in jeopardy." Put in another way, any perceived injustice done to any one of Nigeria's federating units is but a denial of peace; more so when such federating units belong to the minority group. Osodi's photographs presented here in essence have one objective, which is to inspire a positive action targeted at amelioration and the enthronement of justice and peace in Nigeria. The initiative to draw on the power of the pictorial image to remedy derelict conditions of human existence due to wars, neglect of people and the environment by government etcetera, recorded in photographs is not new.

Many commentators on the value of pictures that address derelict conditions of human existence point out how such pictures inspire the emotions of pity; but that what is actually required is how such emotions can be harnessed for positive change. Sontag (2003 & 1977) and Arendt (1973) are concerned that fundamentally such emotional states at best only dampen or diminish the urge to engage requisite political actions that should be targeted at remediation or recuperation. James Fergusson's work entitled 'The Anti Politics Machine' provides a critique on the myth of developmental initiatives when confronted with the guilt of the human conditions sketched above. For him, a bureaucratic power machine that posts, rather, a "depoliticization" of the people has often been responsible for cases of inactions and mere mirage in promises of development. On the other hand, and with some specificity, Uzochukwu J. Njoku (n.d), takes on the derelict situation in the Niger Delta in "Understanding Gen 1, 28 in the Light of Solicitudio rei socialis (numbers 26 and 34)." He appropriates the tenets of "liberation theology" inherent in Solicitudio rei socialis and, latching on the central principle of guilt explicit in negligence, exploitation and mismanagement of wealth accruing from the region by government, suggests a few actions as initial approaches to much needed remediation in the region: In the first place it is a commitment on the part of the government to embrace the "ethics of responsibility," as it leads to the acceptance of guilt, and nurtures a proper conscience for remediation and reparation. Such a stance, he argues, puts collective response into action, in favour of the neglected and wealth mismanaged. Njoku's pointed action plan is but one option. It requires commensurate sensitisation; which Ann Kane (2000), Sontag (2003: 122-3) and James Johnson (2005), complement. For Kane, "narratives explain experience, evoke emotion, engage participation, and normatively evaluate courses of action…" and for Sontag, narratives rather than mere photographic documentations provide immediate drive towards collective response, when relating with photographs that inspire compassion. In the same vein Johnson argues that the work of art proposes a reparation and amelioration agenda, when its sign is read against its historical antecedent and the motivations that inspired the artist to document such scenes.

This documentary of derelict conditions of the Niger Delta, I hope, will engage emotions; and engender positive action towards amelioration of the ills visited on the Niger Delta region. It complements similar work on the travails of the Niger Delta region (Ike Okonta, 2001; Sandy Ciofi, 2007; and Michael Watts, 2008; etc.). Along with these authors the work, which Osodi presents here, are purveyors of the policy's "sacred truth" presented as metaphors, which remain veritable tools for social change. The value of the work presented here is to re-present the morality index of the ruling elite in Nigeria with hope that where "justice" had all the while been offended and "peace" placed in "jeopardy" that there would be achieved an "ethics of nature" that of necessity grooms an "ethics of good life".

Works Cited

Federal Environmental Protection Agency, Guidelines and Standards for Environmental Pollution Control in Nigeria. Lagos, 1991. Johnson, James. "'The Arithmetic of Compassion": Rethinking the Politics of Photography" (other reference details are not available to me). Kane, Anne. "Reconstructing Culture in Historical Explanation: Narratives as Culture Structure and Practice." History and Theory. 39 (2000): 311-330. Lingner, Michael. "Main Aspects of the Concept of nature in Modern Art." Issues in Contemporary Culture and Aesthetics. 5 (1997) 85-95. Michael Lingner, Main Aspects of the Concept of Nature in Modern Art, Issues in Contemporary Culture and Aesthetics, 5 (April, 1997) 85. Njoku, Uzochukwu J. "Understanding Gen 1, 28 in the light of Sollisitudo rel socialis (numbers 26 ans 34)." Unpublished manuscript. Nwachukwu, Orji "Governing 'Ethnicized' Public Sphere: Lessons from the Nigerian Case." Codesria: 12 General Assembly. 07-11/12/2008. Okonta, Ike and Oronto Douglas. 2001 Okonta, Ike. "Where Vultures Feast: SHELL Human Rights and oil," Africa Review of Books, 2006. www.codesria.org/links/publication. retrieved July 13, 2007. Omoweh, Daniel A. Shell Petroleum Development Company, The State and Underdevelopment of Nigeria's Niger Delta: A Study in Environmental Degradation. Trenton NE: Africa World Press, 2001. Paetzold, Heinz. "How to Bridge the Gap between the philosophy of Art and the Aesthetics of Nature: A Systematic Approach." Issues in Contemporary Culture and Aesthetics, 4 (1997): 4-15. Palti, Elias Jose. "The Nation as a Problem: Historians and the National Question" History and Theory, (October 2001) 324-346. Peel, Michael, A Swamp Full of Dollars: Pipelines and Paramilitaries at Nigeria's Oil Frontier. London: I. B. Tauris and Co., 2009. Plotinus, Enneads 7 Volumes, trans. A. H. Armstrong, Cambridge Massachusetts: Harvard University Press, 1989, 1, p. 261. Mcfague, Sallie. The Body of God: An Ecological Theology. Minneapolis: Fortress Press (1993). Senghor, Leopold Sedar. Prose and Poetry. Selected and translated by John Reed and Clieve Wake, London: Oxford University Press, 1965. Sontag, Susan. On Photography. London: Penguin 1977 Sontag, Susan. Regarding the Pain of Others. New York: Picador, 2003. Tagg, John The Burden of Representation: Essays on Photographies and Histories. Houndmills: Macmillian, 1988. Ugiomoh, Frank A. O. "Pale Reflections and Fables of Life: George Osodi's 'Real People' of the Niger Delta in Nigeria." Camera Austria International. 106 (2009) 23-34. Ugiomoh, Frank A. O. "Musicalising Trash: Osahenye Kainebi's Rhythmic Trump" Exhibition Brochure, Trash-ing: An Exhibition of New Works By Osahenye Kainebi, Sept 12 – Oct 10, 2009 CAA, Watts, Michael. Curse of the Black Gold: 50 Years of Oil in the Niger Delta. New York: Power Books, 2008

BEWAILING A BLESSED LAND

by Chinwe Azubuike

Permit my thoughts if I am not slain,
my own pen inflicting pain upon myself,
to question your mission as you relish your tenure on that throne
why you wallow in the demise, caused by you,
of a nation fortified with riches
why you rejoice at being a menace to humanity
Permit my thoughts if I am not slain,
to question why you choose to transform us into a looting booth
eager to grab every penny from our sweats
before you vacate that seat
The likes of you and your predecessors,
so peculiar are your characters
I wonder if it isn't an inherited craze?
Permit my thoughts if I am not slain,
to question why you blow false alarms to line your safes
and that of your generation with foreign bootieswhat
robbery is worse than this?
Make a monkey of us and leave us in debts
Yet, these we swallow in silence for fear of losing our heads
heads we've lost from trigger happy hooligans
who invade our lives, homes and freedom
in desperate attempts to wipe us off our soil

and that of our ancestors.
Permit my thoughts if I am not slain,
to question your chant of "Democracy" in our ears
when we still writhe in pain under the hammer of your reign
why we cringe in disgust at the sight of your presence
What is leadership without the love of your subjects?
How long must we live in this tyranny?
The thought of death and violence on our land
Mangled bodies like roses squashed underfoot
Vivid imaginations of chaos tearing us apart
If I have a wish, it is to relive the reign of those leaders
history spoke of in my unborn days.
Let our scholars, who witnessed the better days of our nation
congregate to boost the morals of the layman;
and if they decline,
Let our parents uphold this task;
and if they decline,
Then let we, their children uphold this task
as we witness perpetual days of a land
bathed in the orange glow of burning oil
May we, with every ink, create teardrops
as vivid imaginations tear us apart.
Let it be ticked in history
that our generation,
and generation to come,
was wiped out in the course for justice.

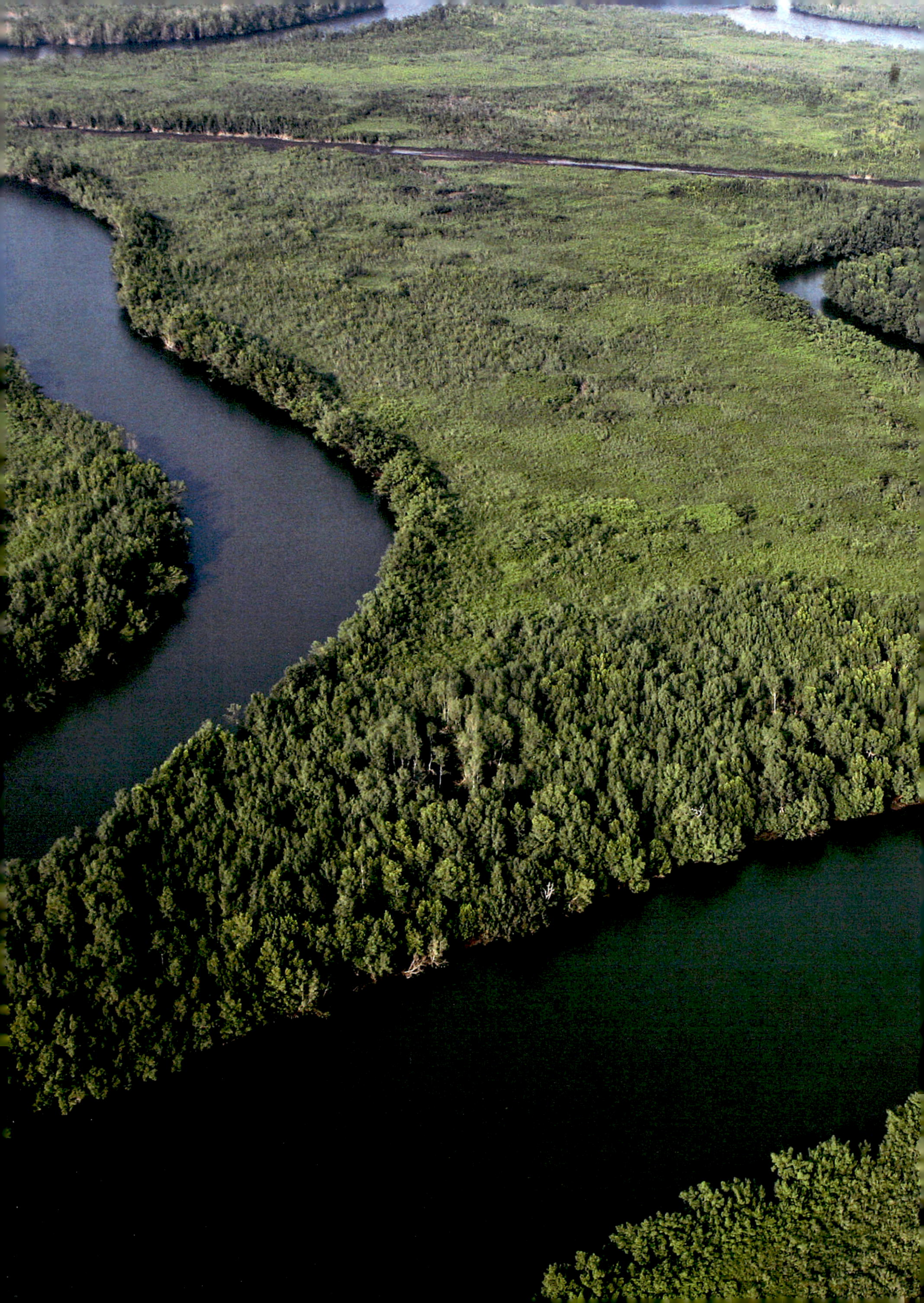

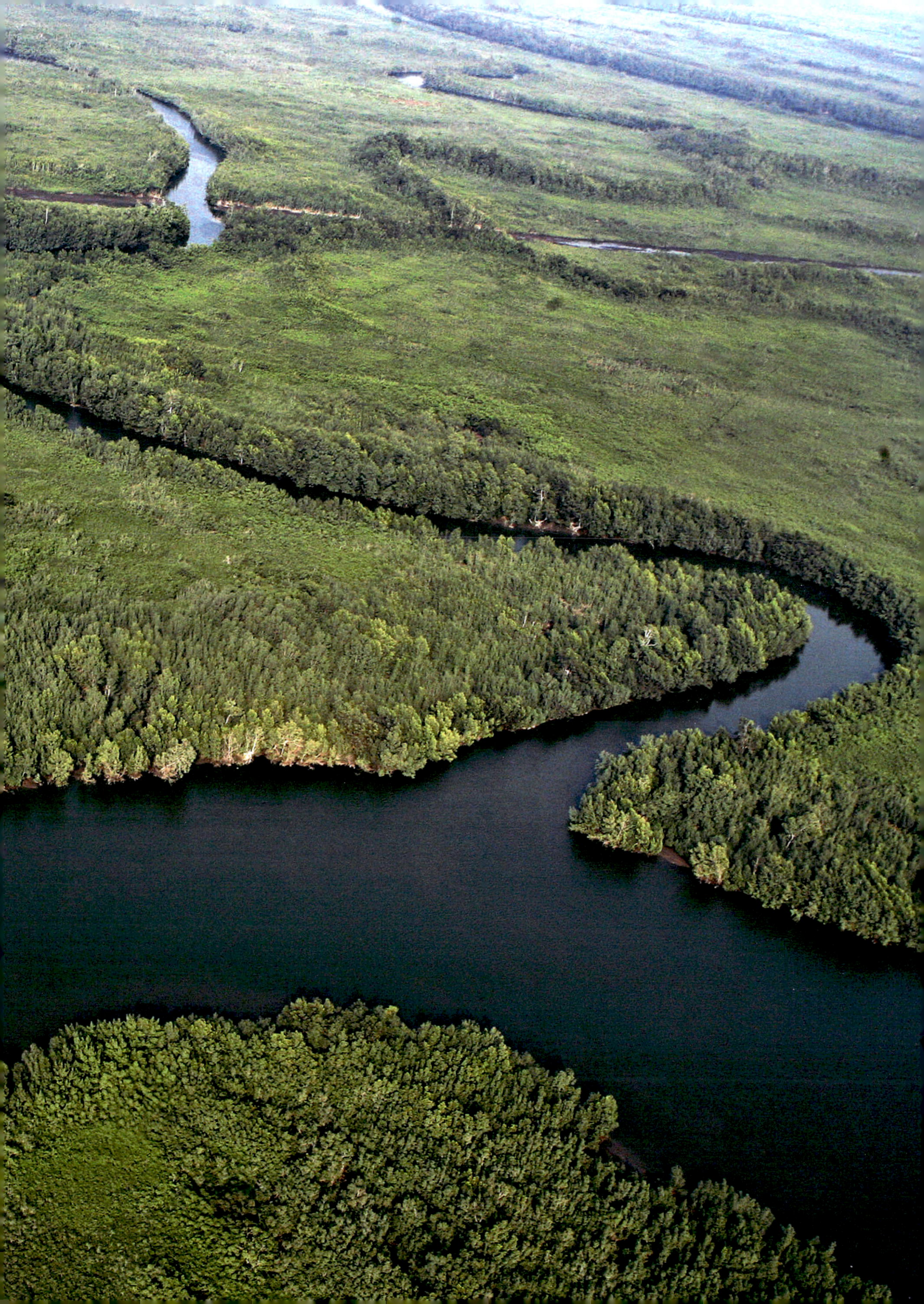

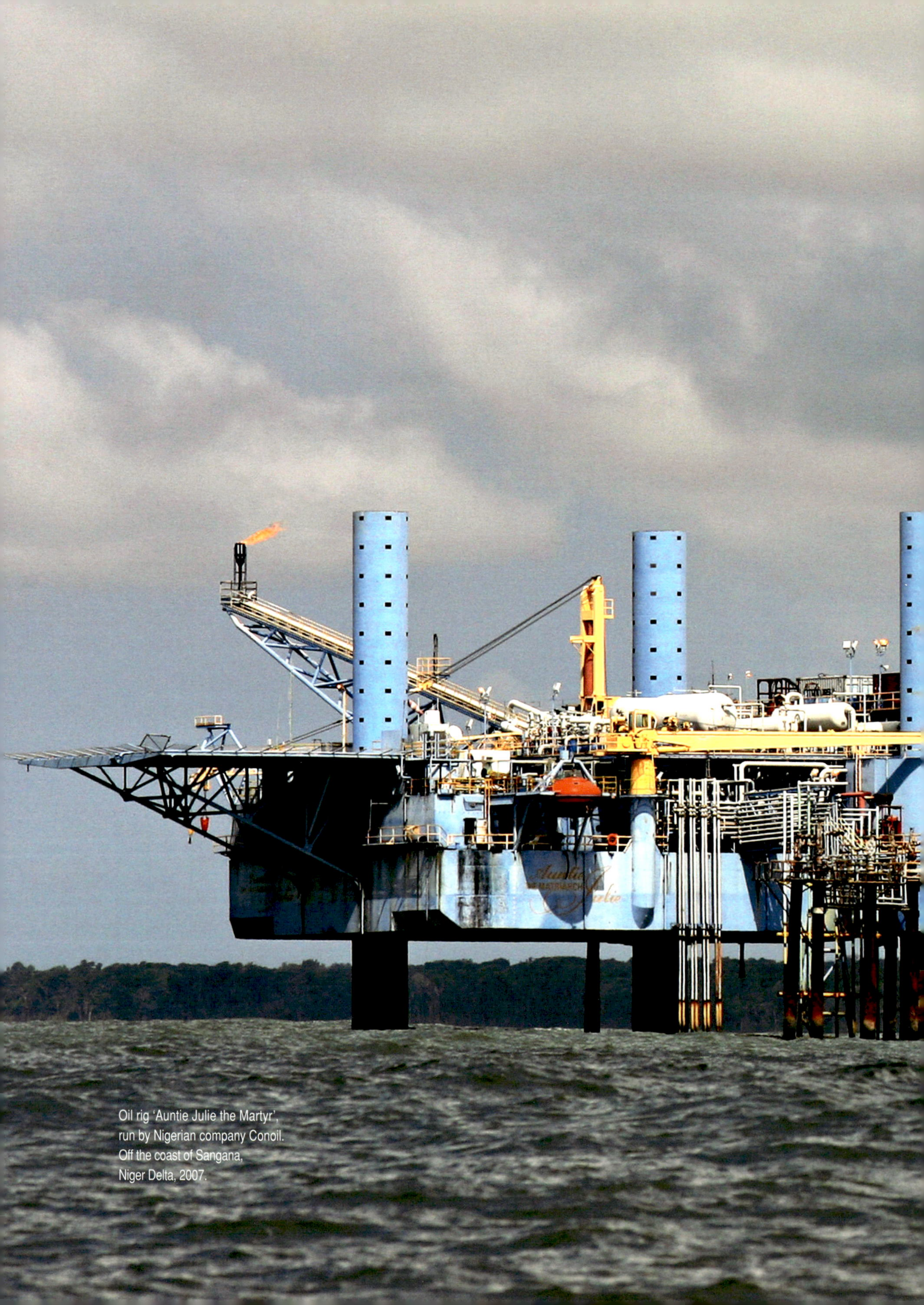

Oil rig 'Auntie Julie the Martyr',
run by Nigerian company Conoil.
Off the coast of Sangana,
Niger Delta, 2007.

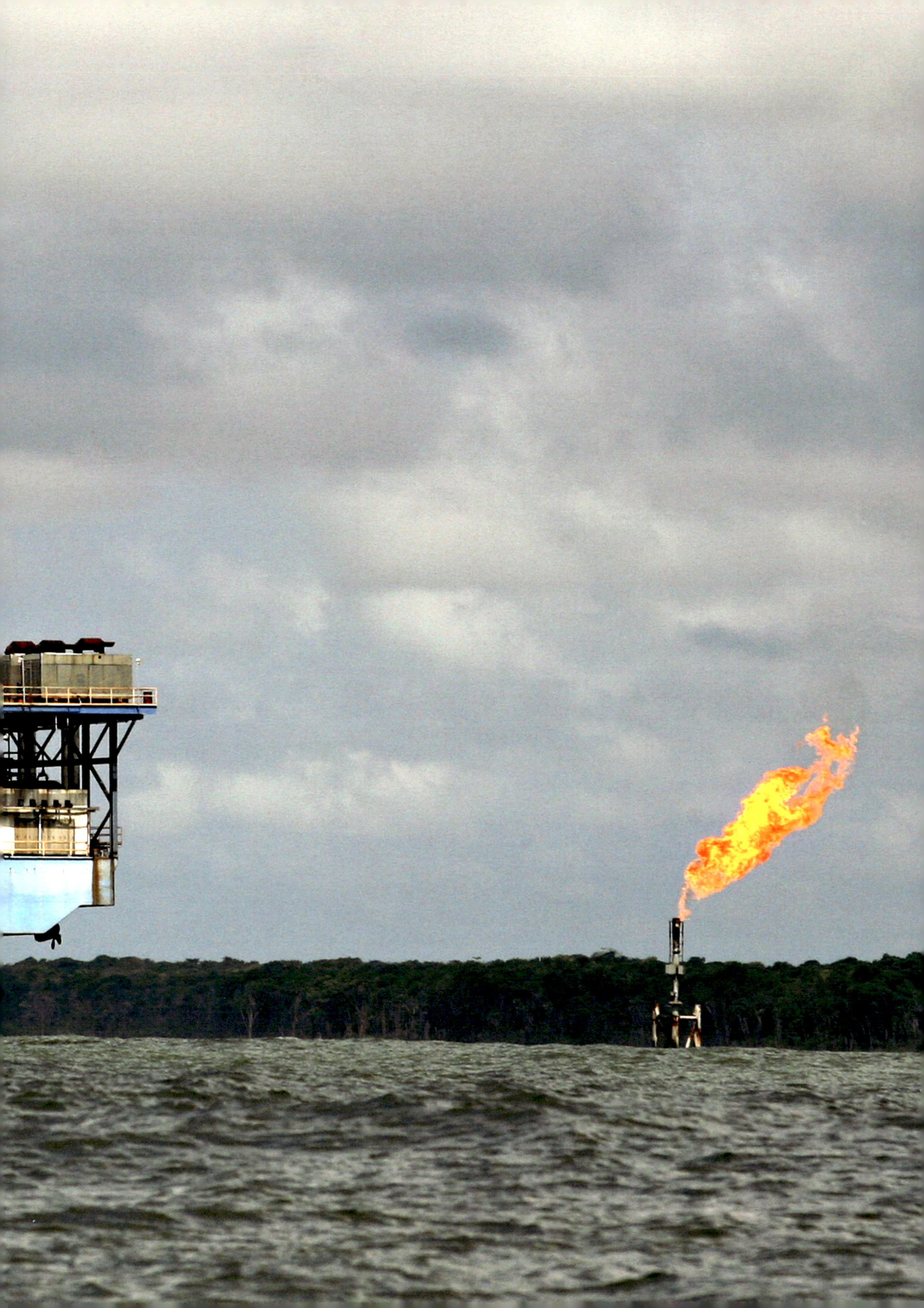

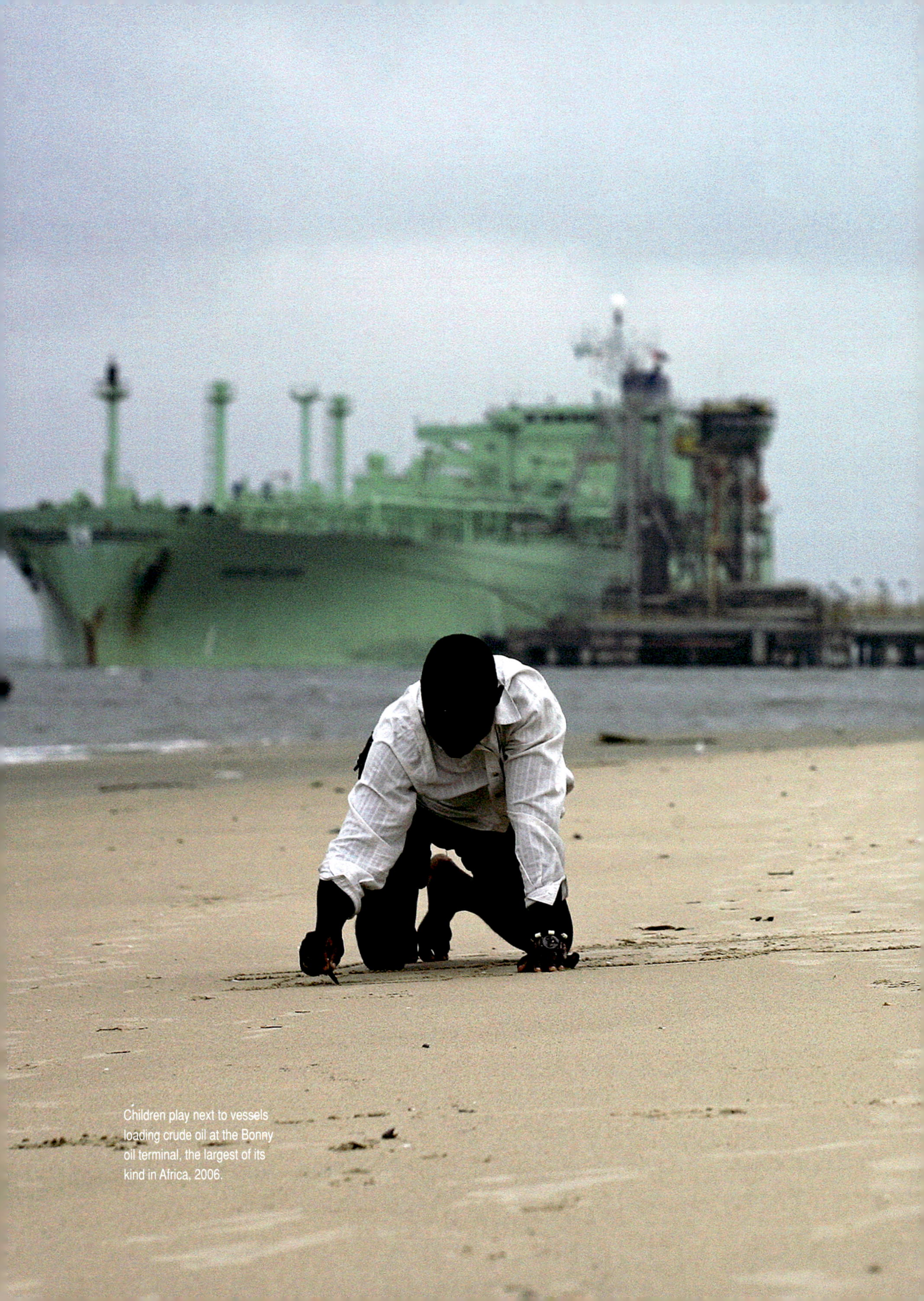
Children play next to vessels loading crude oil at the Bonny oil terminal, the largest of its kind in Africa, 2006.

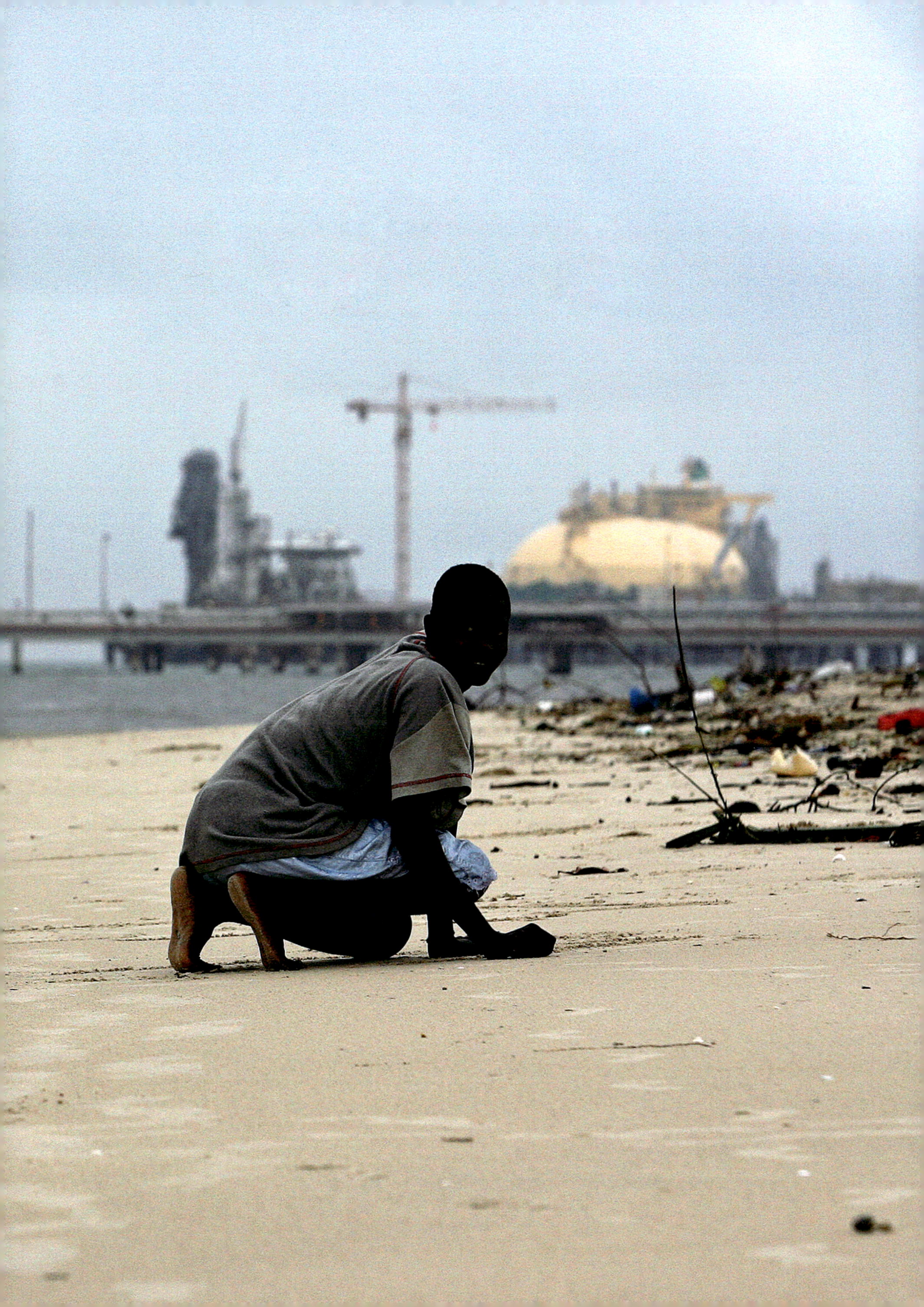

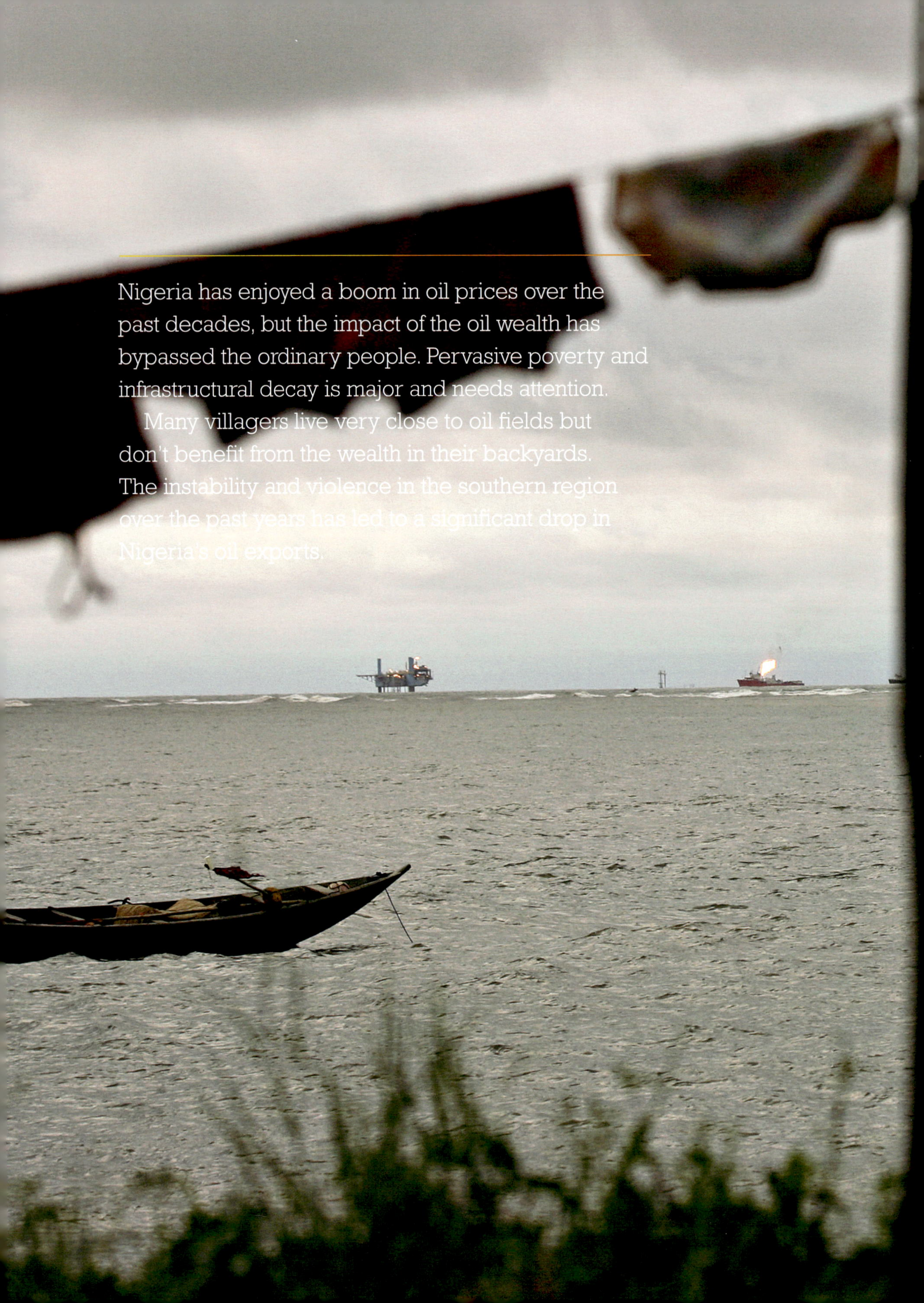

Nigeria has enjoyed a boom in oil prices over the past decades, but the impact of the oil wealth has bypassed the ordinary people. Pervasive poverty and infrastructural decay is major and needs attention.

Many villagers live very close to oil fields but don't benefit from the wealth in their backyards. The instability and violence in the southern region over the past years has led to a significant drop in Nigeria's oil exports.

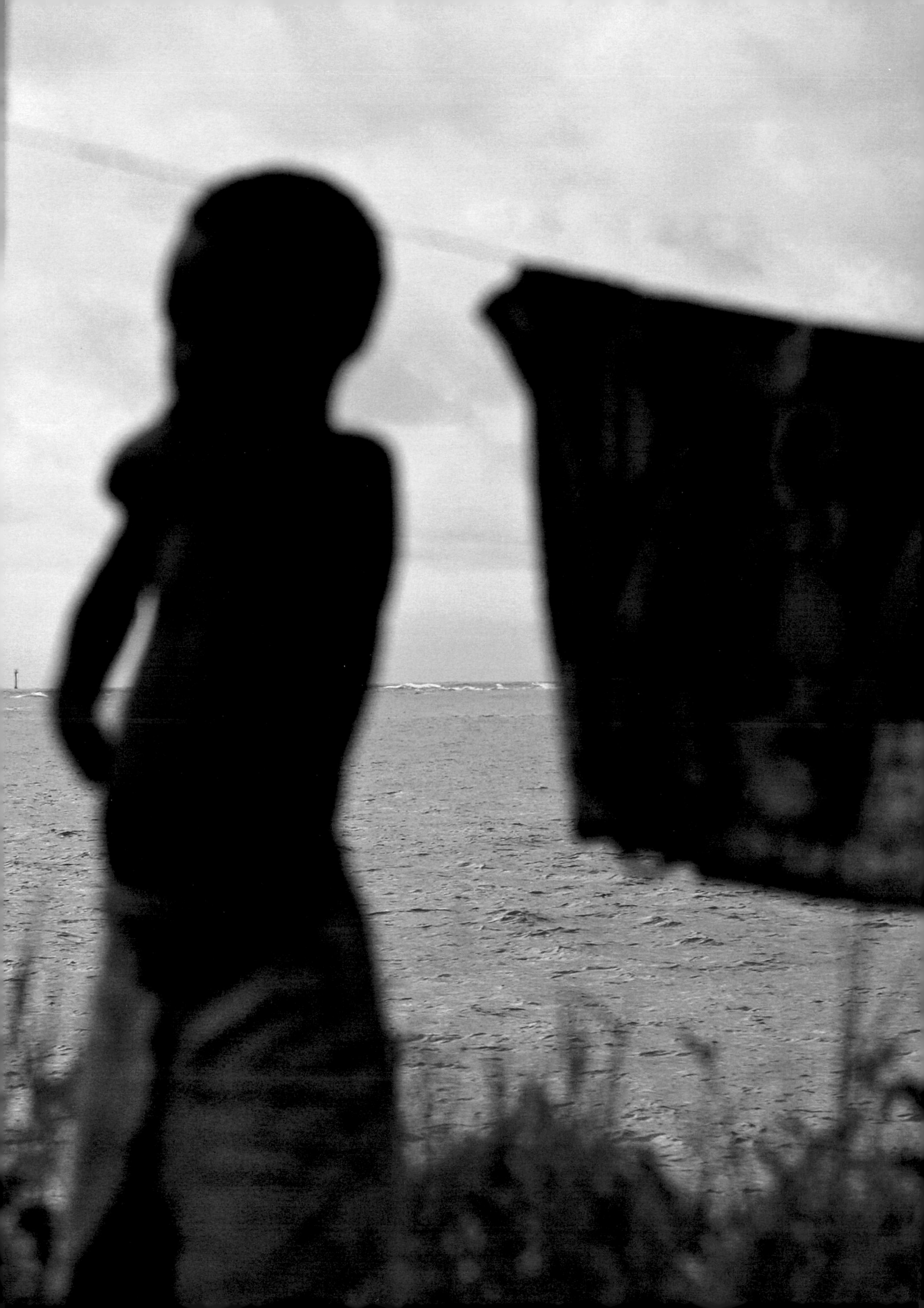

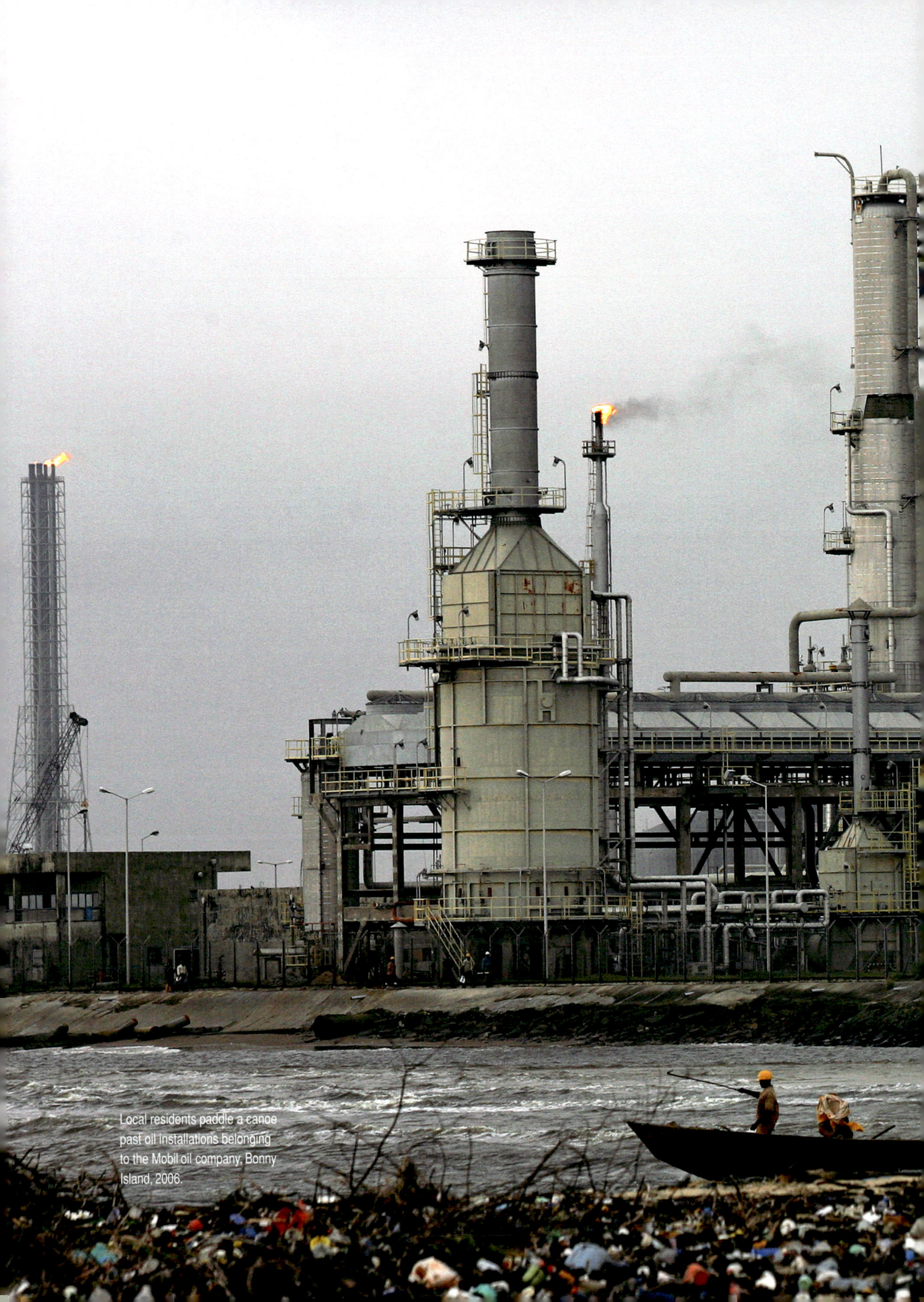

Local residents paddle a canoe past oil installations belonging to the Mobil oil company, Bonny Island, 2006.

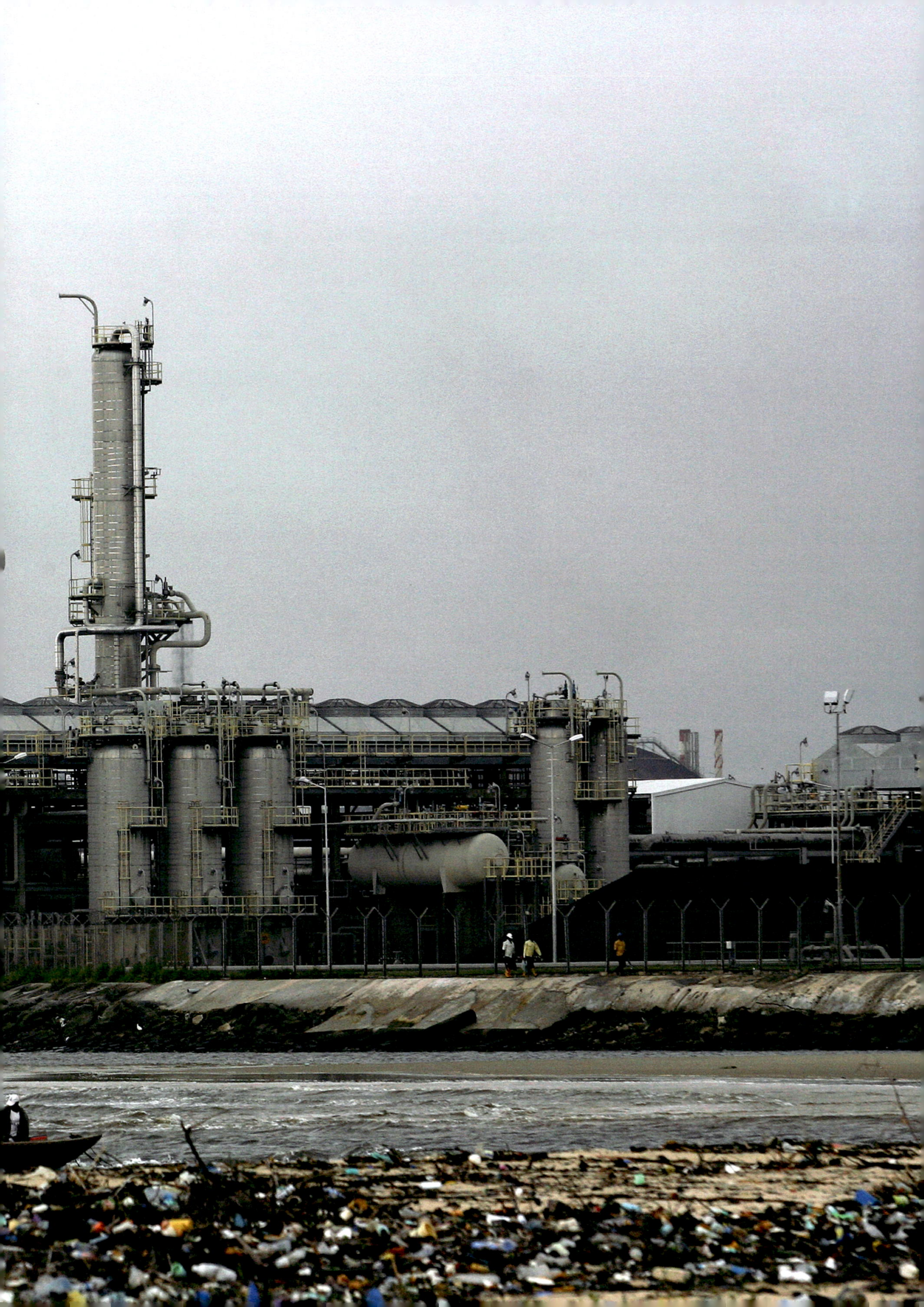

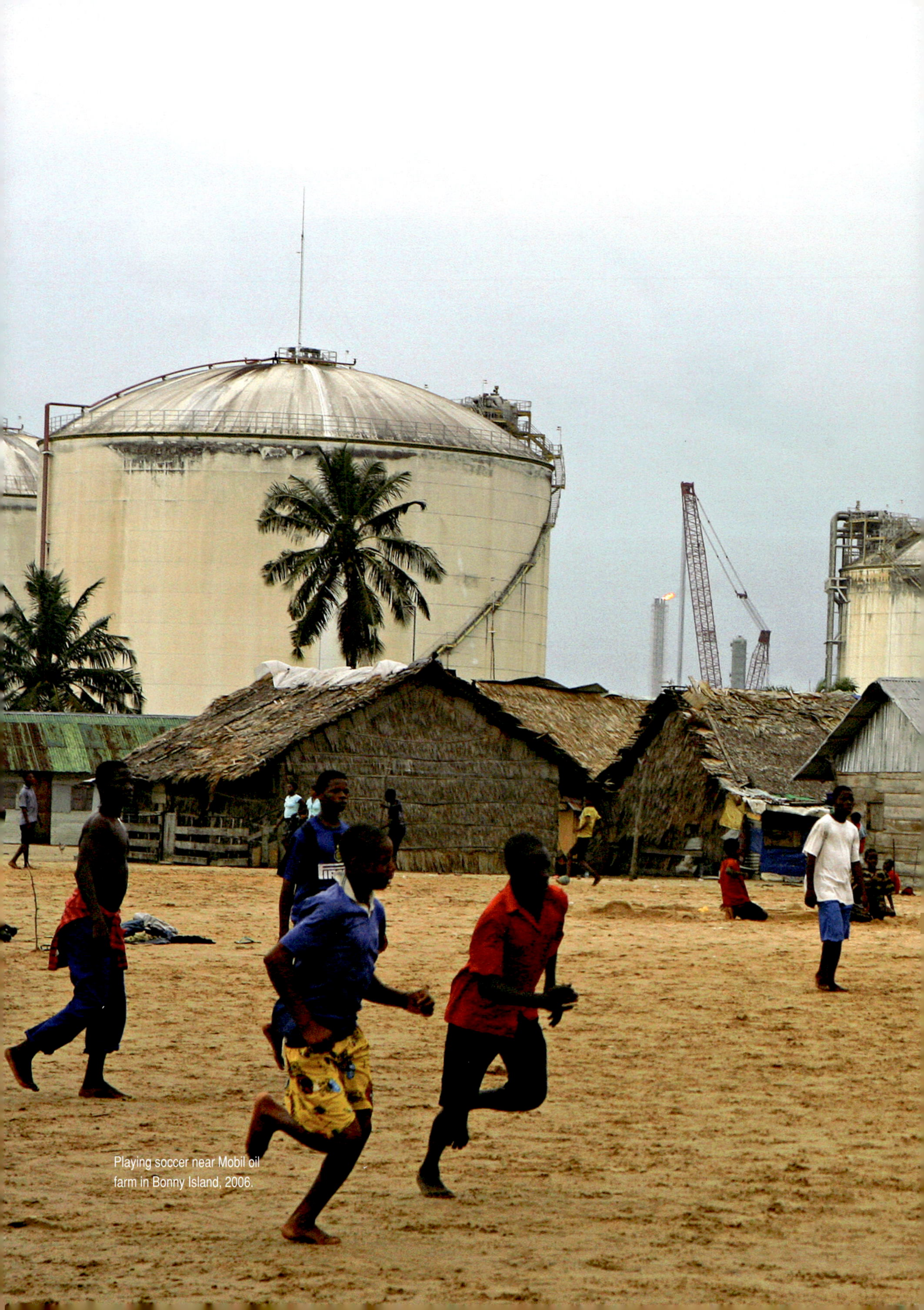

Playing soccer near Mobil oil farm in Bonny Island, 2006.

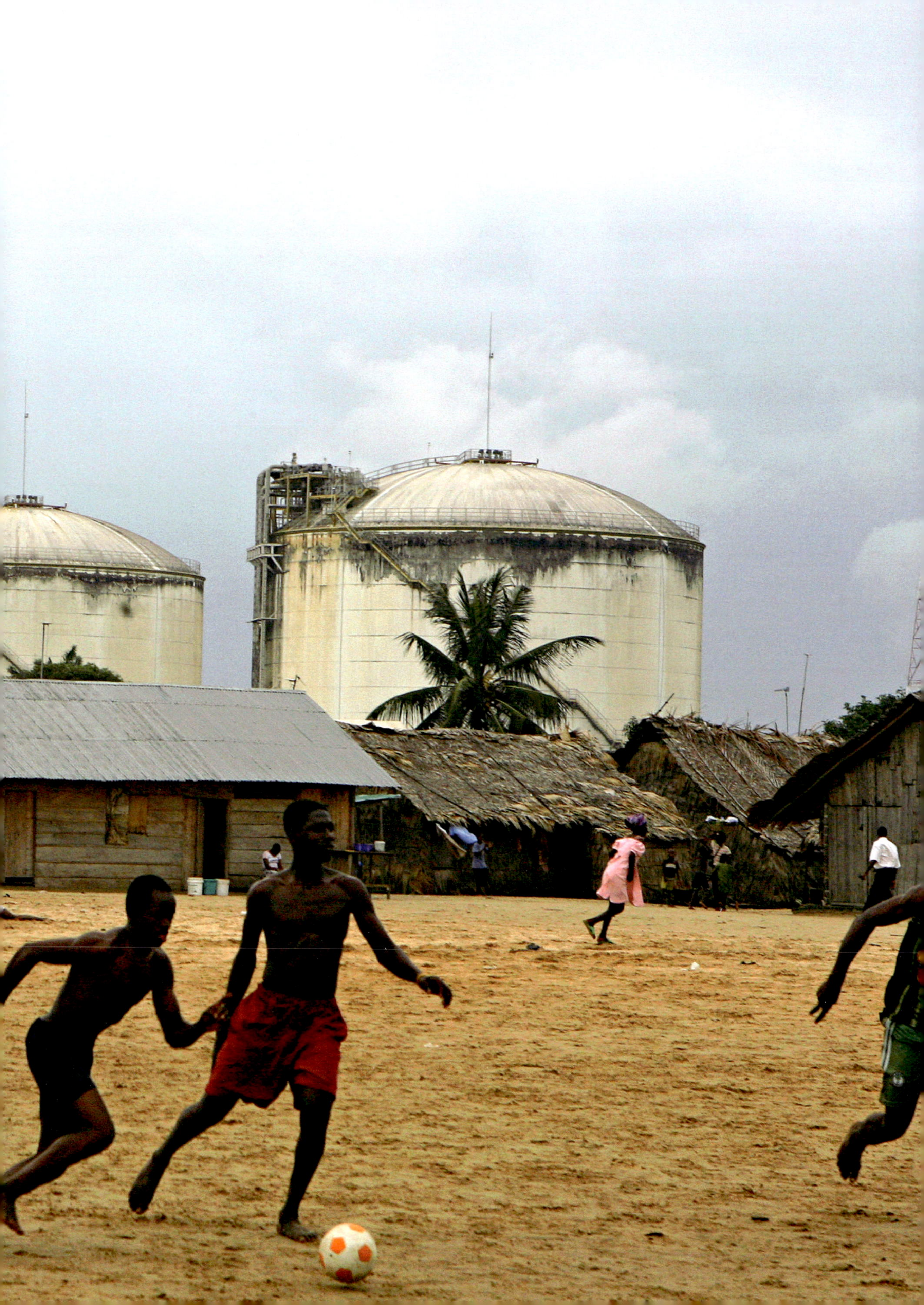

GEORGE OSODI IN CONVERSATION WITH BISI SILVA

BISI SILVA You have been documenting the Niger Delta for over five years and developing it into an ambitious project. What was the impetus for this important body of work?

GEORGE OSODI I became involved in the project because I noticed that there was so much injustice being perpetuated in the region. So much had been heard about the situation in the Delta but very little had been seen in images, presumably due to the hostile topography of the region. In 2003 I took my camera with me to the Delta and began the project. I started talking to people, asking questions, going into the creeks and to the villages to document the reality of what was happening. I felt that photography was the most appropriate tool to communicate what I was seeing, hearing and experiencing. I felt the urge to share this information, to create some form of document in the hope that it would be useful, not only for the present but also to future generations irrespective of the attendant risks or the duration involved.

BS The risk of going to the Niger Delta due to the activities of the militants as reported by the local media and especially the international media such as CNN has reached almost hysterical proportions. It makes it difficult for the average person to decipher what the actual situation may be like. What was your experience with them and how were you able to gain access?

GO Gaining access to the militants is all about risk taking, between life and death, as you really do not know what is going to happen in the creeks.My first encounter with the militants was during a visit to take pictures of oil installations. After 50 minutes driving in the creeks, a group of 7 men - two of them with guns - appeared out of nowhere and surrounded my boat. They asked questions about my identity, my activities and my reason for being in the creeks. In spite of my efforts to explain that I was a photojournalist doing a documentary on the Delta, they remained suspicious, accused me of being a government spy, took my cameras and equipment and promised to deal with me accordingly. I was taken to the camp of their 'boss'. The next few hours were terrifying, being interrogated and watching the militants dancing, drinking and smoking all kinds of stuff. It was a situation in which anything could have happened. In the end, I was able to convince the group of my mission, but they warned that if I gave out incorrect information, I would be tracked down and killed. I was given access to their village, to talk to the locals and to take pictures. At night they put me on a boat and took me out.

My second meeting with militants was with Asari Dokubo's Niger Delta People's Volunteers Force (NDPVF) after the 2003 presidential election. This was during Asari's early days in the creeks. We met with him thanks to a contact and he spoke to us about the group's unsavoury role in the 2003 elections, about the questionable political activities going on in the region and in the country. He also talked about the way in which his group's conflict with another group led by Ateke Tom's Niger Delta Vigilante (NDV) has brought untold hardship and distress to the local communities. The fighting between these two rival groups has led to the destruction of lives and property in some areas of the Delta.In the end my feeling is that meetings with militants is all about people playing 'god' and they are everywhere in the Delta, accounting for a lot of the social unrest and underdevelopment of the region.

BS Your response highlights the pervasive thinking about militants in the collective minds of Nigerians.

GO I think that is a comfortable compromise for me because nothing can be done at this stage to deny their centrality within the contemporary Deltan landscape. However there is more to the region and I also wanted to focus on that aspect, on the interweaving stories that constitute the region. Very little of the Delta seemed to be known by Nigerians in general until the widespread visibility of the militants parading themselves commando-style in the creeks. The Niger Delta was generally a peaceful place with great people, traditions and environment. The region contains the world's third largest wetland. Its rich soil has supported the cultivation of produce like timber, cocoa, yams, rice and palm trees. For centuries they were a great and prosperous trading port and came in contact with many different peoples and cultures. They have a rich and fascinating culture that manifests in all its splendour during their numerous boat regatta festivals. All these aspects are unfortunately ignored and neglected.

BS Taken from the layman's perspective, what do you think? Are we near to finding a lasting solution?

GO My experience of visiting and documenting the Delta in the past decade reveals that no one seems to be fighting the Delta cause, only their own personal interests at the peril of the ordinary people. The Delta problem is the Nigeria problem and the Delta solution could be the Nigerian solution. These problems can be corrected through transformation in terms of leadership, sincere governance and respect for the rule of law. A leadership that truly has the commonwealth of the people at heart. This could signal a beginning for other things to fall into place bringing about a drastic change in Nigeria. In the meantime, resistance against the oil companies, government officials and the military has started at local level with the emergence of various militant groups who claim to be fighting for the people, demanding that the people get a greater share of the region's oil wealth. But the question being asked is how sincere are the rising number of militant groups surfacing in the Niger Delta today?

BS During a recent visit to CCA, Lagos by one of the secondary schools, I asked what they knew about the Niger Delta and the collective chorus was militants, guns and killing but few understand why. This to a certain extent symbolises the collective ignorance of many Nigerians about the complexity of the Niger Delta situation, a key region of Nigeria. How does your work in this exhibition deal with this?

GO Whilst this exhibition shows my works that deal with the consequence and effects of economic stagnation, environmental degradation and political upheaval, it was also important to portray this region from a different perspective. I wanted to show the duality of life in the Delta regions, the beauty of ugliness, the children playing football in a green field with gas flaring high in the background, the women in their traditional attire waving their symbolic white handkerchief as they danced, the greenness and abundance of the mangroves, the men at their aluminium windows, the water glistening as the sun sets and the women fishing in polluted waters. It is amazing how people carry on with their lives, with their daily routines with a smile. I wanted to put a human face on this paradise lost. I think the portrait and landscape images contribute to a different vision of the Niger Delta. The Niger Delta was and is a beautiful place. If the independence vision had been carried through today it would rank high in the world as a leading eco-tourism destination. But unfortunately oil was discovered over fifty years ago.

BS You work mainly in a documentary style for international press agencies. This has allowed you to build up an impressive body of work on the different facets of contemporary African society. However beyond earning your livelihood you have also created an important body of images in which the documentary and the artistic coalesce. How do you negotiate this delicate pendulum between what you are assigned to do and what you want to do or show from your perspective?

GO I see the world as a stage and a studio so everyday is work for me especially in Africa where almost every aspect of our daily life is worth photographing. However, as a documentary photographer, it is not just what I want to show that matters, but how I want to show it taking into consideration the aesthetics and visual power of the image. I want to carry the viewer along so as to appreciate the message I am trying to communicate. I have never really compromised my assignments with my personal projects. The two are fairly separate but naturally impact on each other.

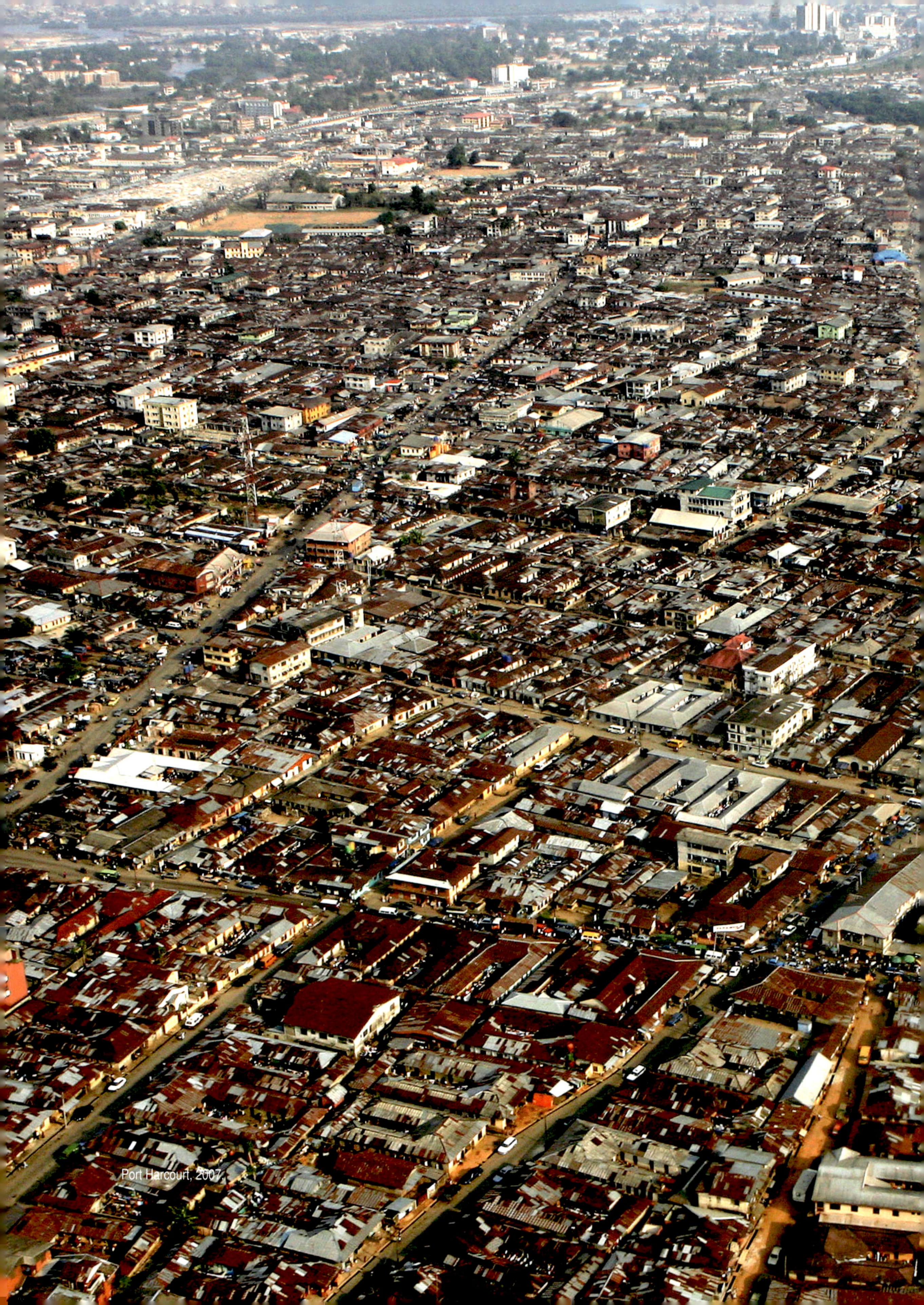

Port Harcourt, 2007

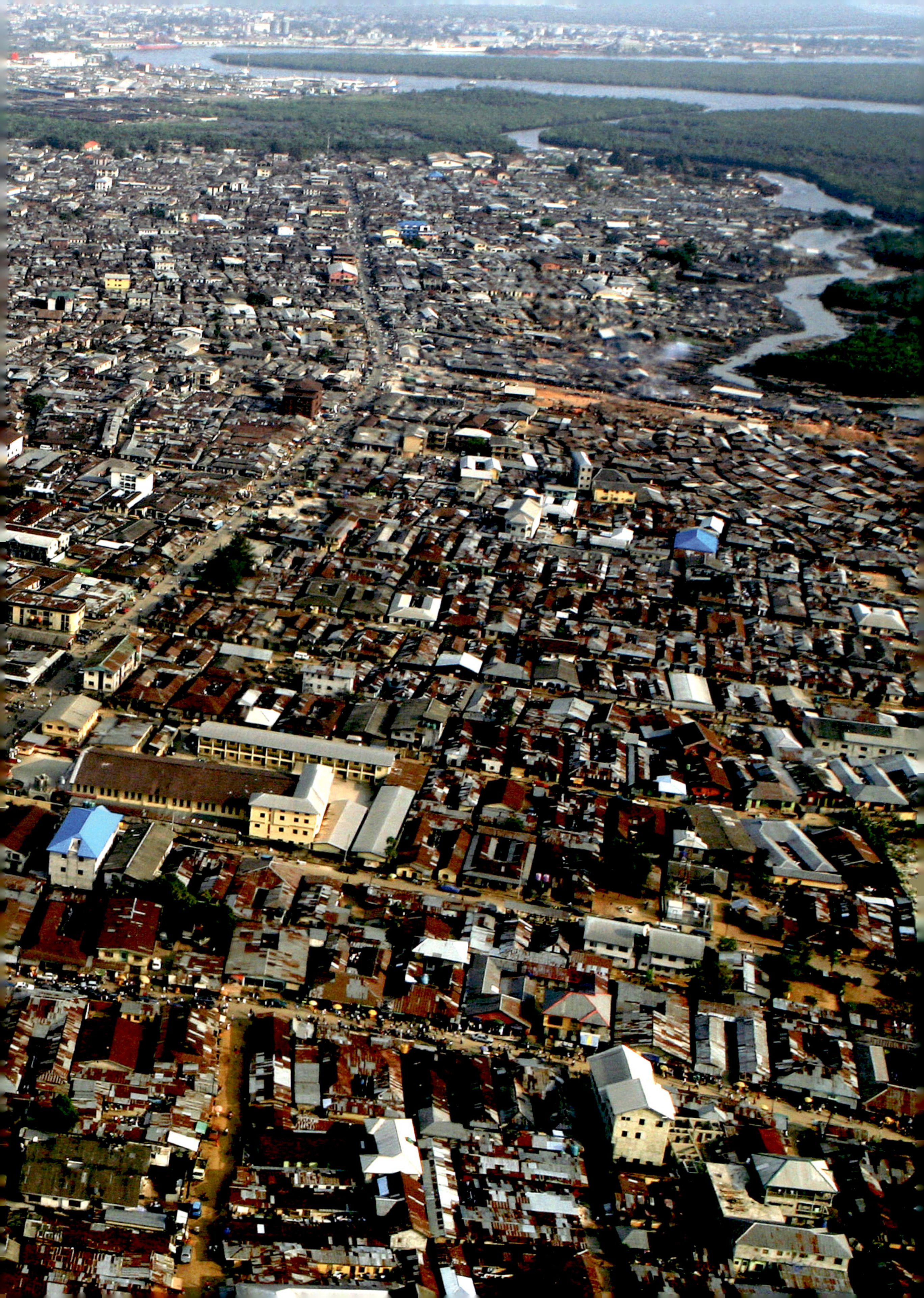

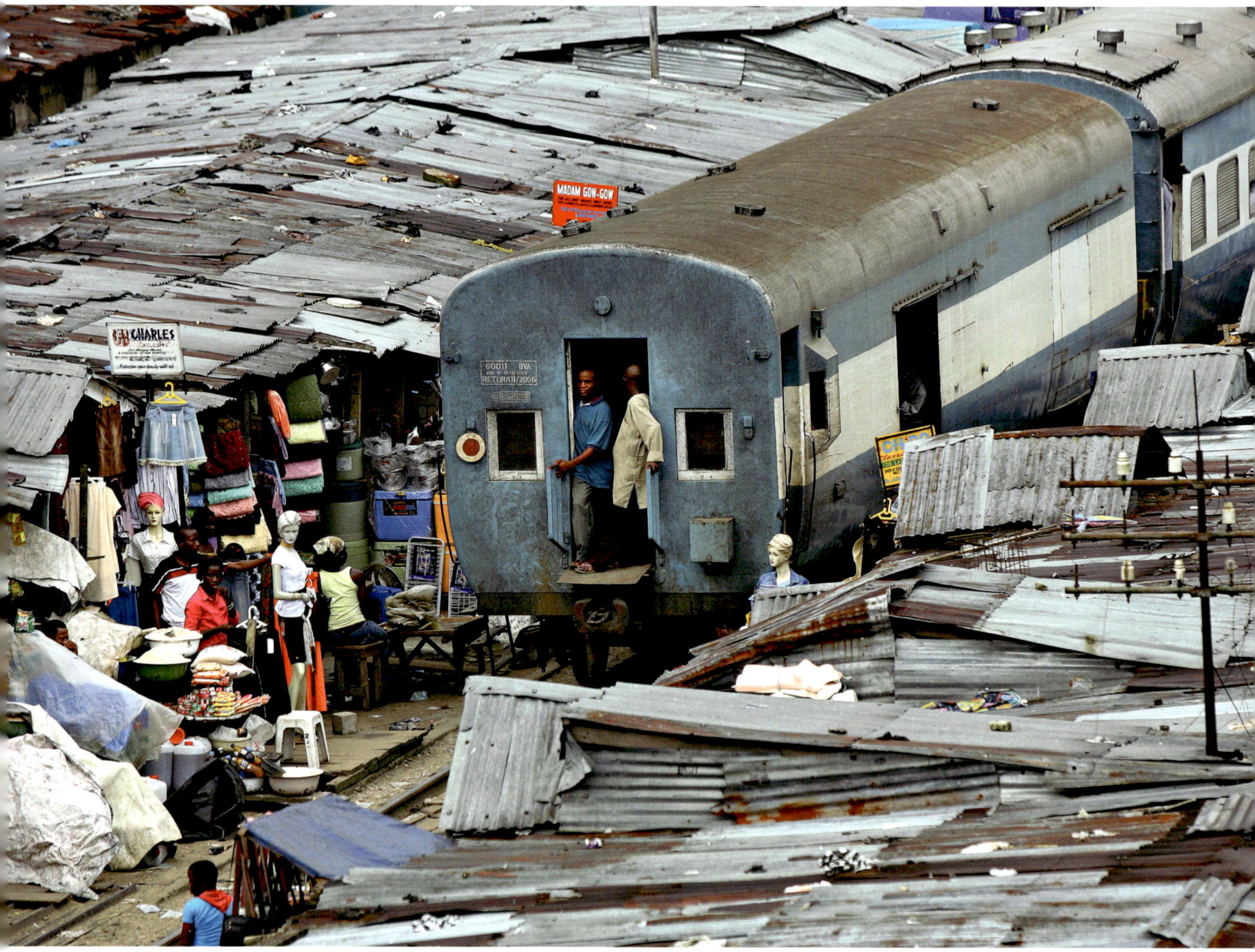

Above and right: Mile One Market, Port Harcourt, 2006.

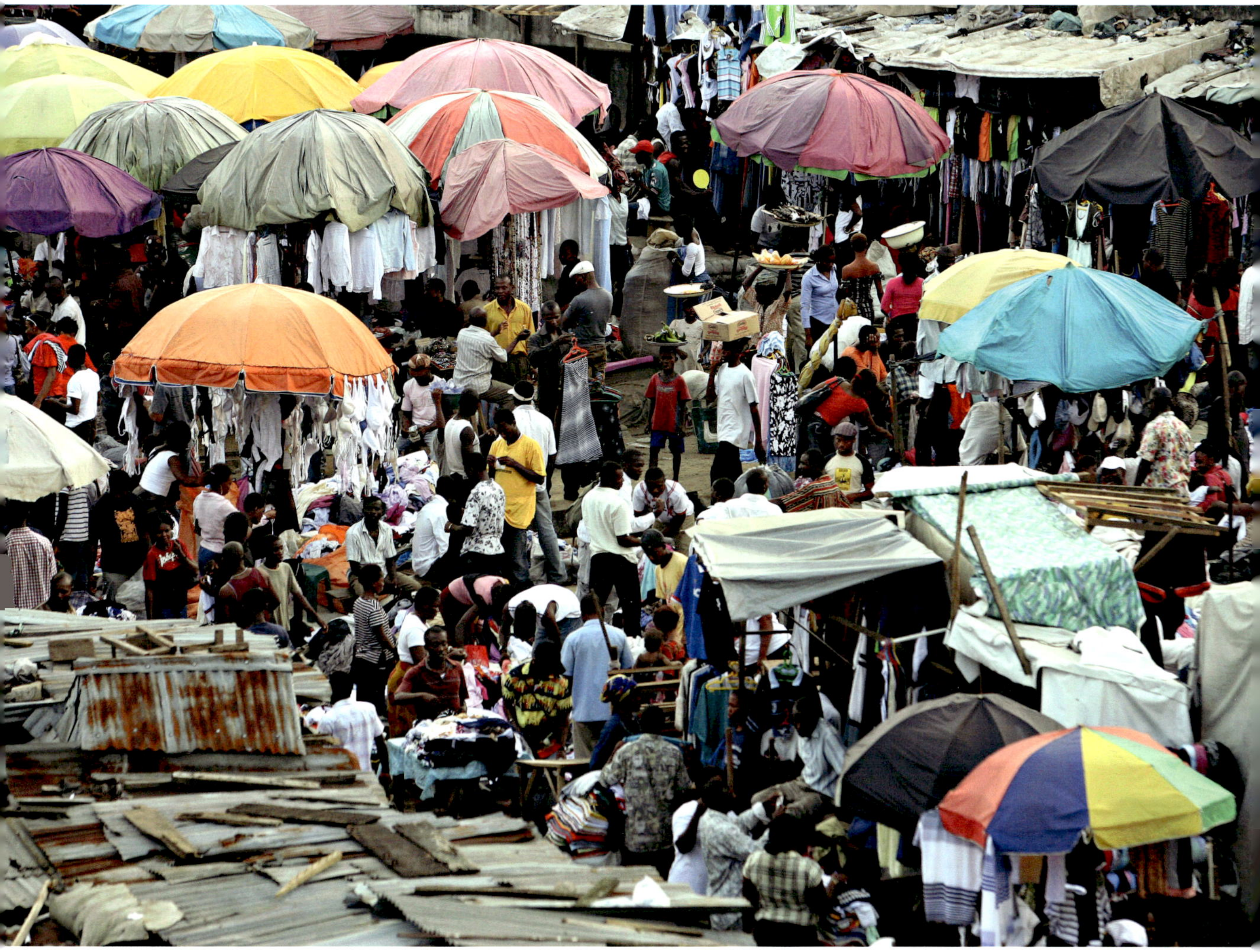

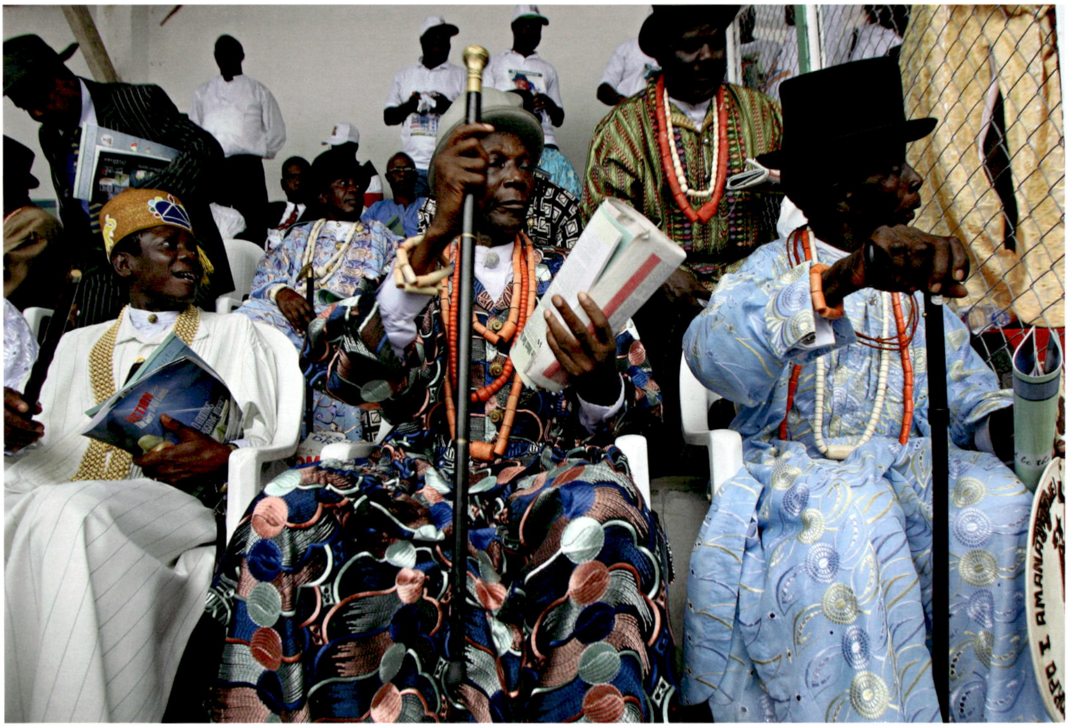

Above: Ijaw chiefs attend a function in Yenagoa.

Right: Ijaw women dance with their symbolic handkerchiefs in Kaiama.

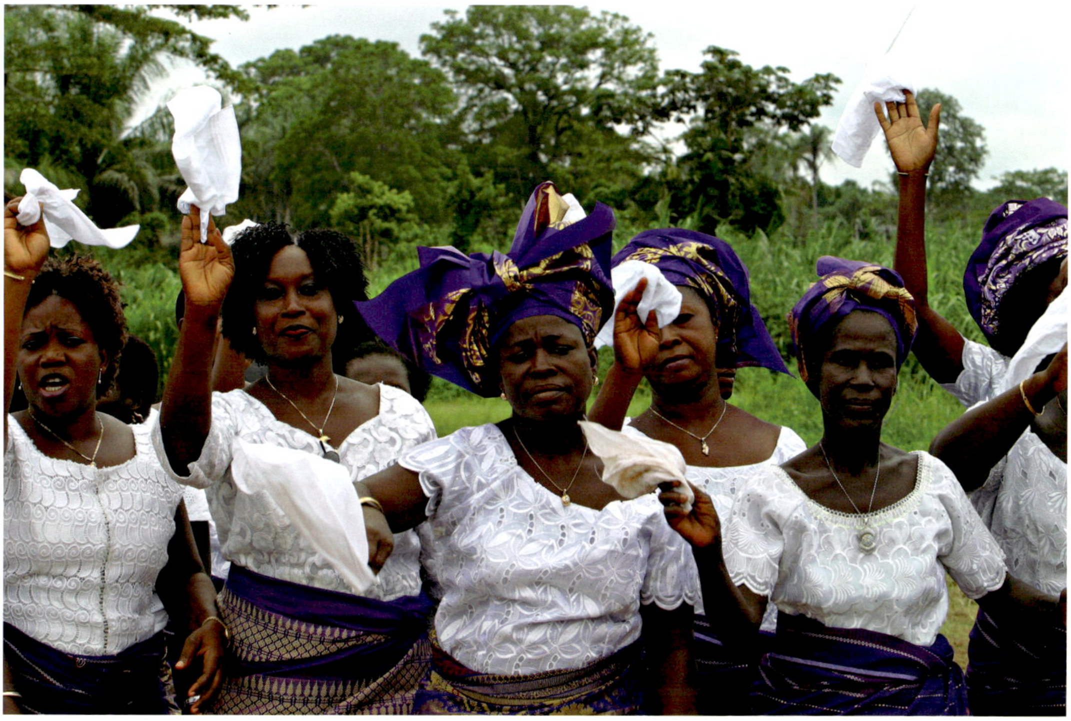

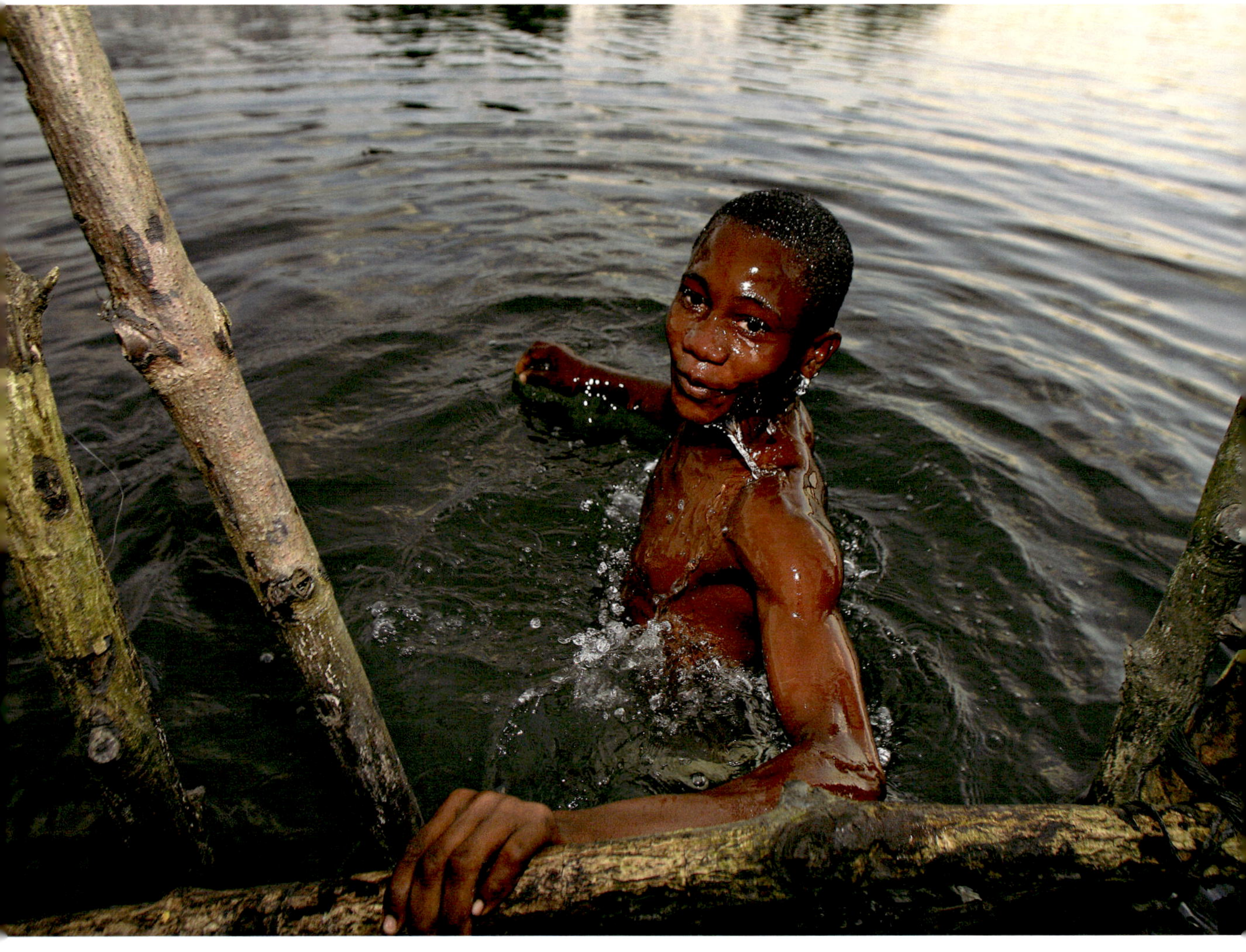

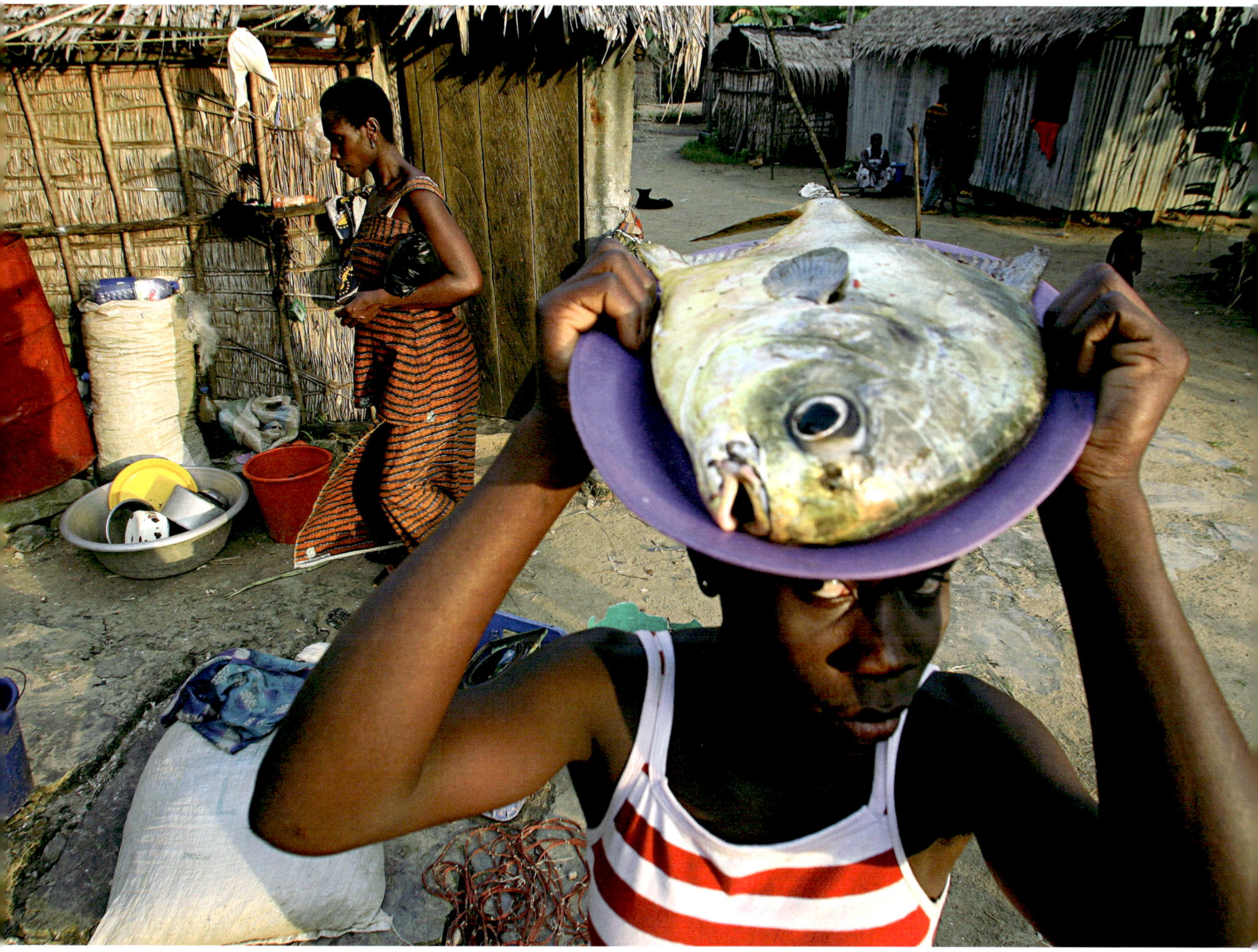

Ogberemedaye Pius, who lost his left leg while fighting for the Nigerian Army during the Nigerian civil war, is left abandoned with no care in Ughelli while his wife Joy, who works in the farm to provide for their welfare, stands outside their house. Delta residents accuse successive Nigerian governments and international oil companies of depriving them of the wealth pumped from their backyards.

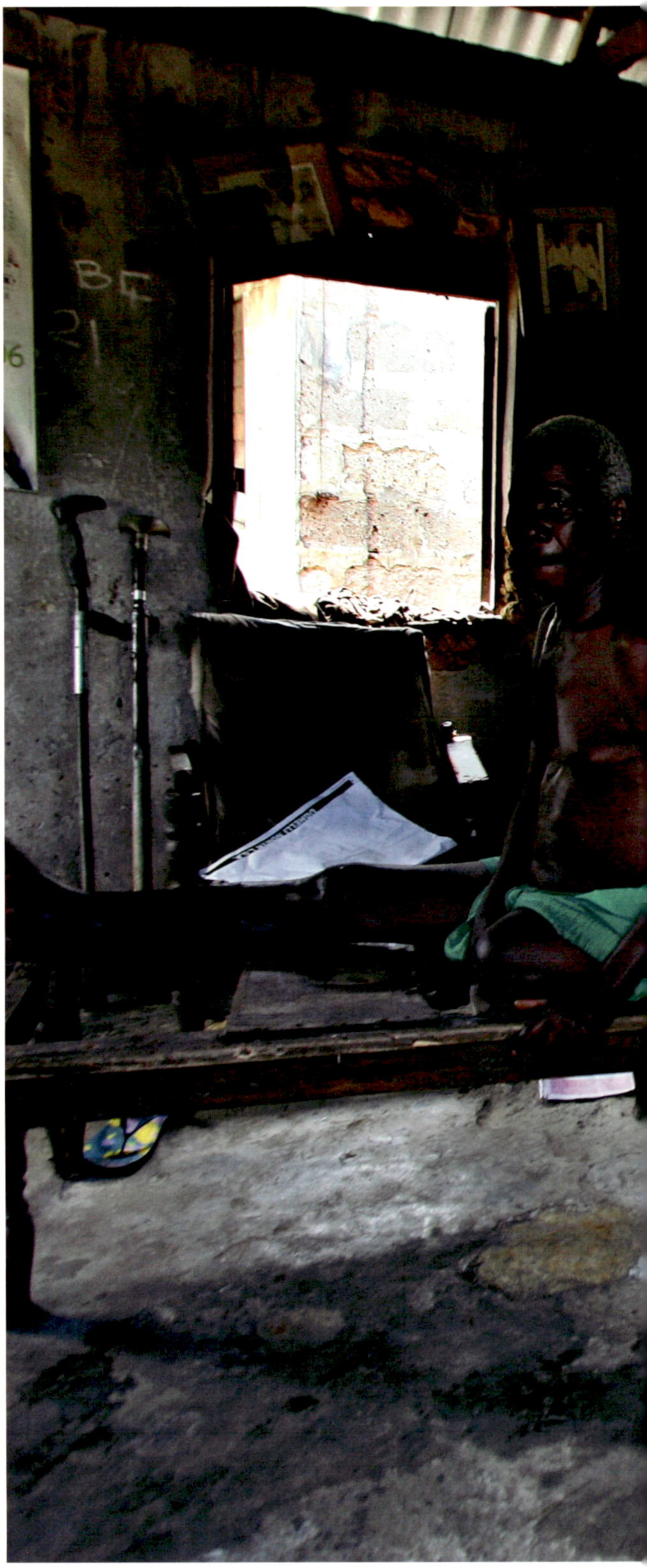

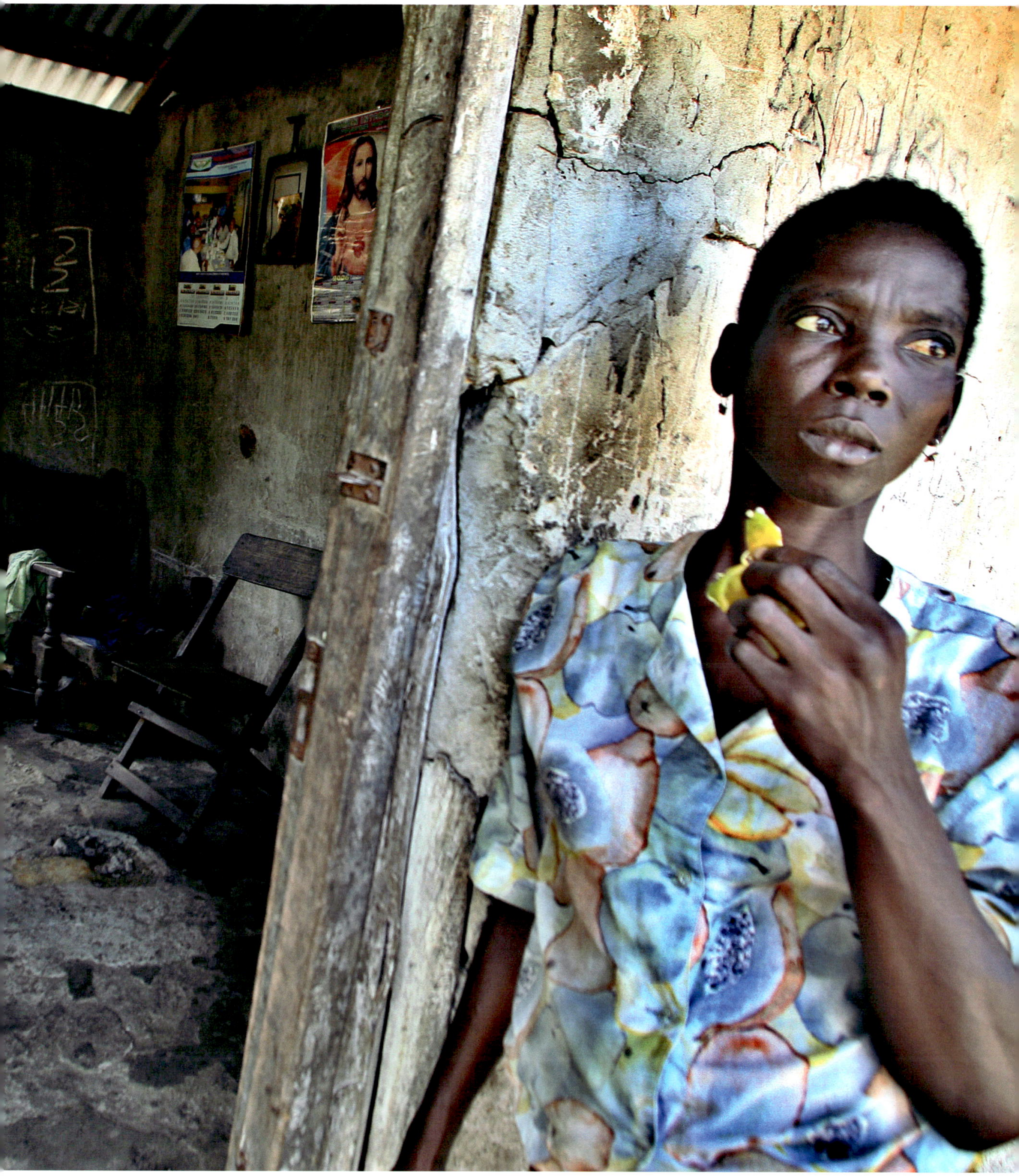

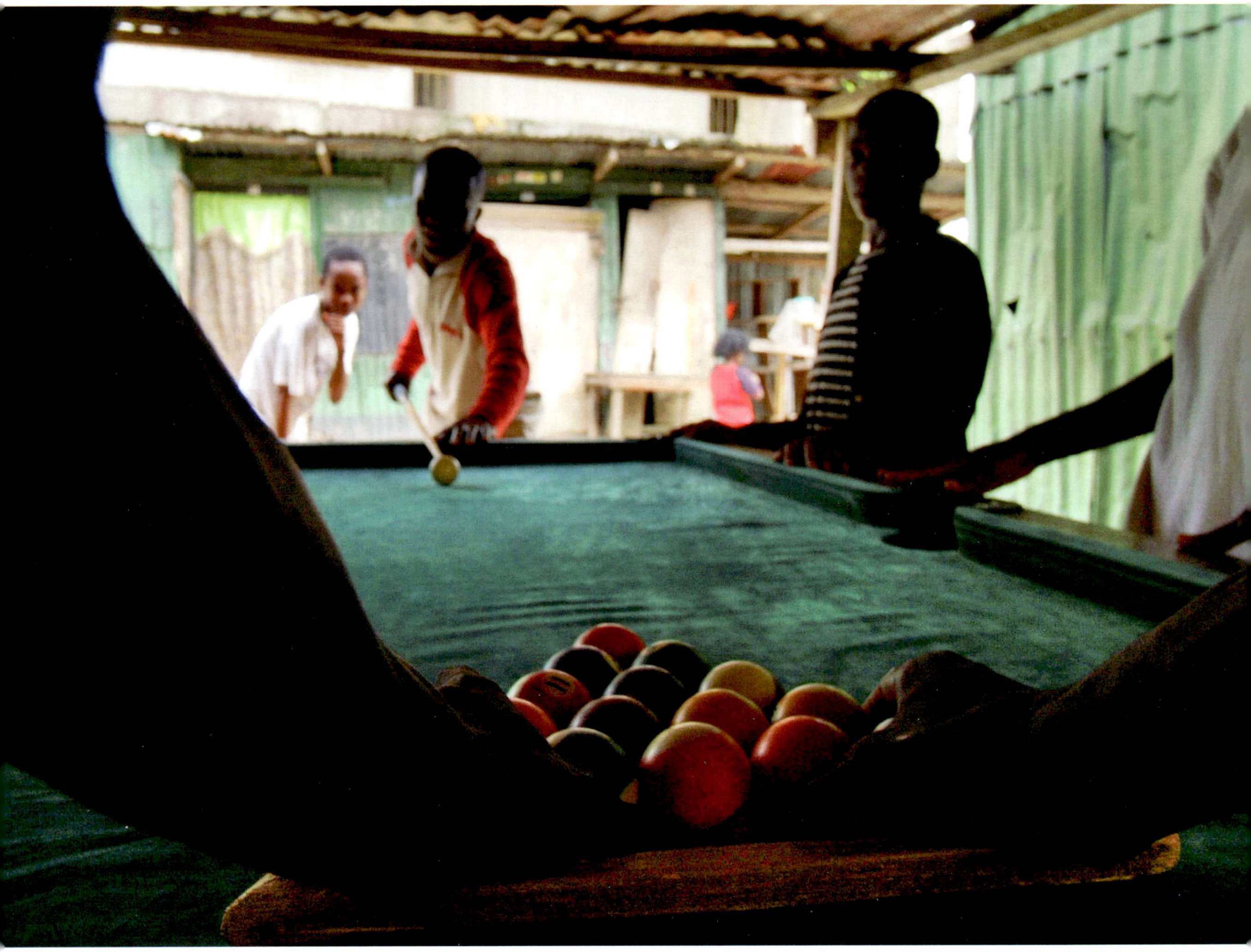

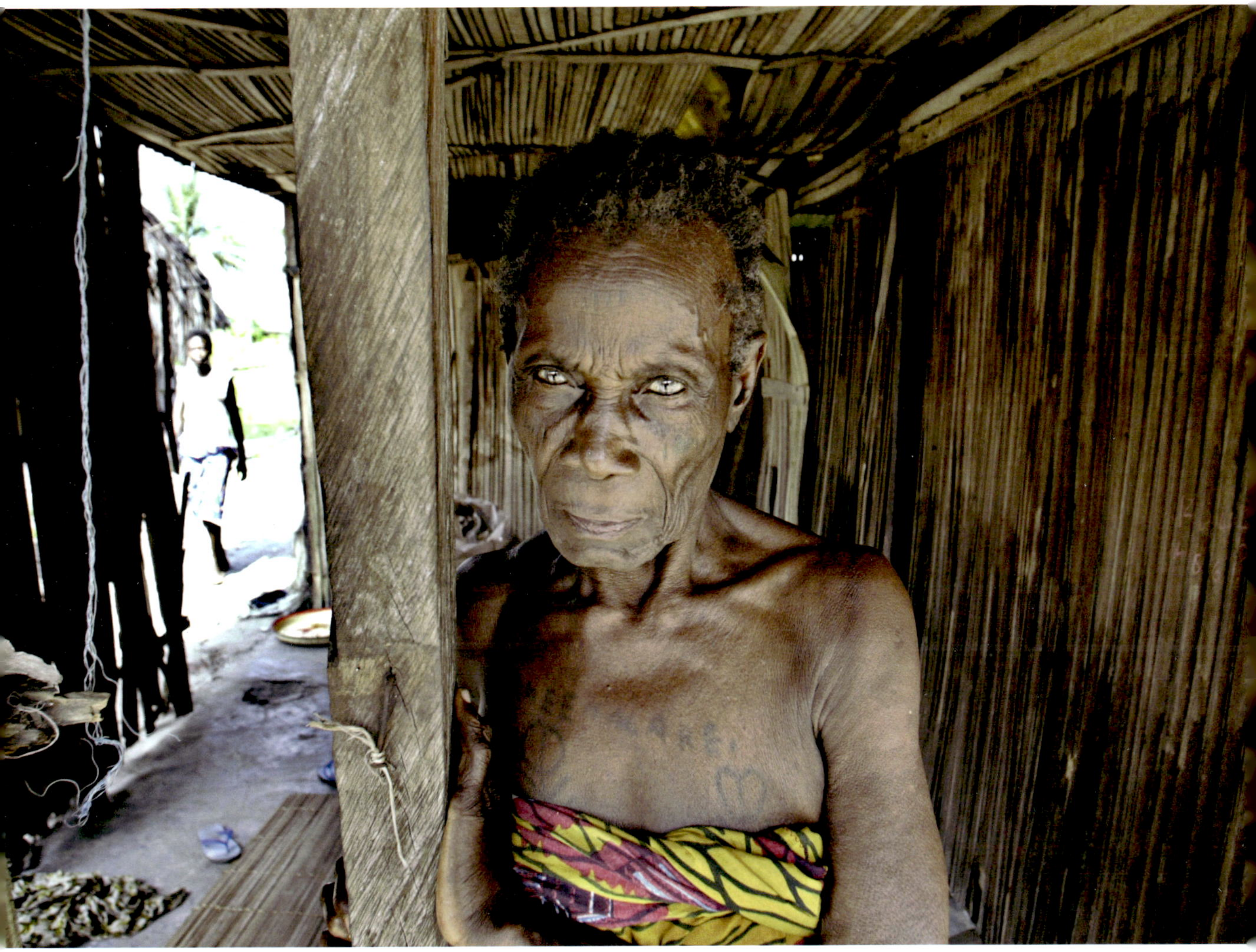

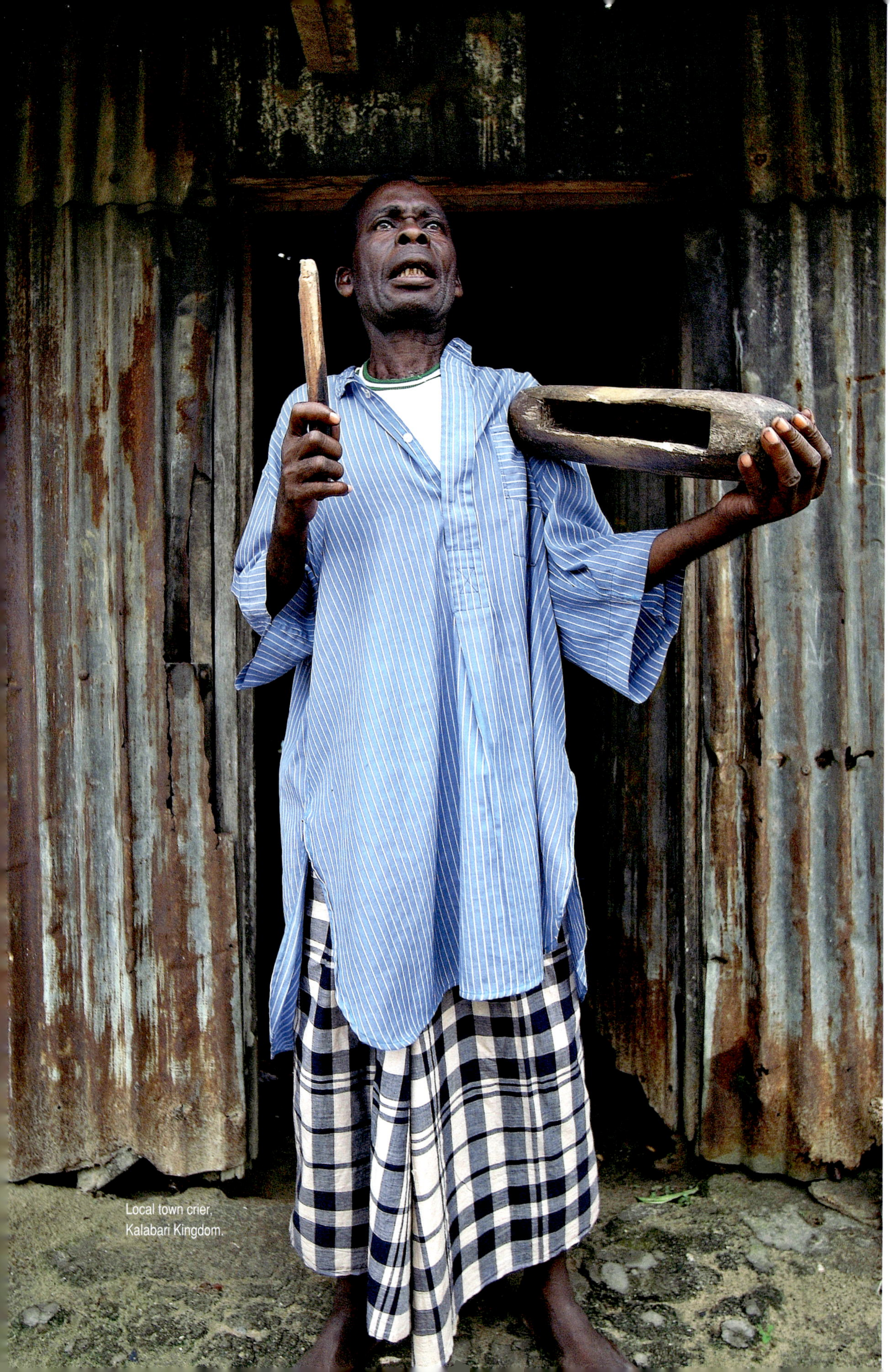

Local town crier,
Kalabari Kingdom.

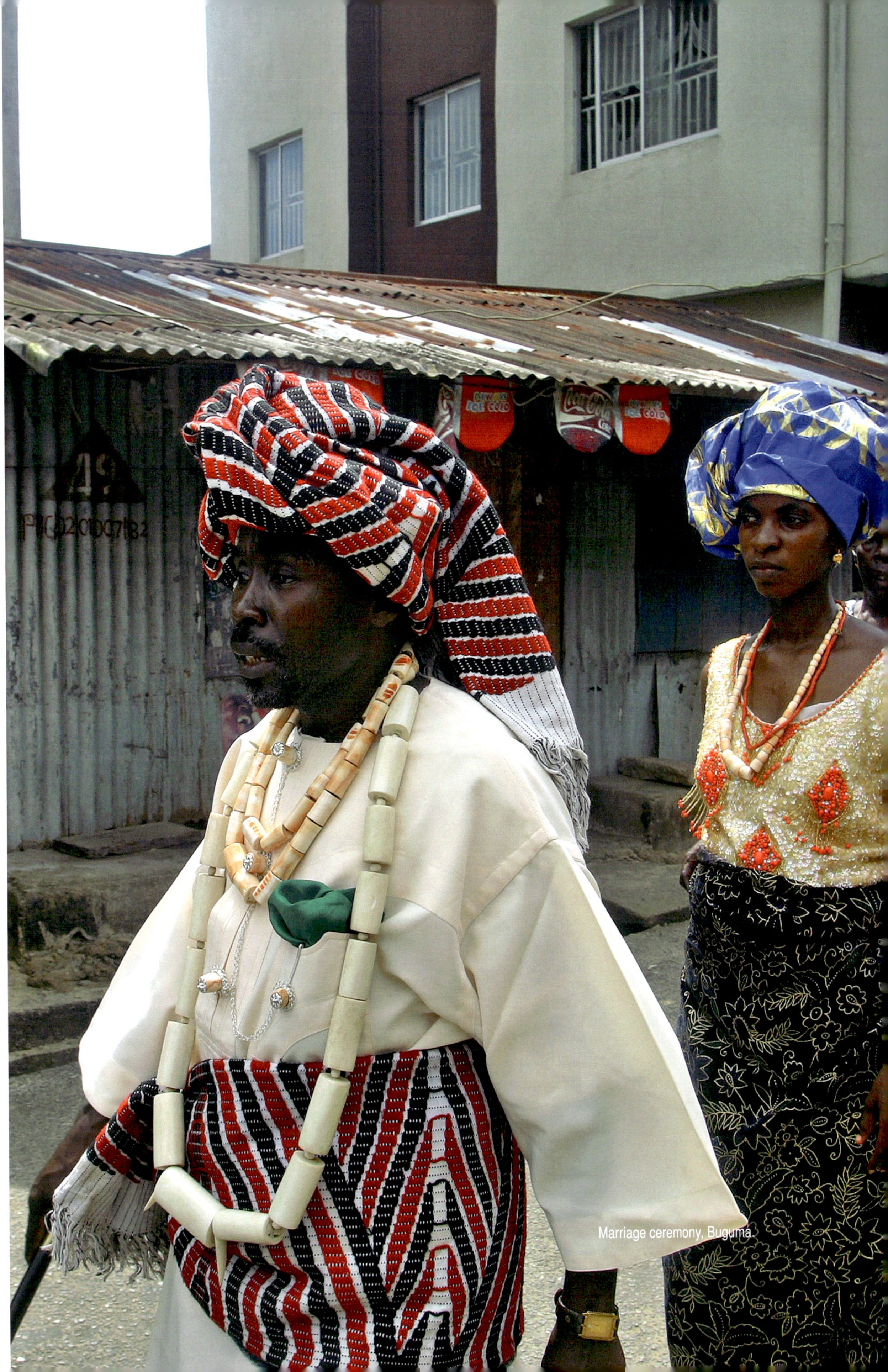

Marriage ceremony, Buguma

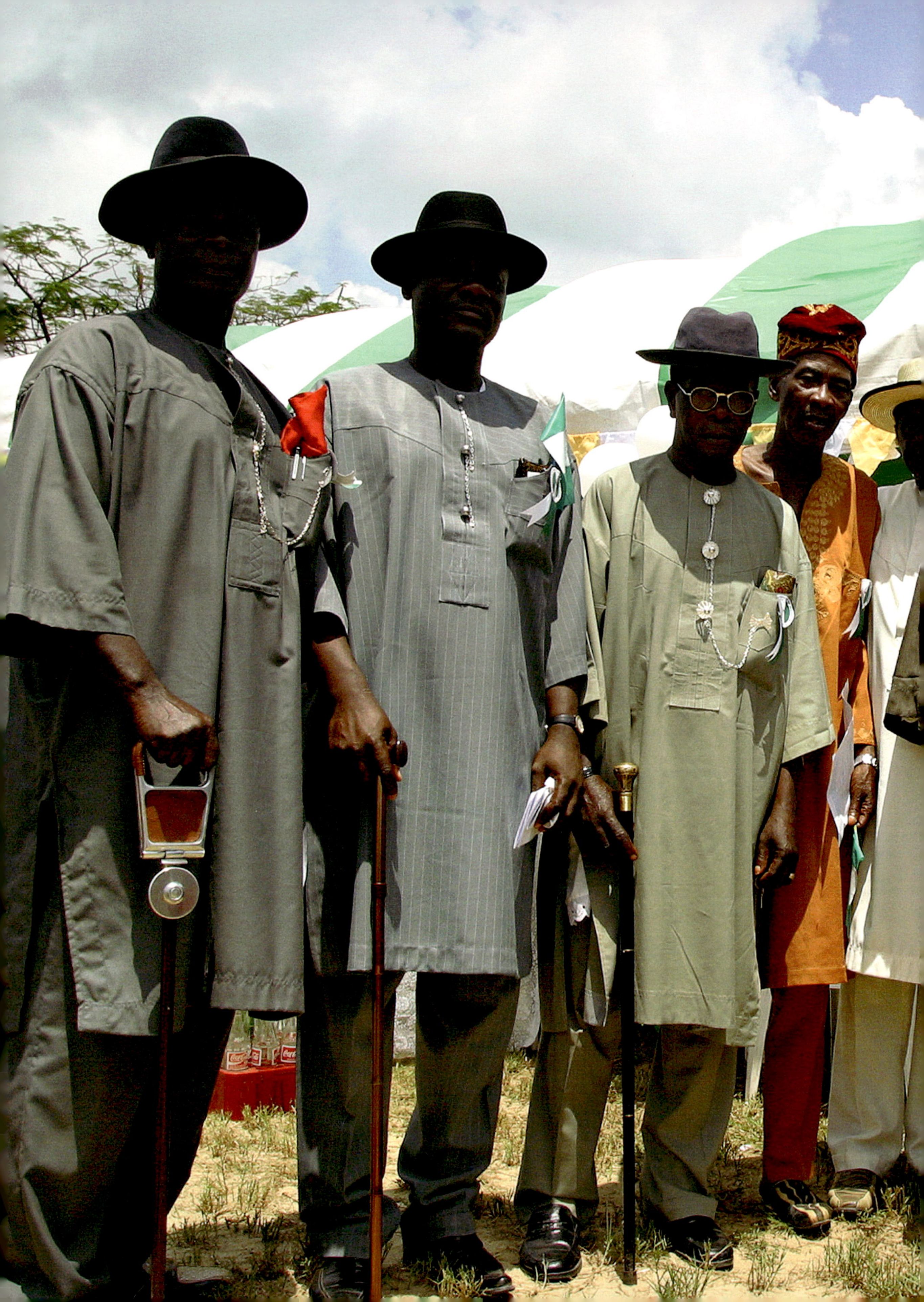

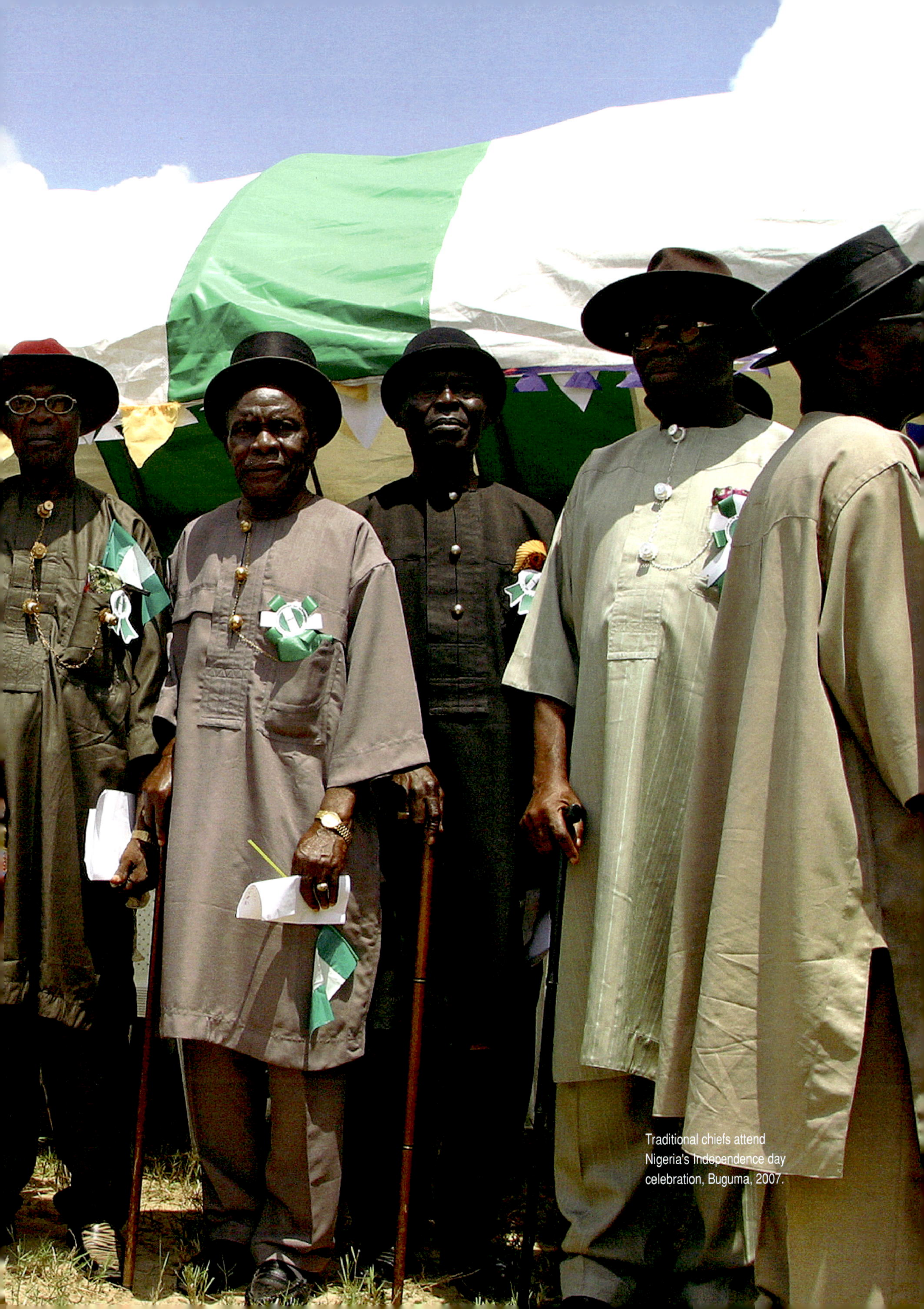

Traditional chiefs attend Nigeria's Independence day celebration, Buguma, 2007.

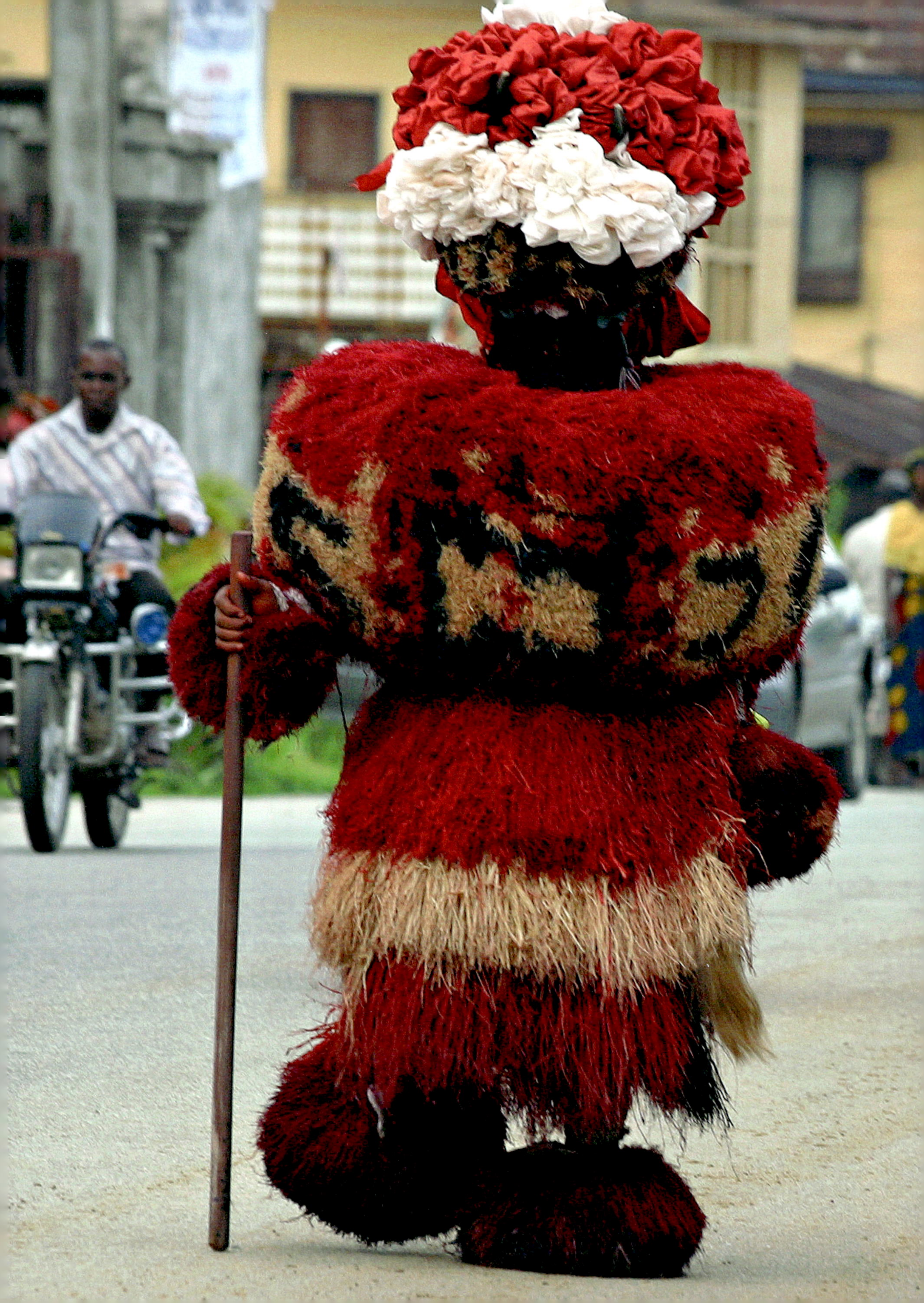

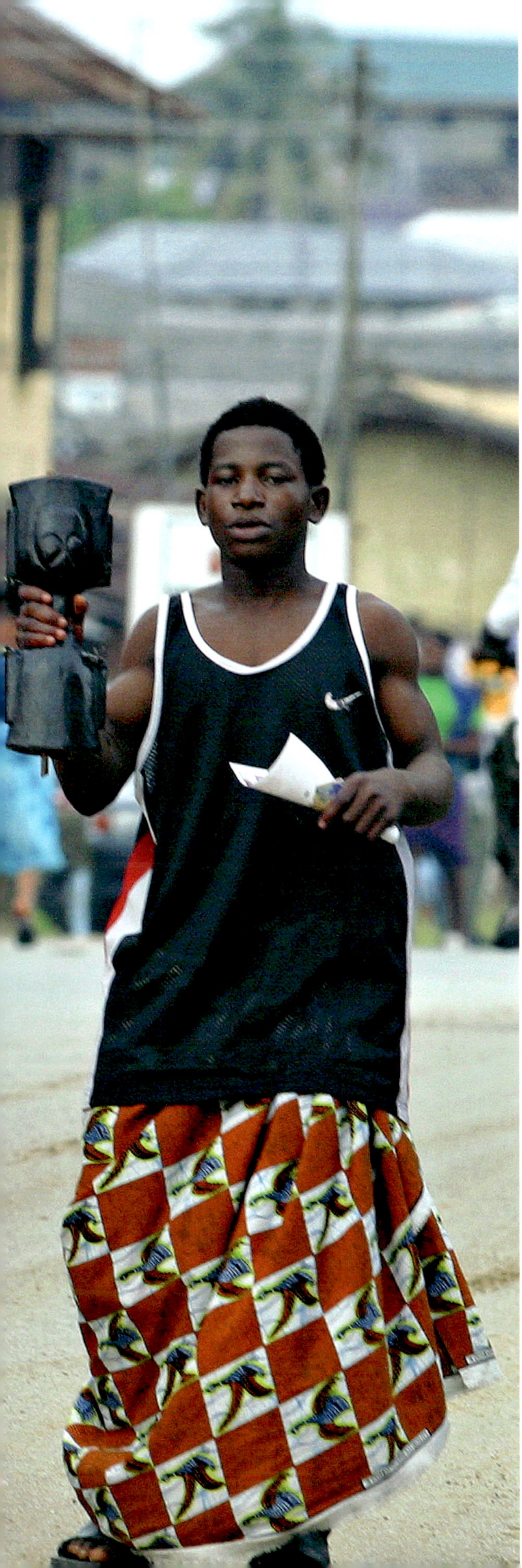

Ekpe masquerader from Calabar, Cross River State, South East Nigeria. Ekpe is a secret society whose beliefs were based on the invocation of the leopard spirit.

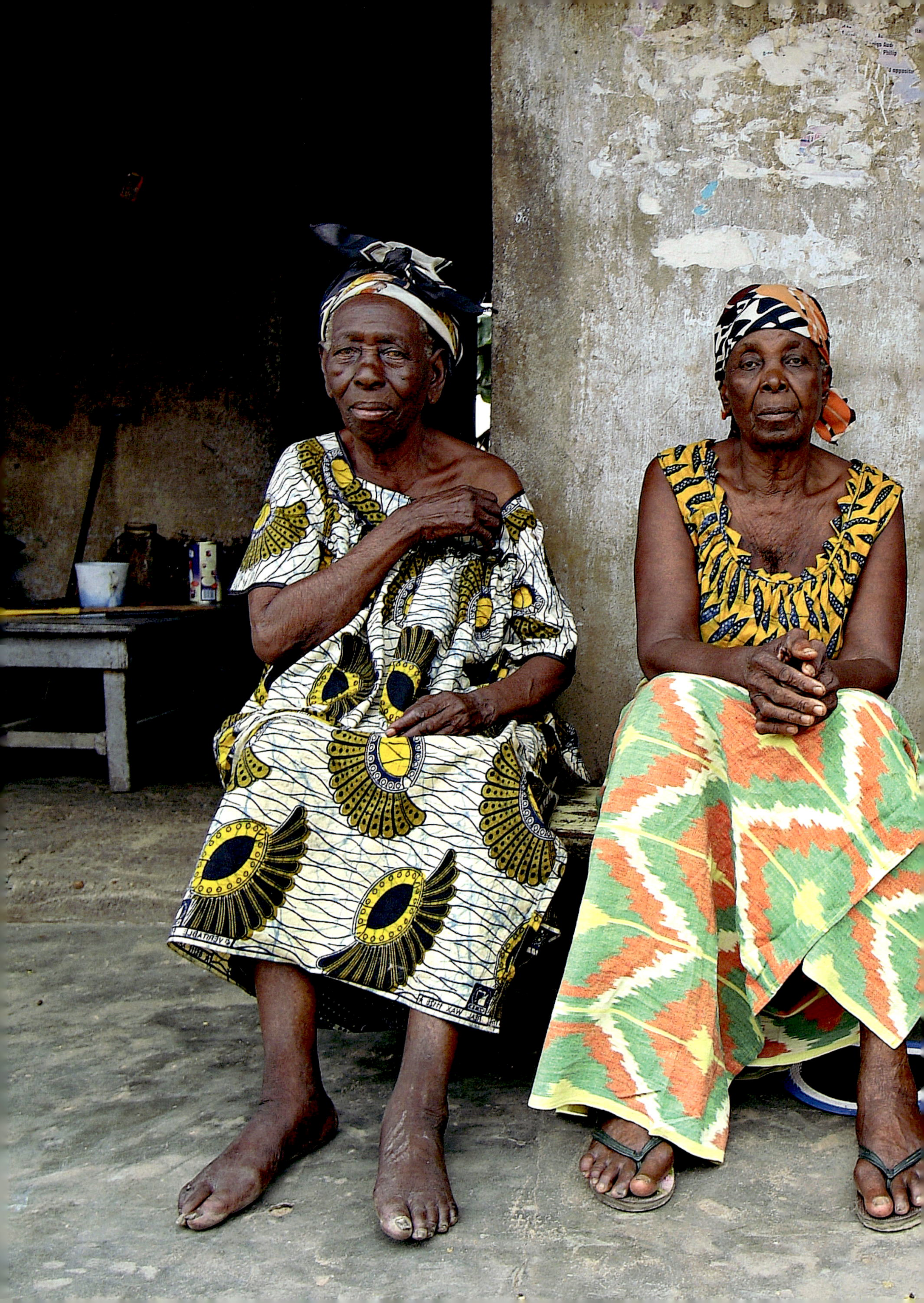

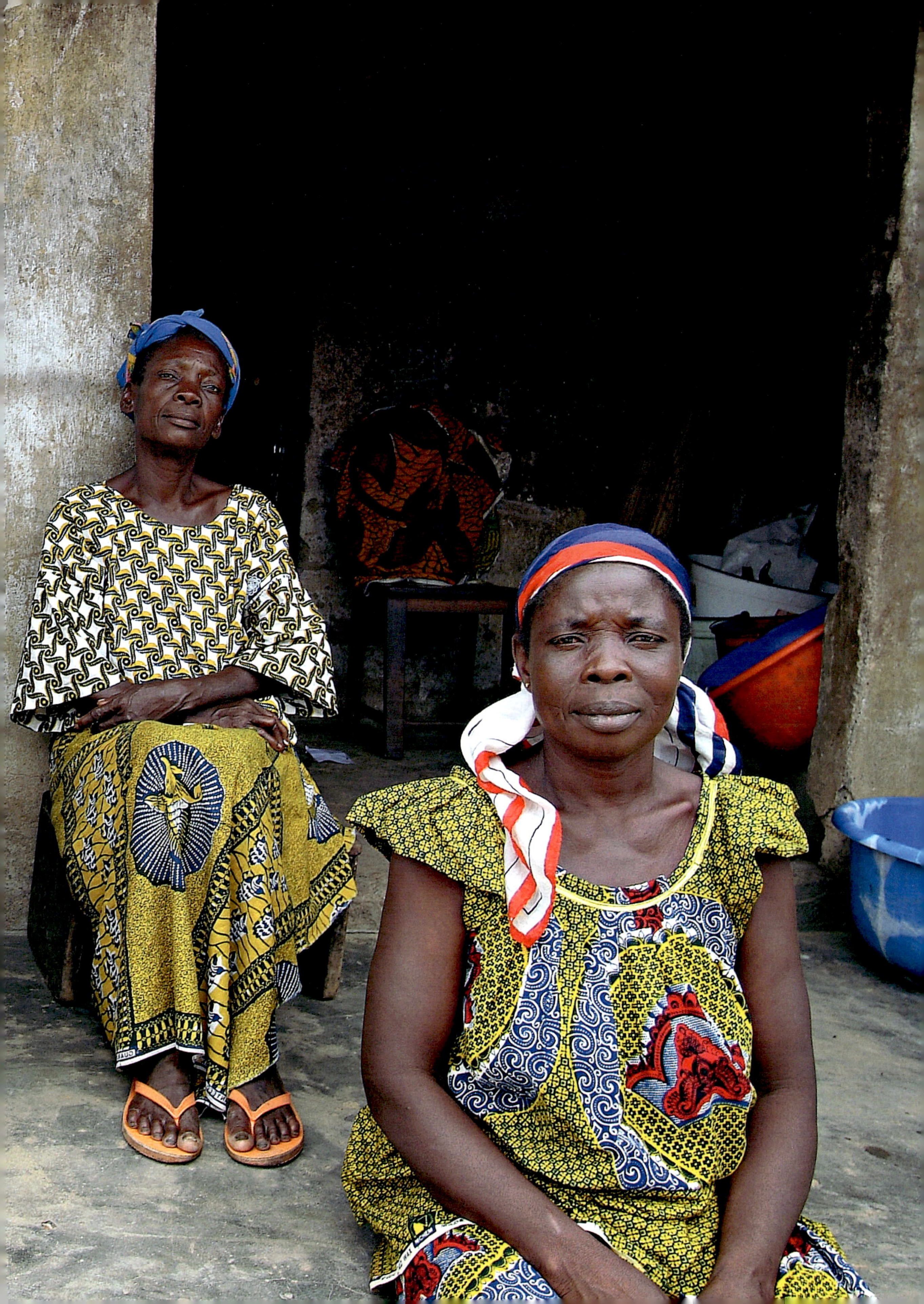

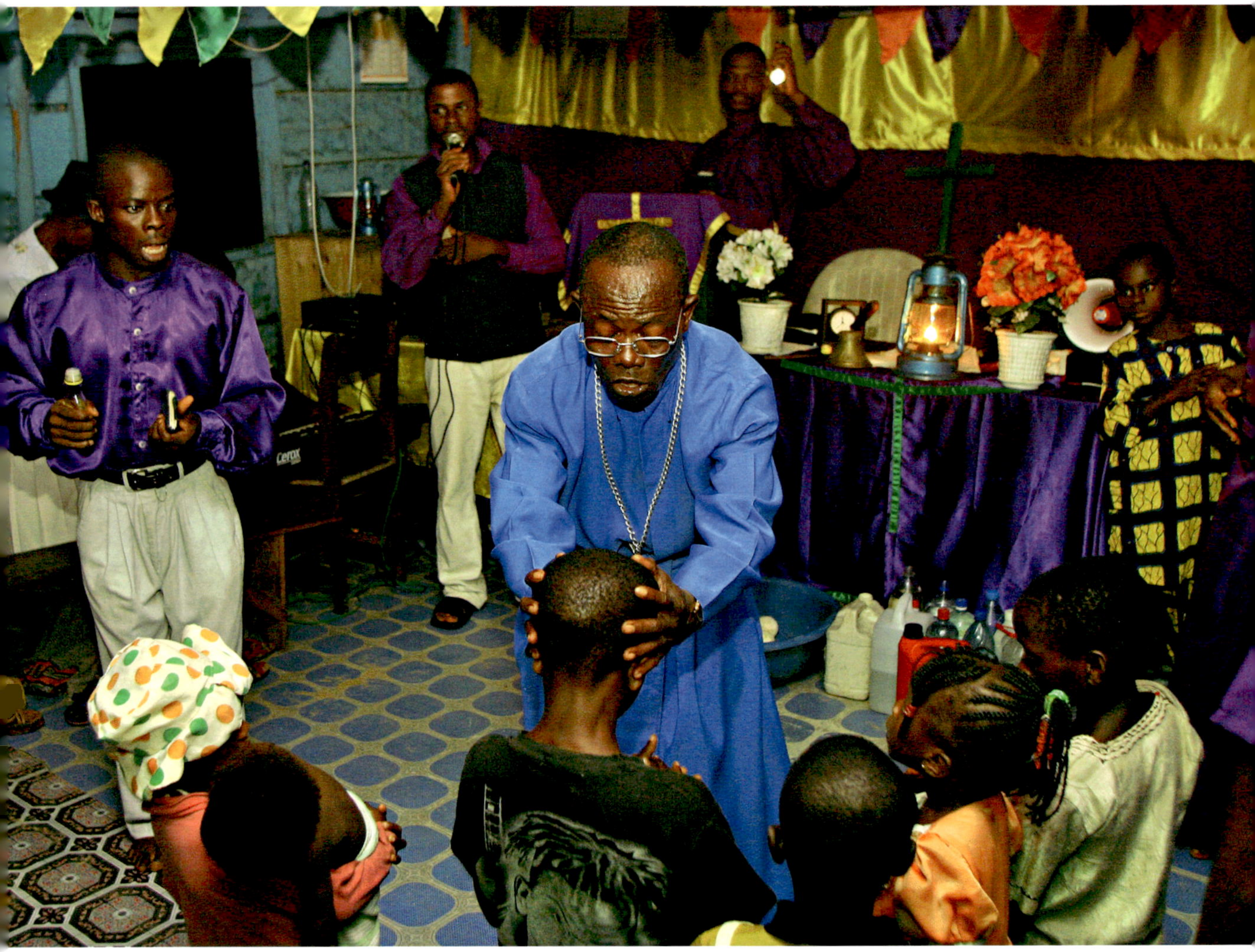

Above and right: Night vigil service, Finima.

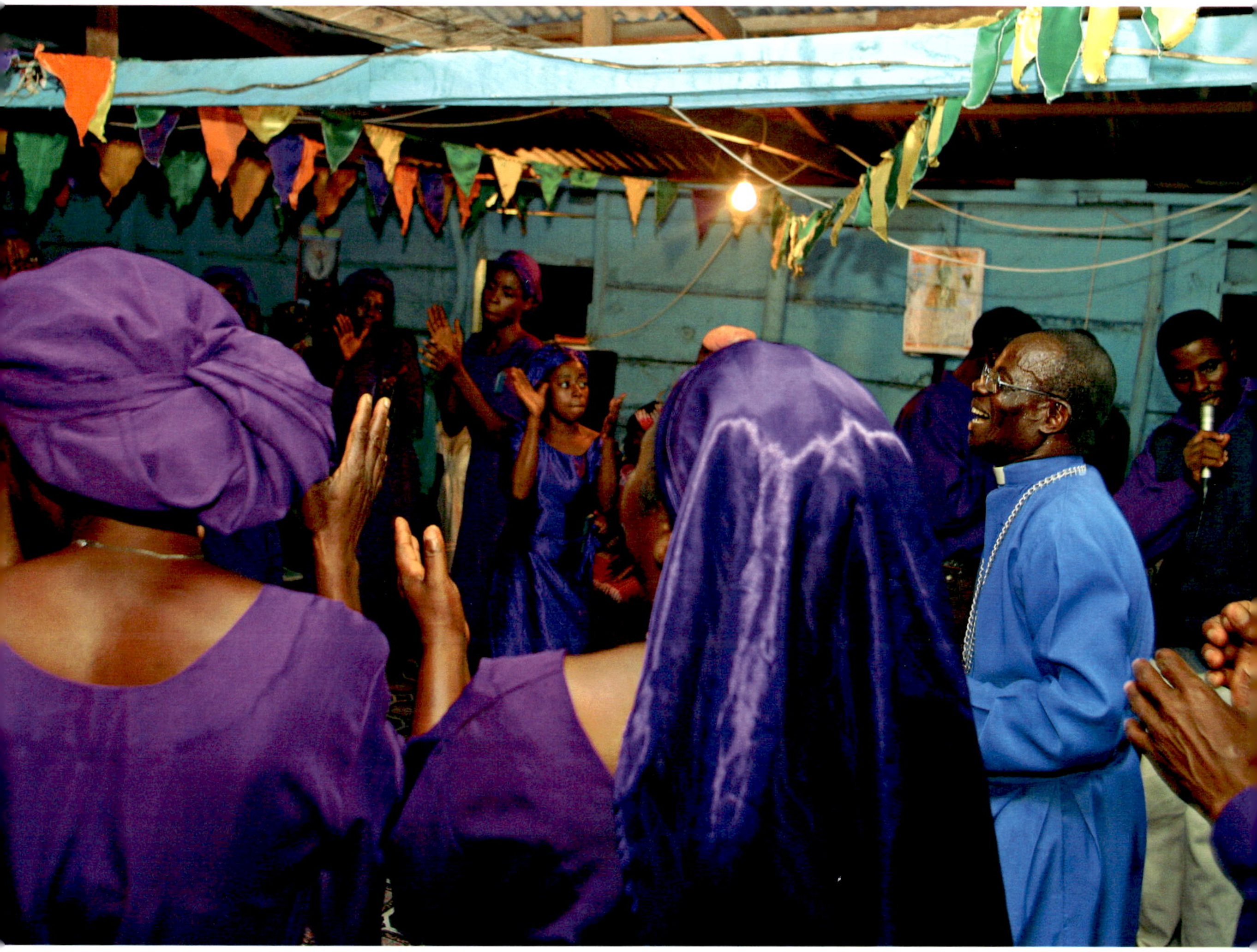

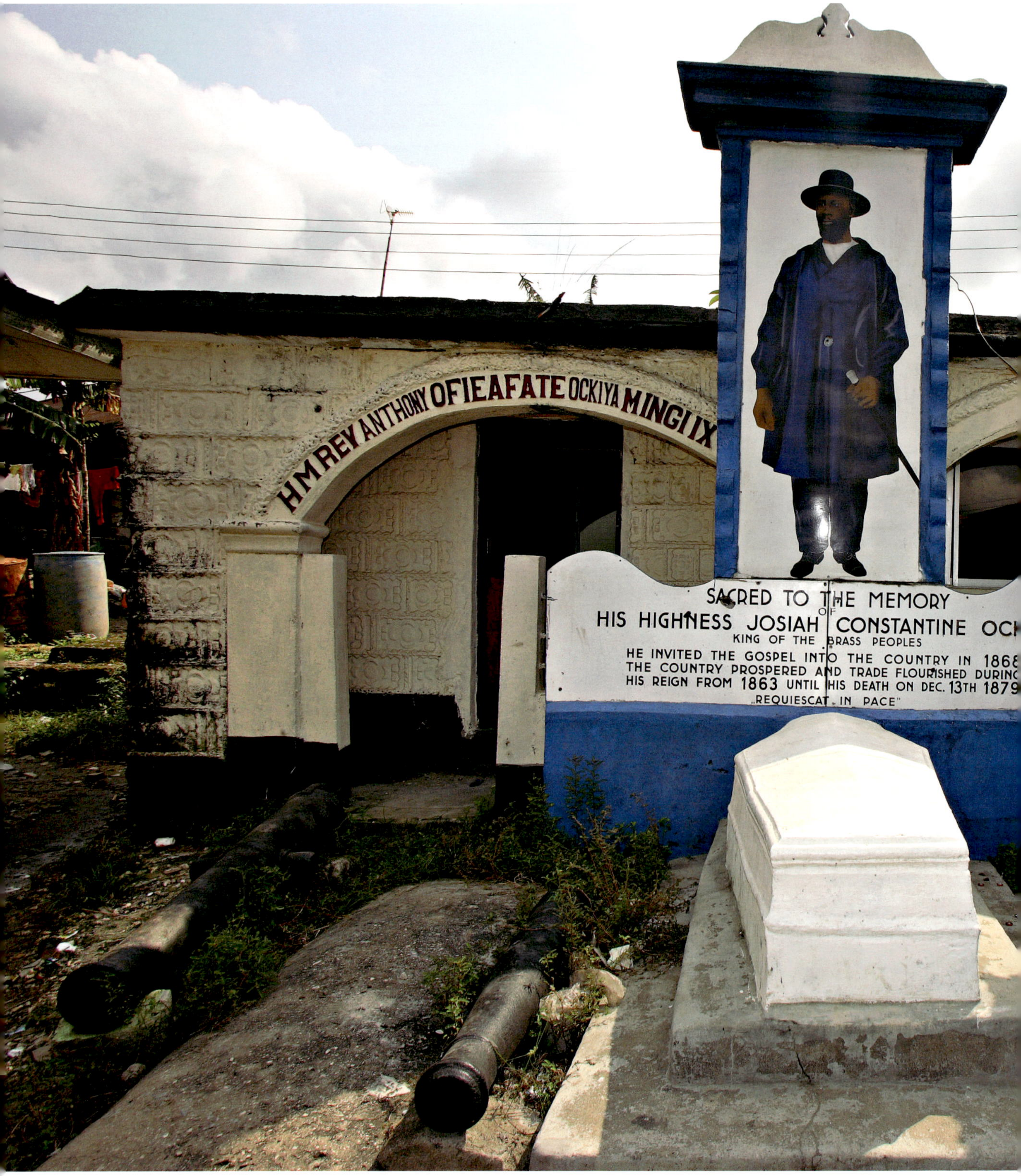
HM REV ANTHONY OFIEAFATE OCKIYA MINGI IX
SACRED TO THE MEMORY
OF
HIS HIGHNESS JOSIAH CONSTANTINE OC
KING OF THE BRASS PEOPLES
HE INVITED THE GOSPEL INTO THE COUNTRY IN 186
THE COUNTRY PROSPERED AND TRADE FLOURISHED DURIN
HIS REIGN FROM 1863 UNTIL HIS DEATH ON DEC. 13TH 187
"REQUIESCAT IN PACE"

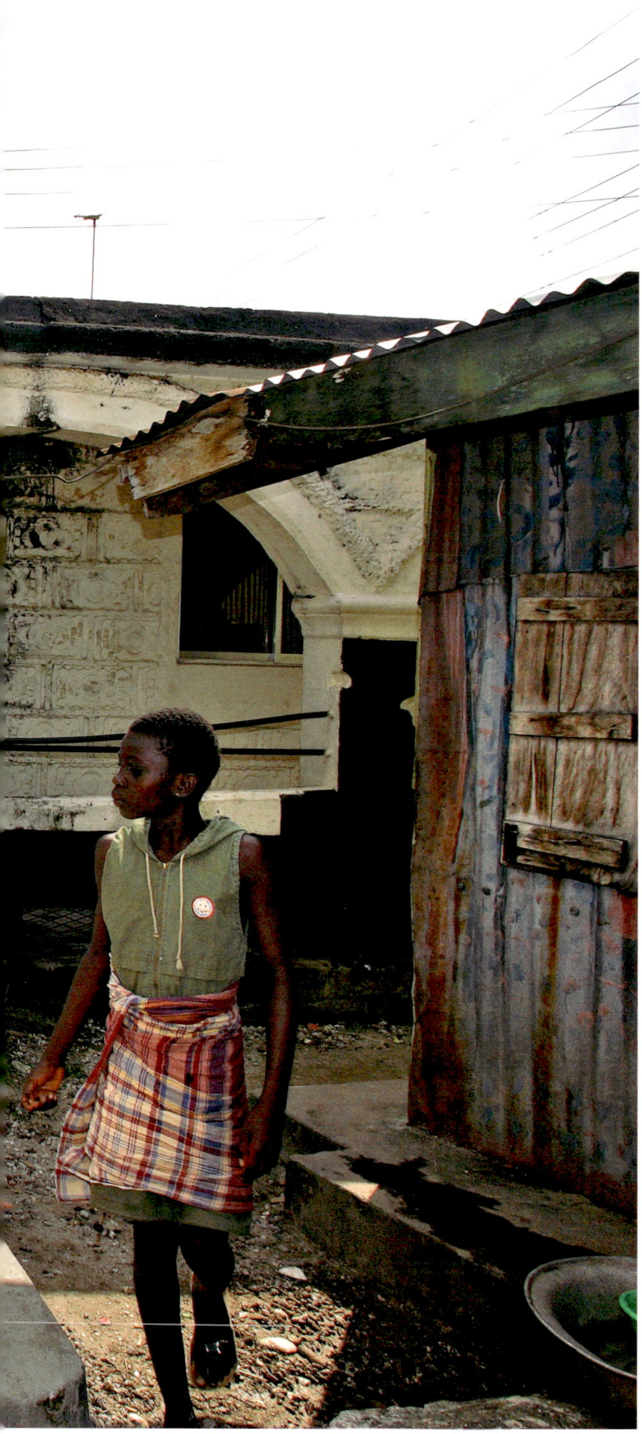

Tomb of King Josiah Constantine Ockiya, Mingi VII, Nembe, Bayelsa state, 2007.

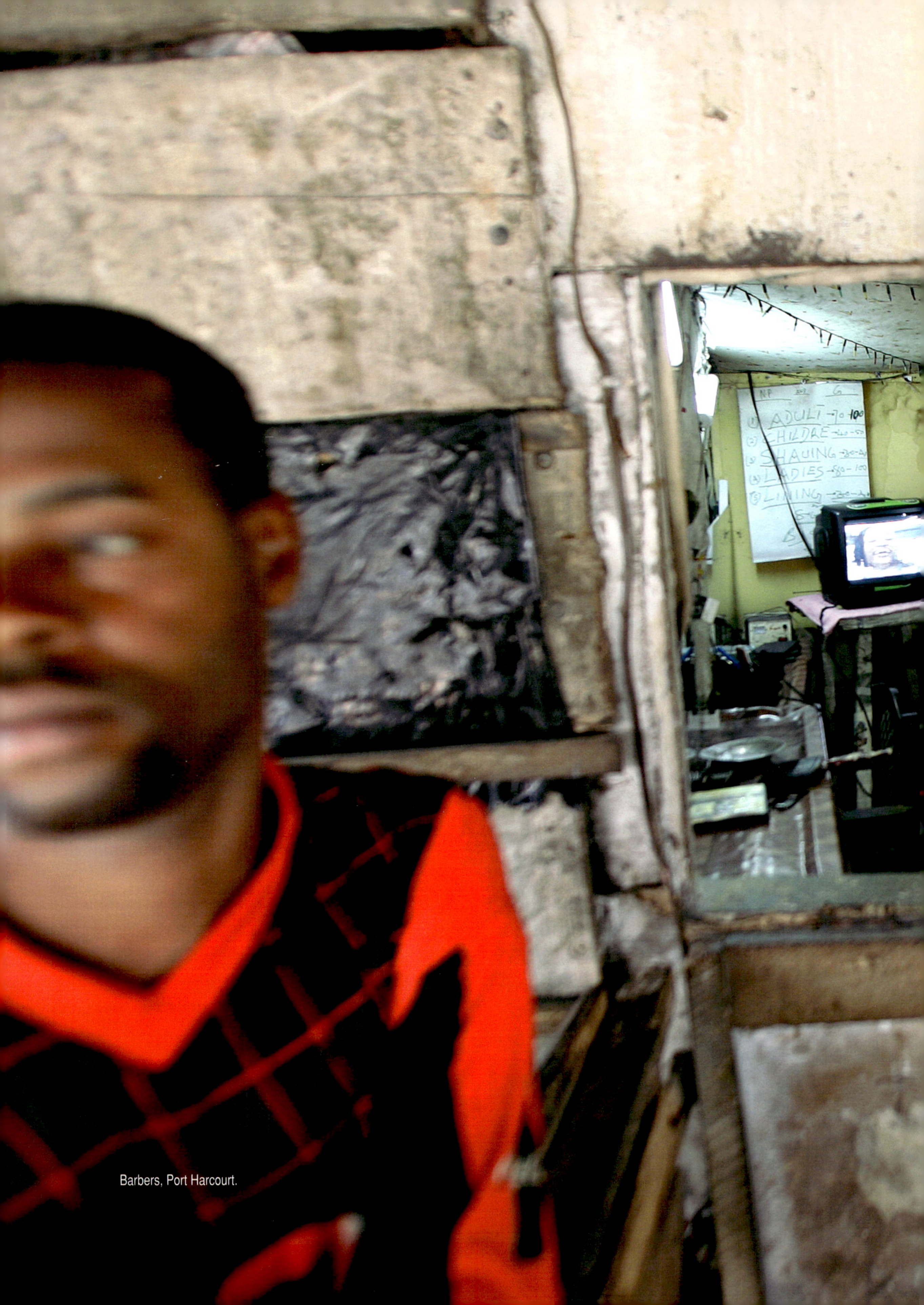

Barbers, Port Harcourt.

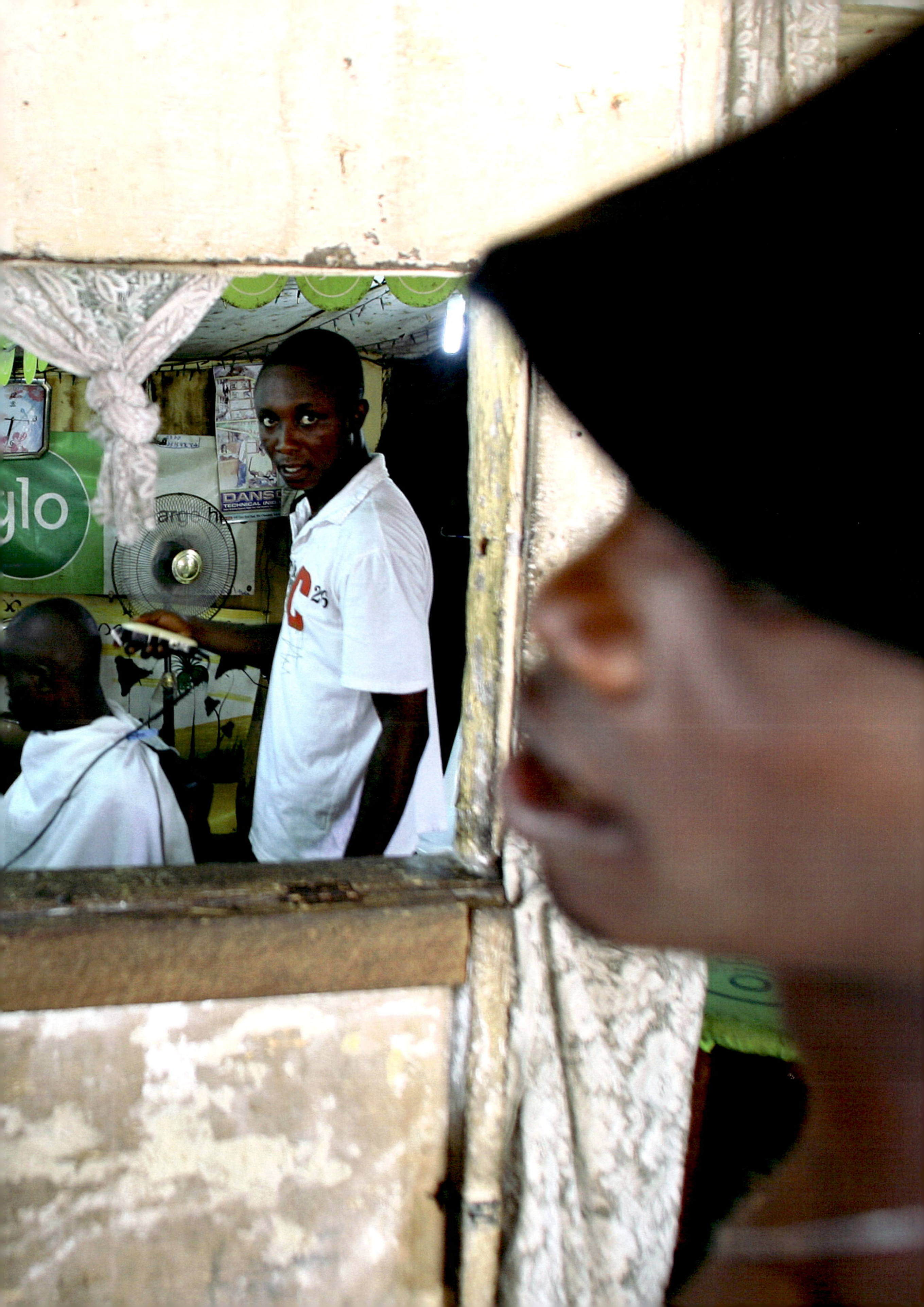
glo

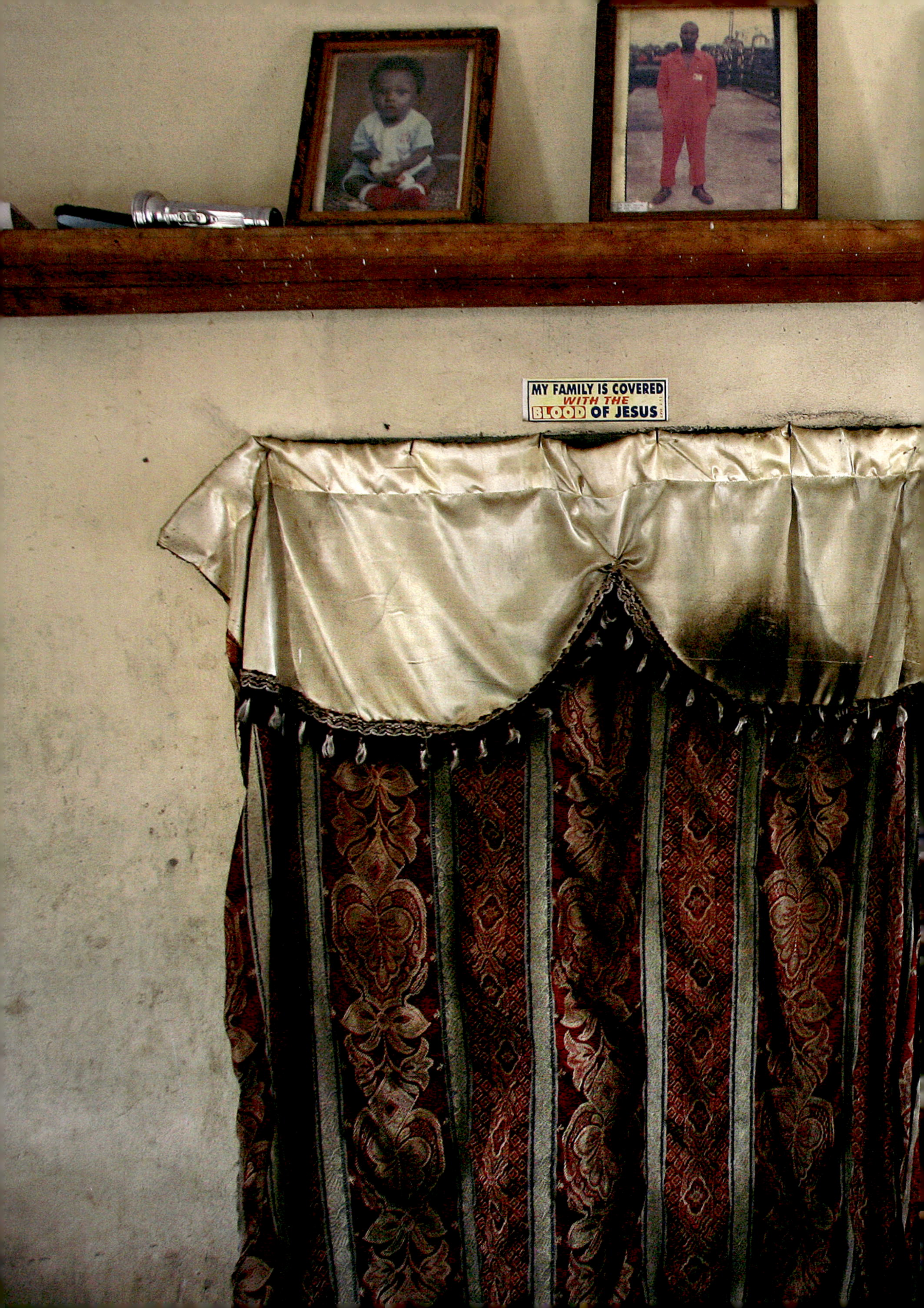
MY FAMILY IS COVERED
WITH THE
BLOOD OF JESUS

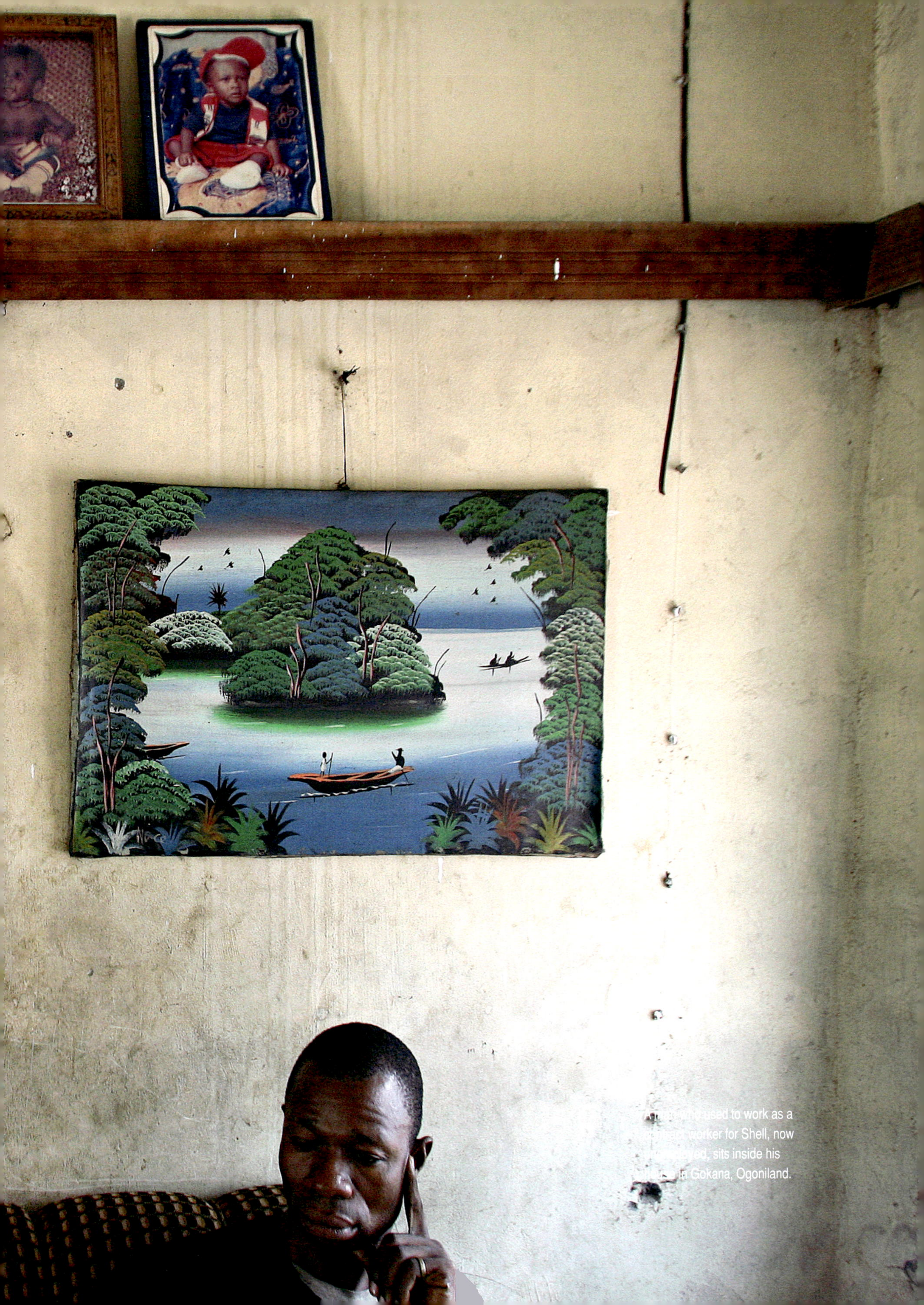

A man who used to work as a [illegible] worker for Shell, now unemployed, sits inside his [illegible] in Gokana, Ogoniland.

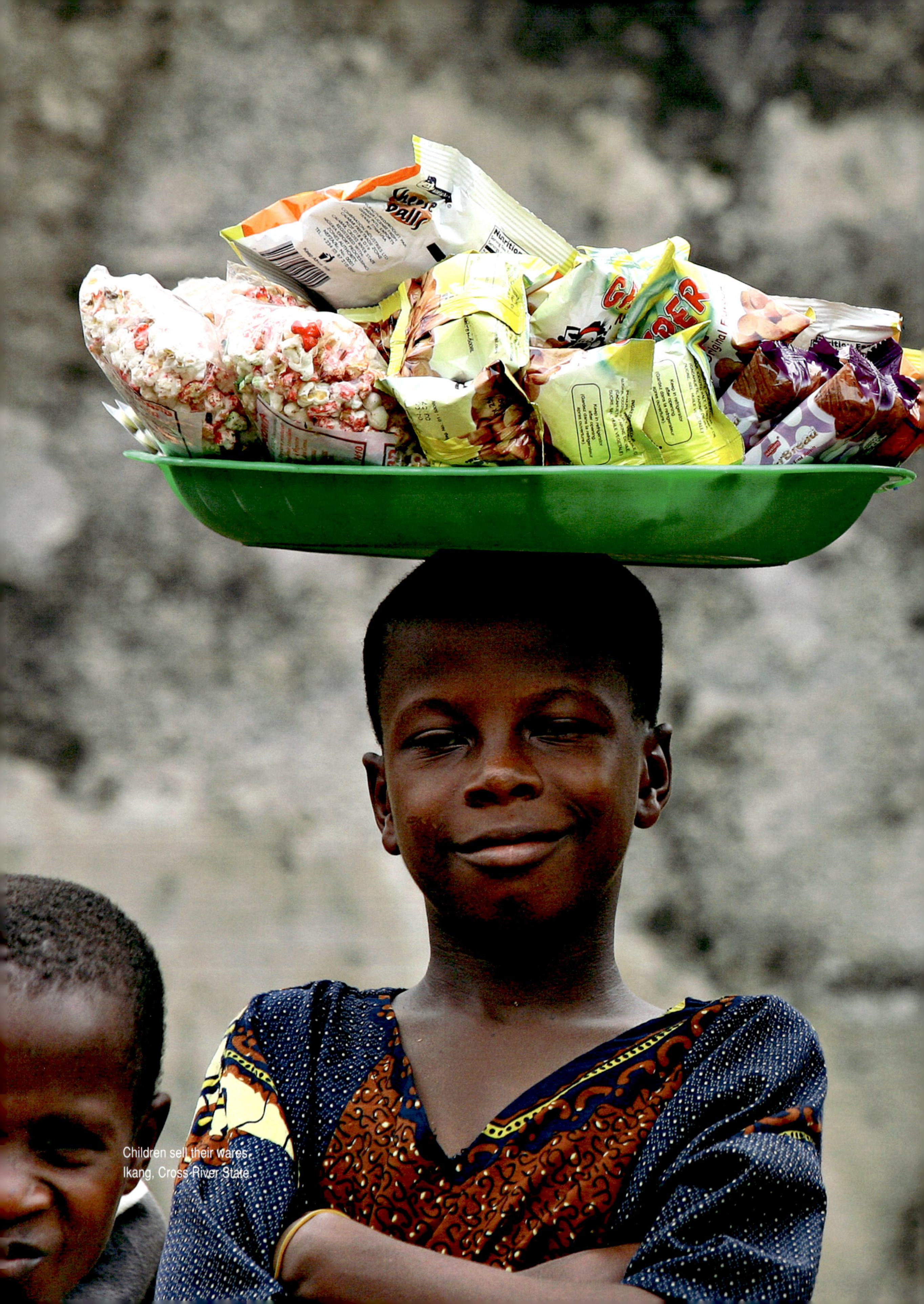

Children sell their wares,
Ikang, Cross River State

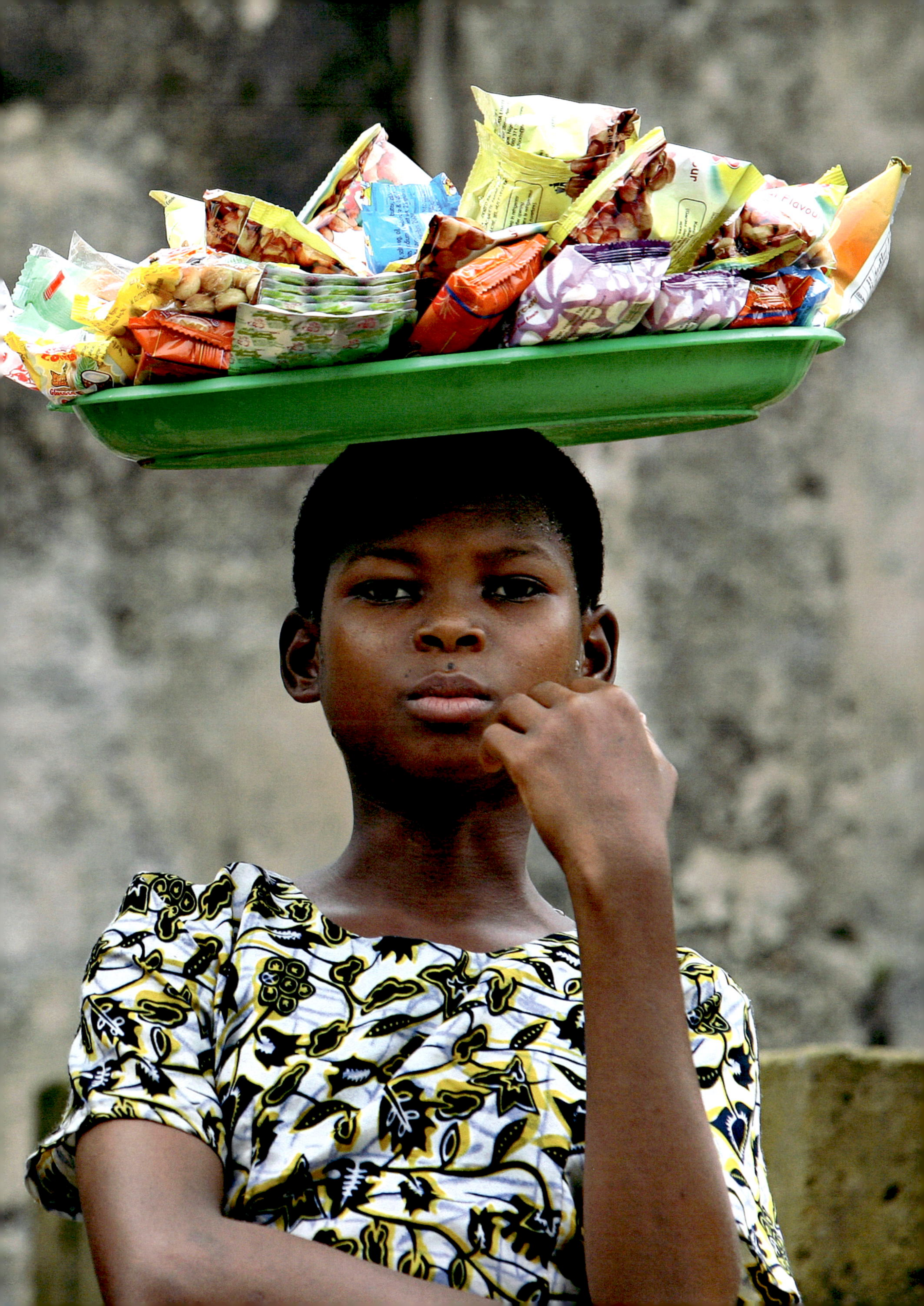

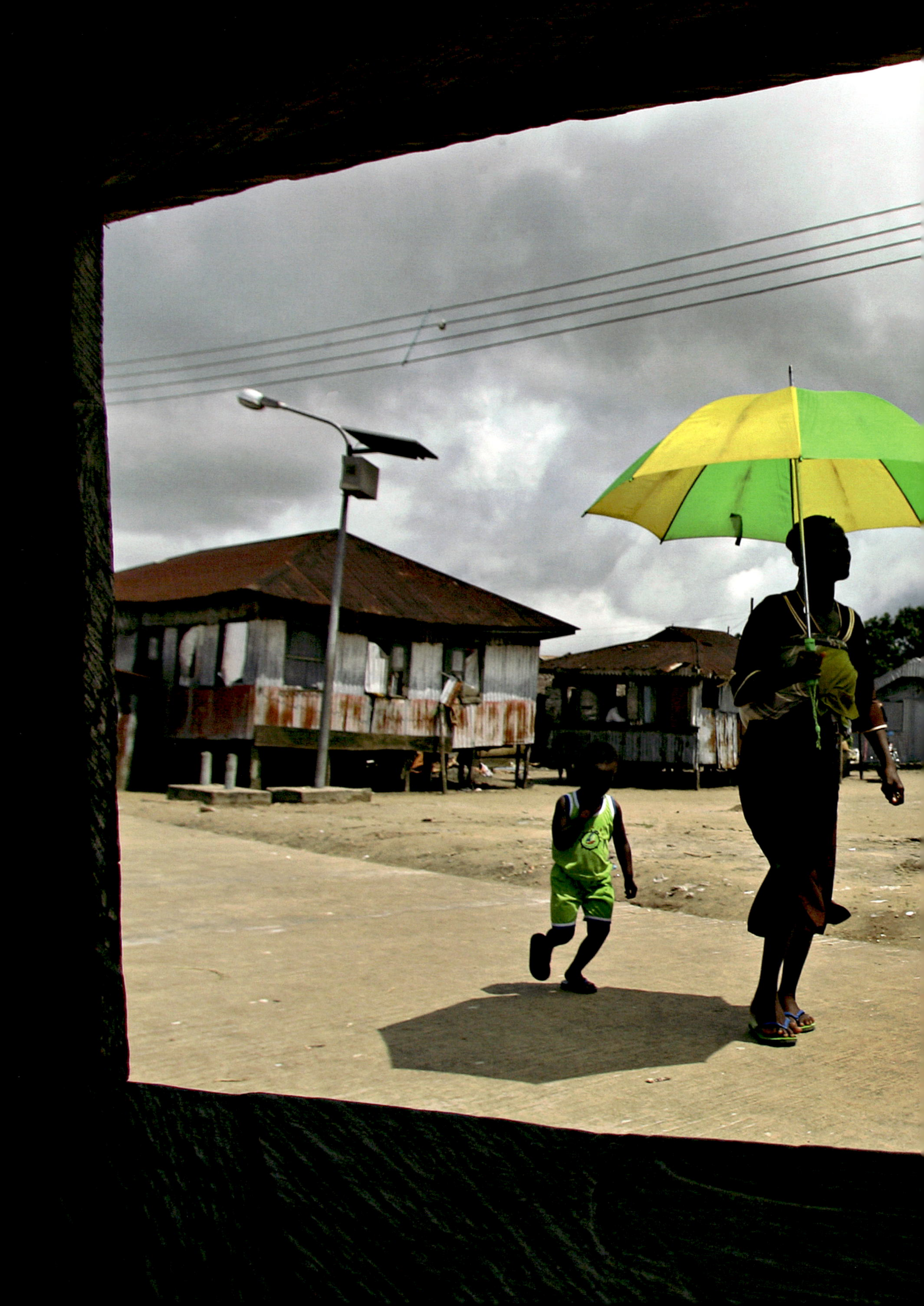

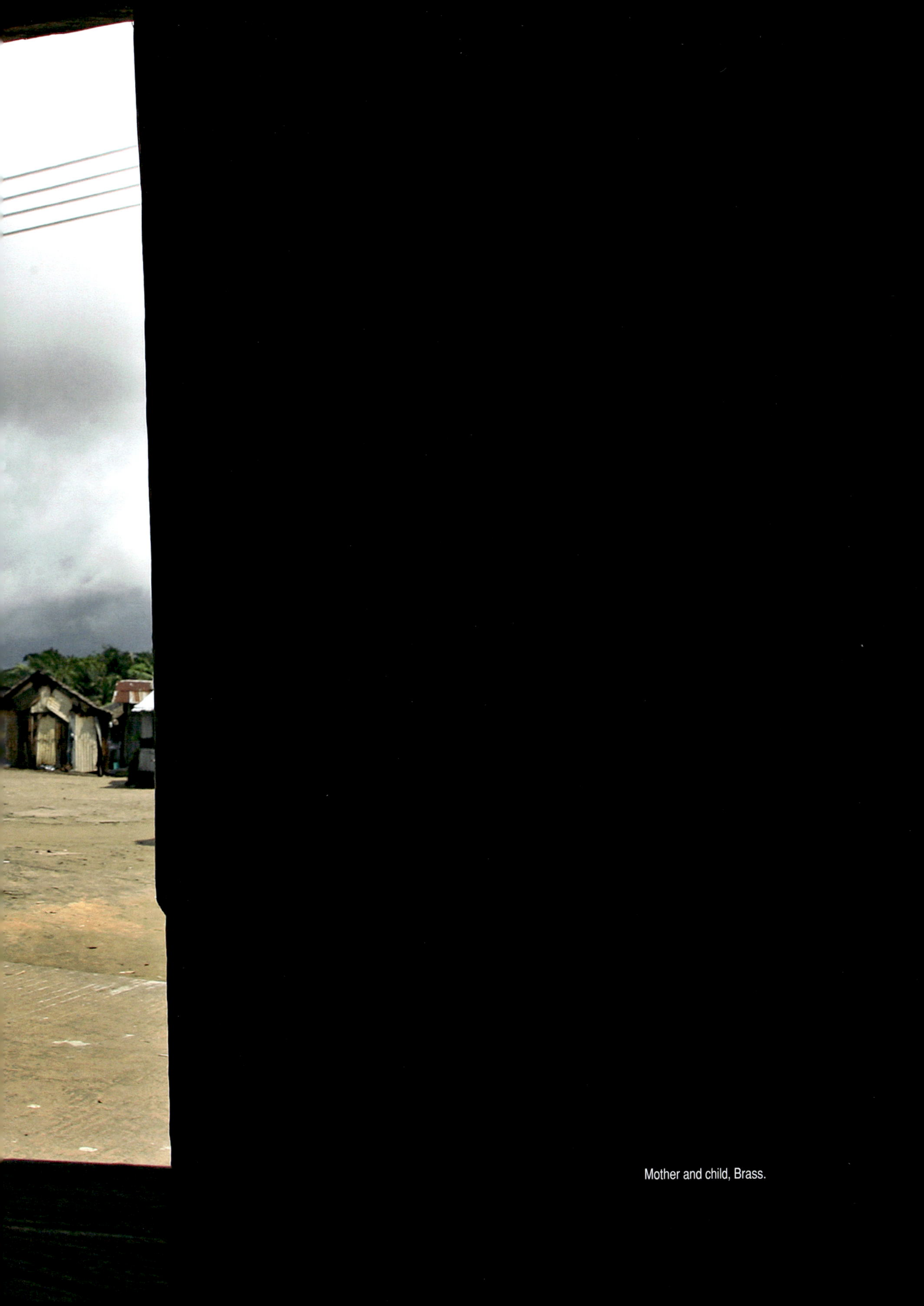

Mother and child, Brass.

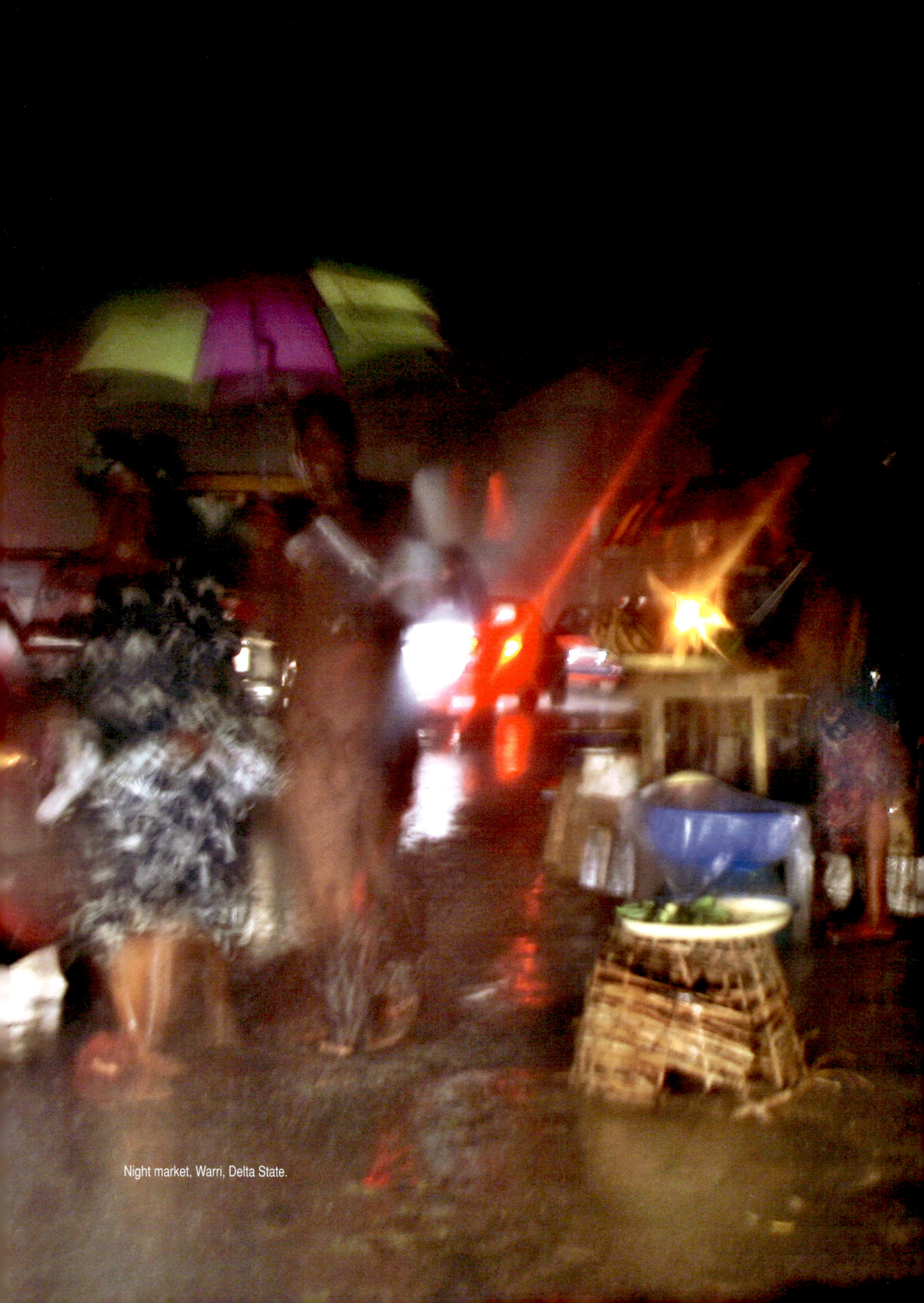

Night market, Warri, Delta State.

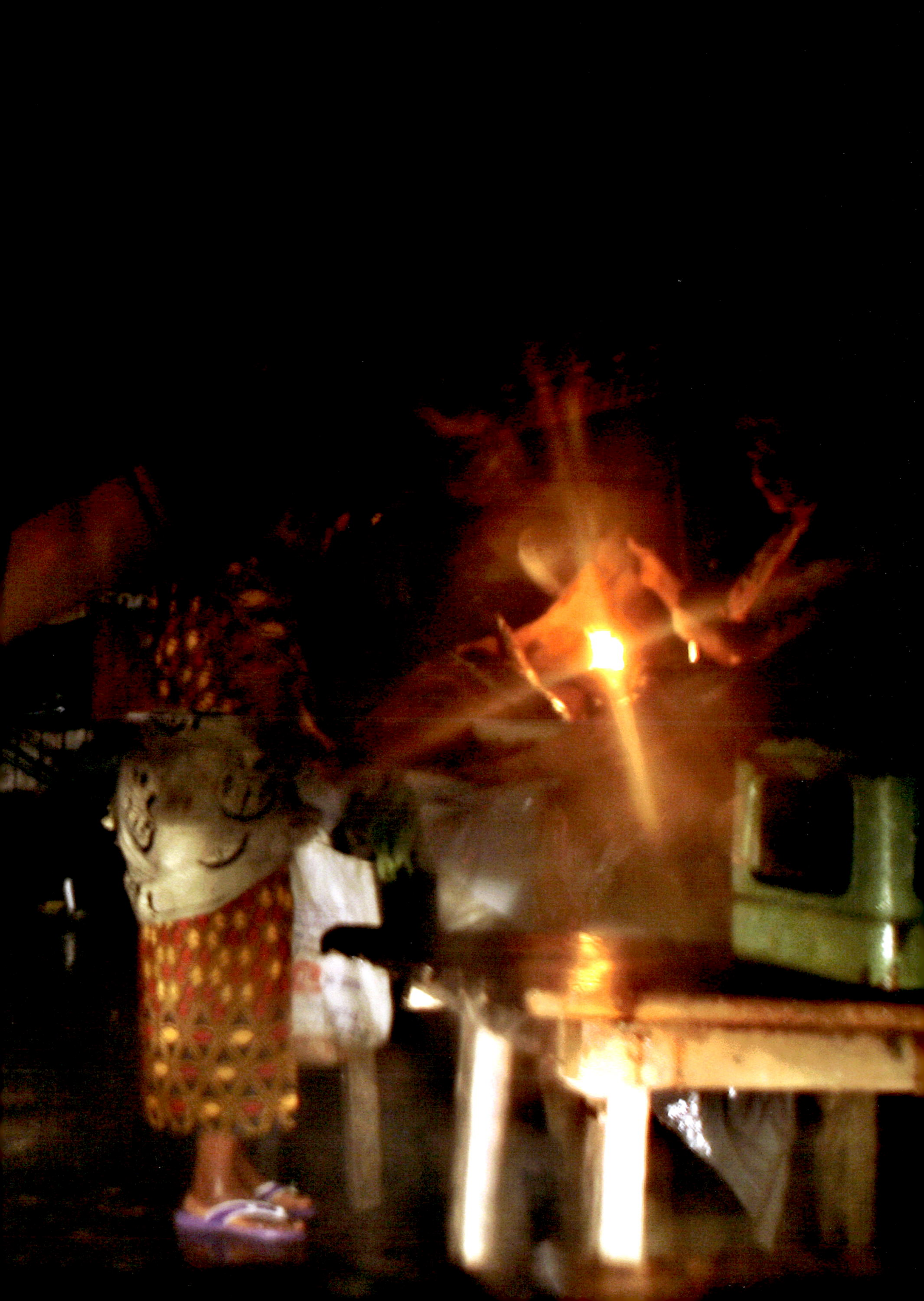

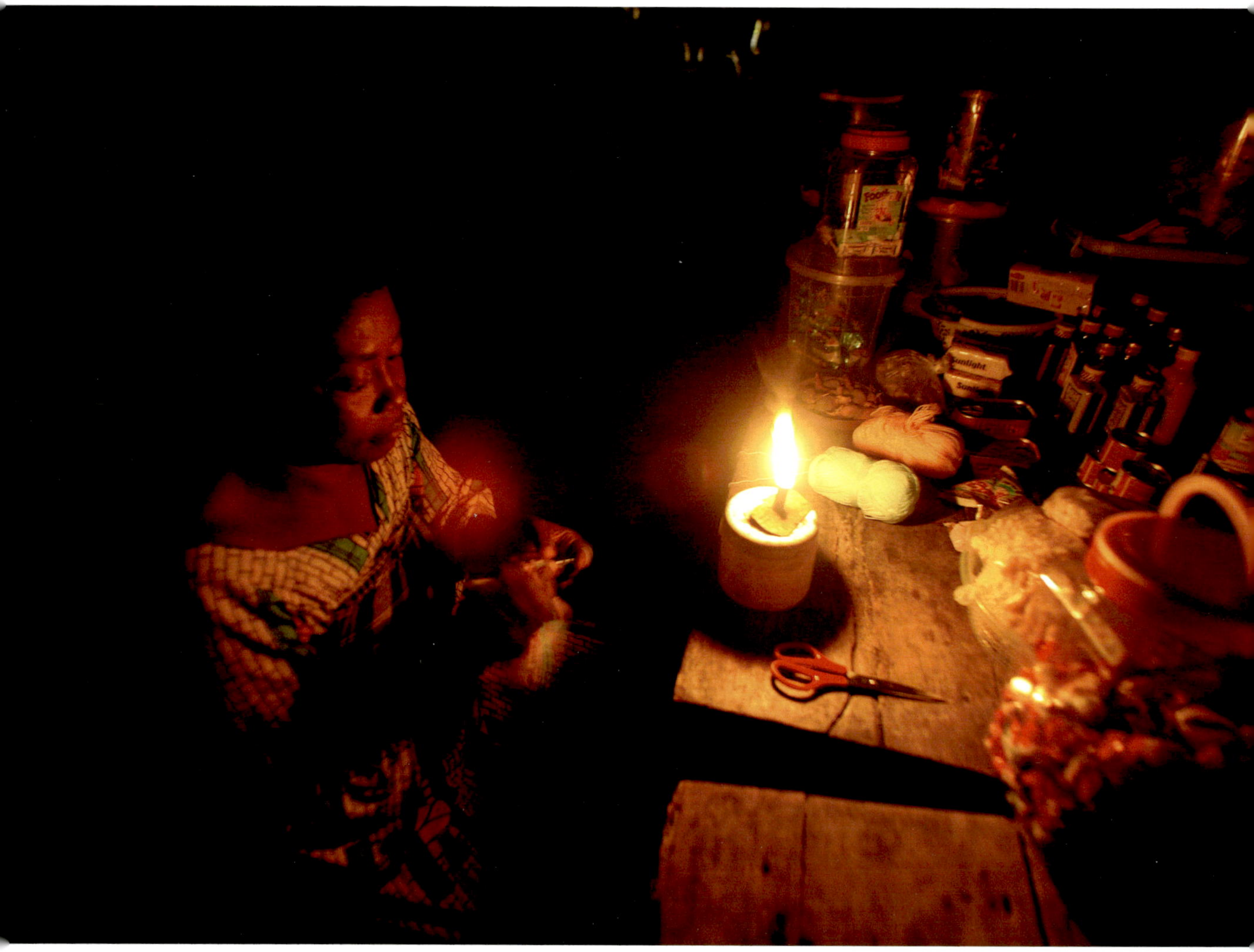

Above: A woman sells her wares at night in Nembe.

Right: House in Timber slum area, Port Harcourt.

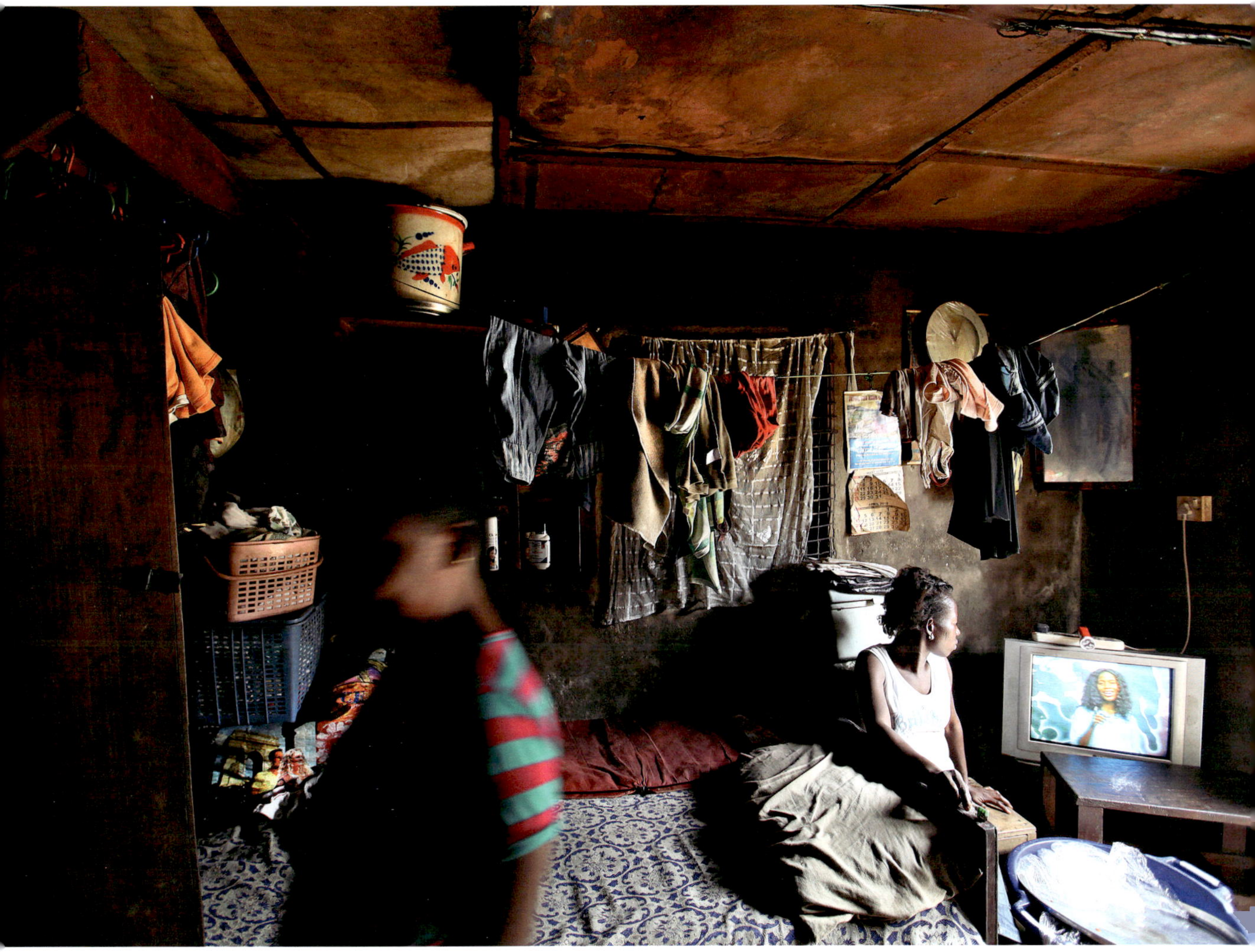

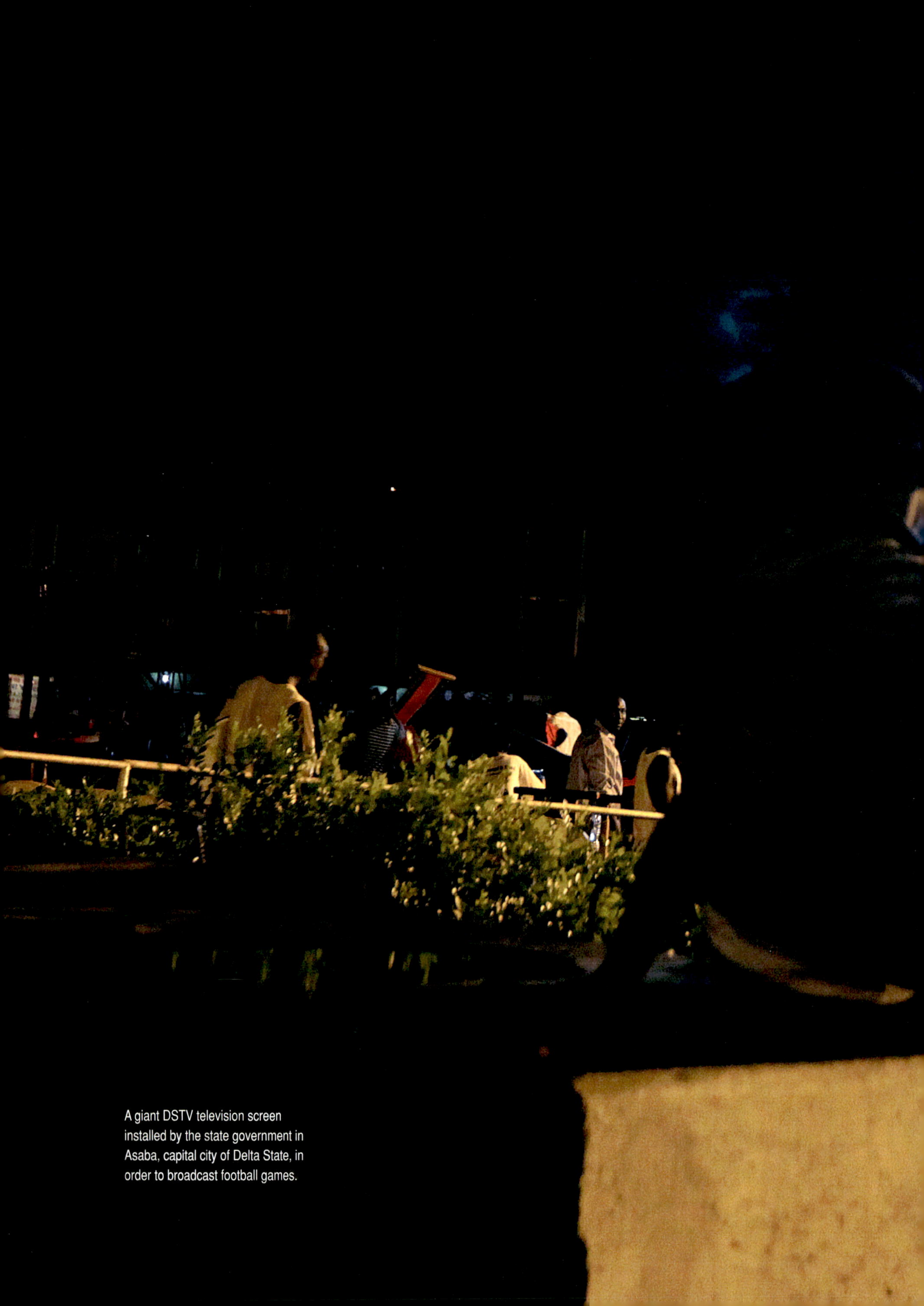

A giant DSTV television screen installed by the state government in Asaba, capital city of Delta State, in order to broadcast football games.

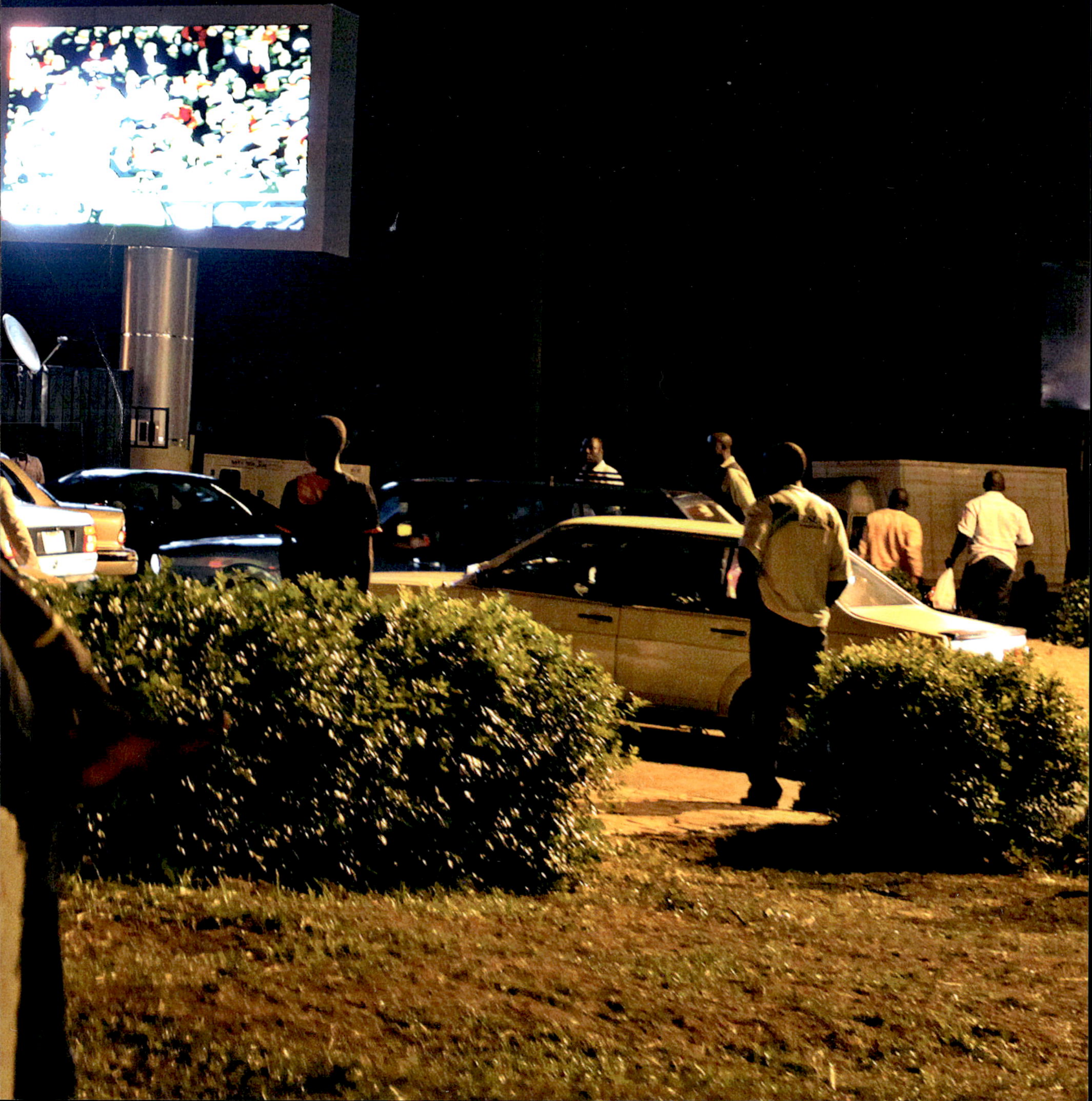

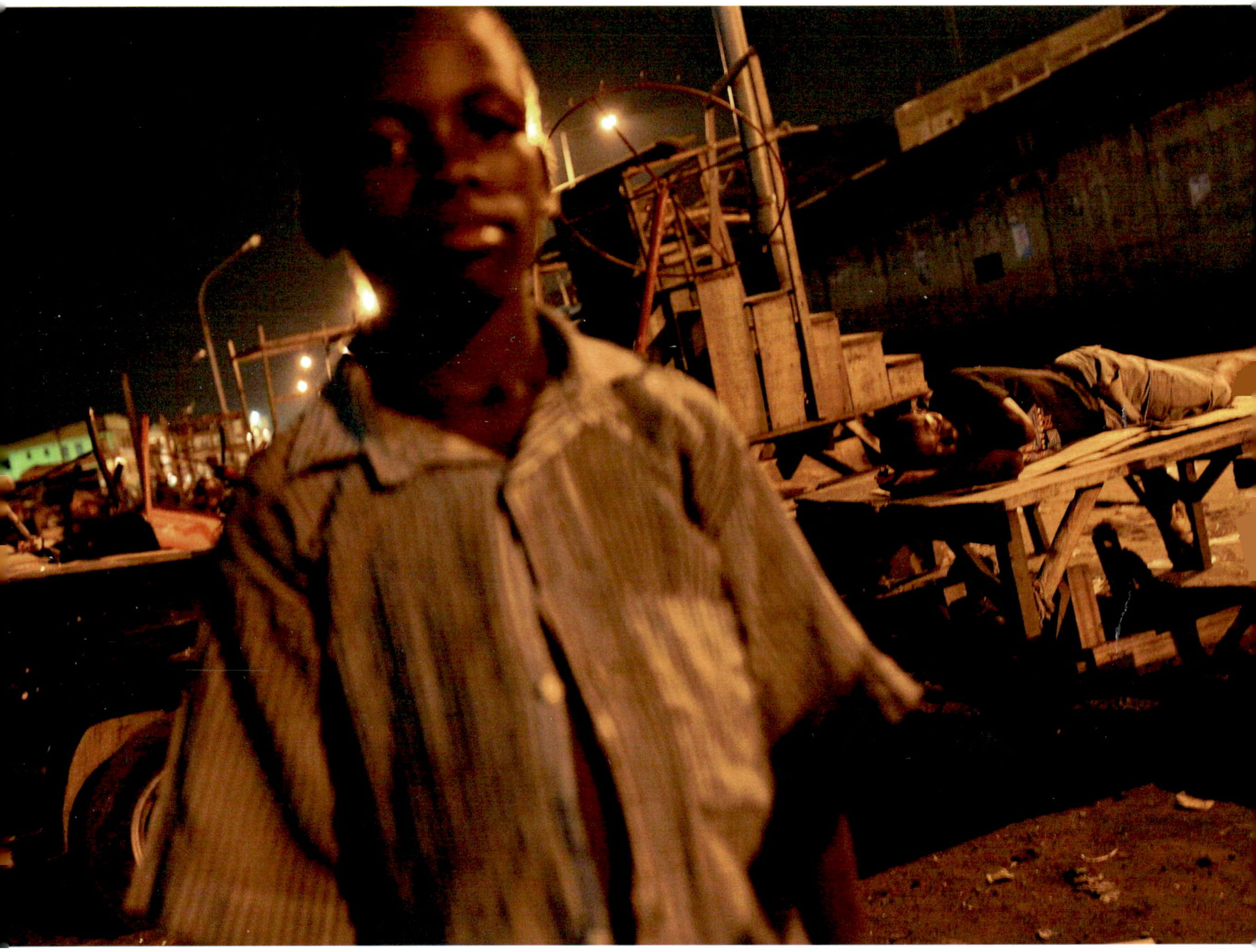

Above: Samson Jacknoble, 11 year old homeless street kid, stands near a wooden table where he sleeps at the Mile One Market in Port Harcourt.

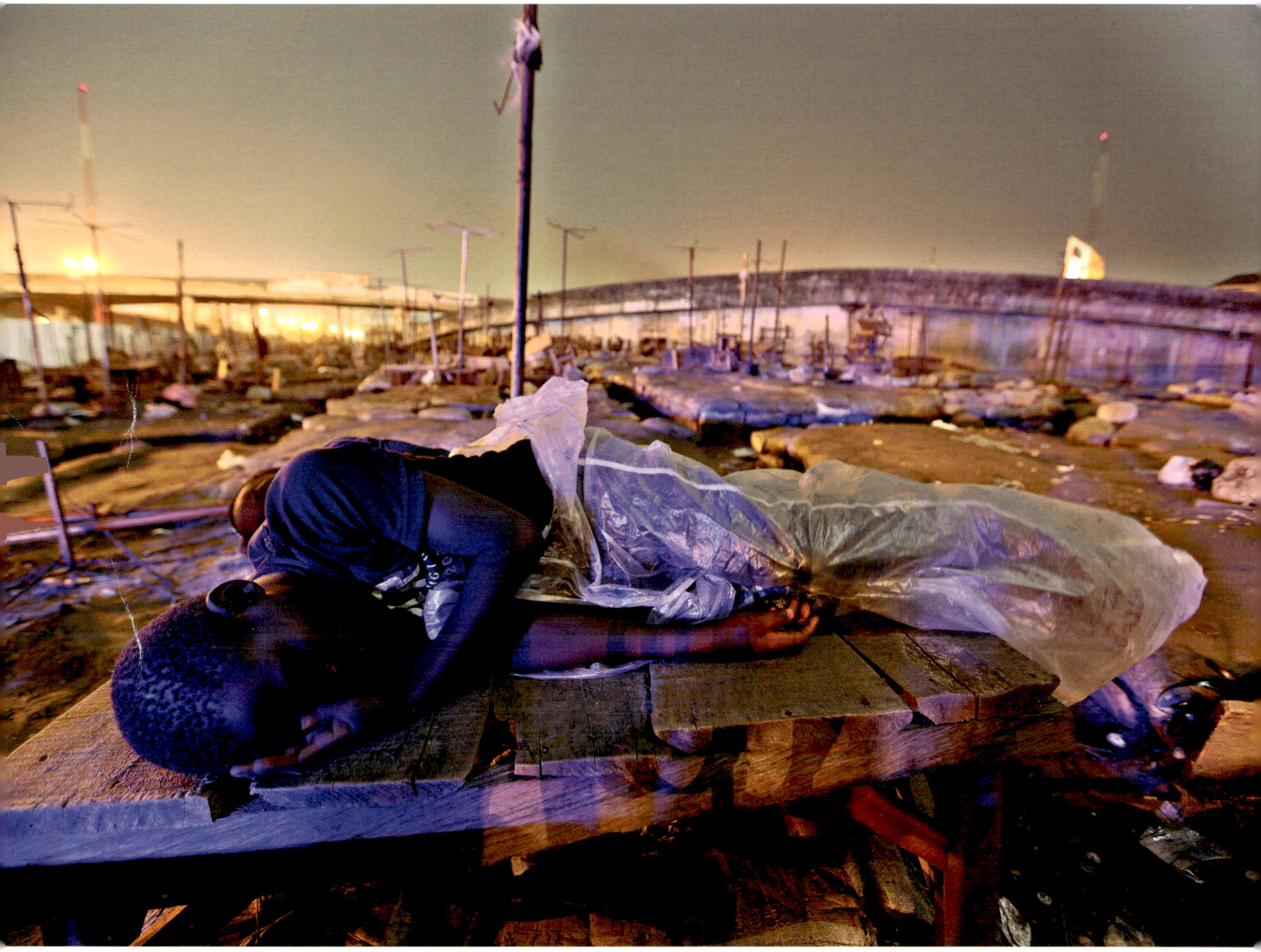

Above: Aniekan Etim, who used to live in Njemanze before it was demolished, sleeps on a wooden table at the Mile One Market where he and the other street kids live.

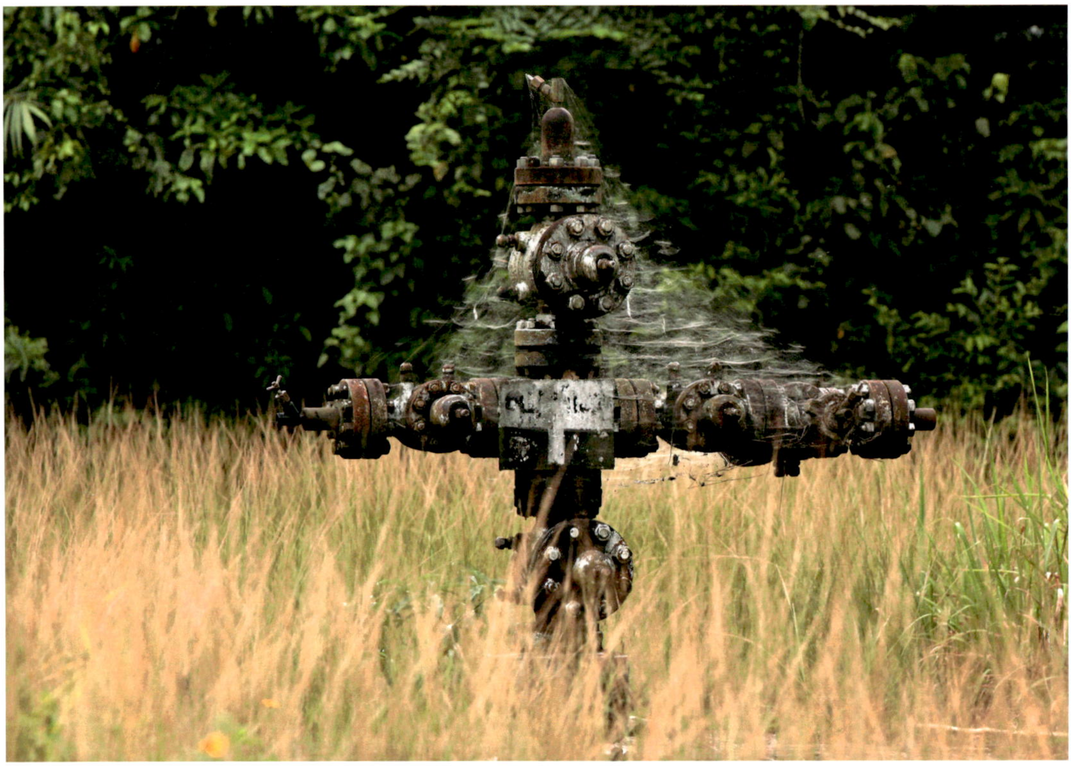

THE MASQUERADES

"This community has not benefited from Shell's presence here in the past 20 years. It was only in June last year that Shell awarded even a one kilometre road contract", says the traditional ruler of Otabagi community, one of the communities where oil was first struck. "Infrastructural development by Shell in Oloibiri community is nothing to write home about here." These words are echoed by the residents of the three communities where Shell Petroleum Development Company (SPDC) has been involved in oil exploration and exploitation activities in the '50s to '70s, before milking the area and leaving it dry. The most common features in the communities now are the relics of oil activities dotting the landscape (above), abandoned oilwells nicknamed 'masquerades', due to their appearance as traditional tribal masks. The older men and women in the communities have been left impotent by the damage done to their farmlands by the oil activities, whilst their children and grandchildren have no respect for them for allowing it to happen in the first place, and for believeing the empty promises of the oil companies. They call the old 'masquerades' too, comparing their lives to the abandoned oil wells.

Above right: Augusta Amadukua inside her home, Otabagi.

Below right: 70 year old Karibo Fuah, Otabagi.

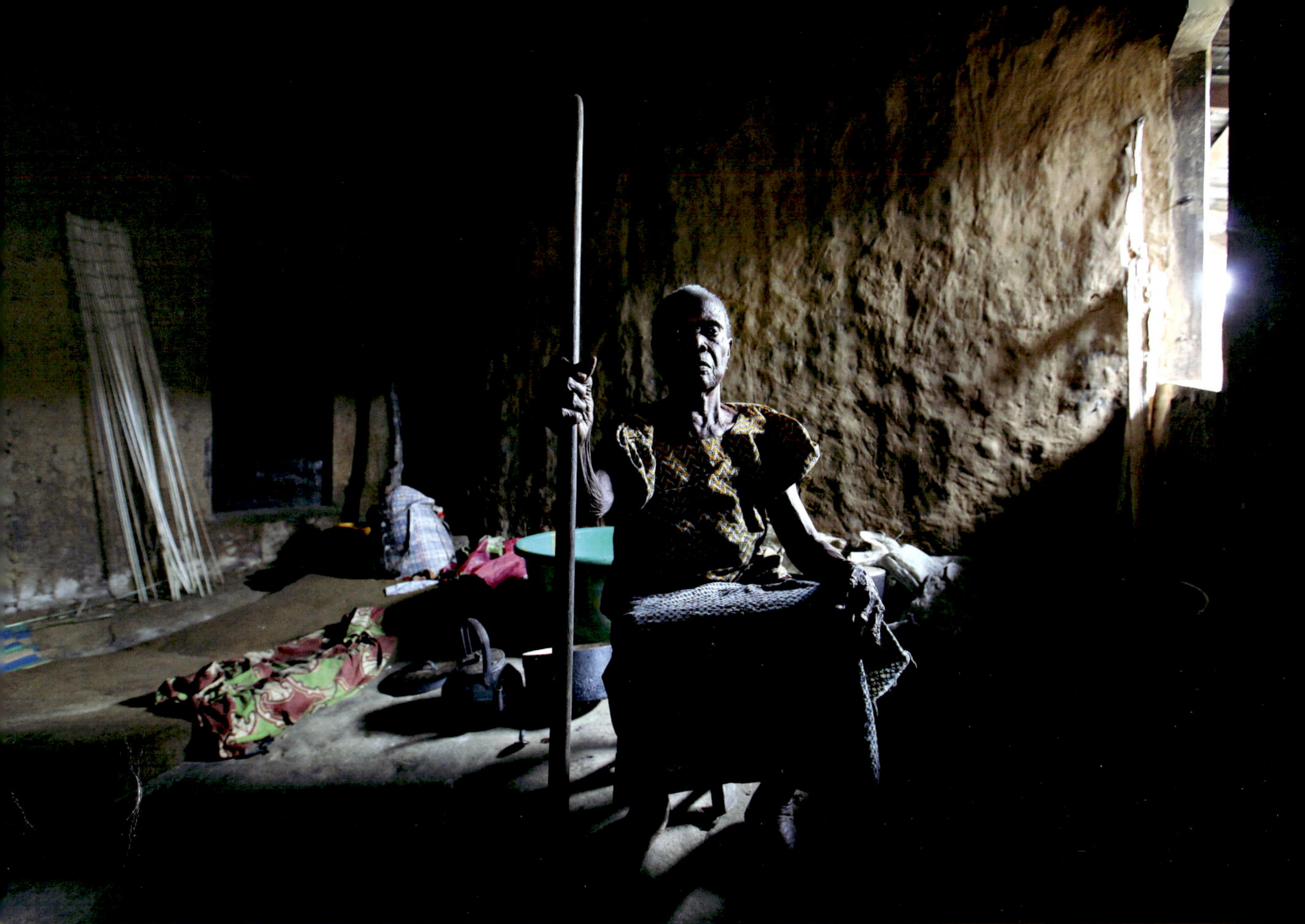

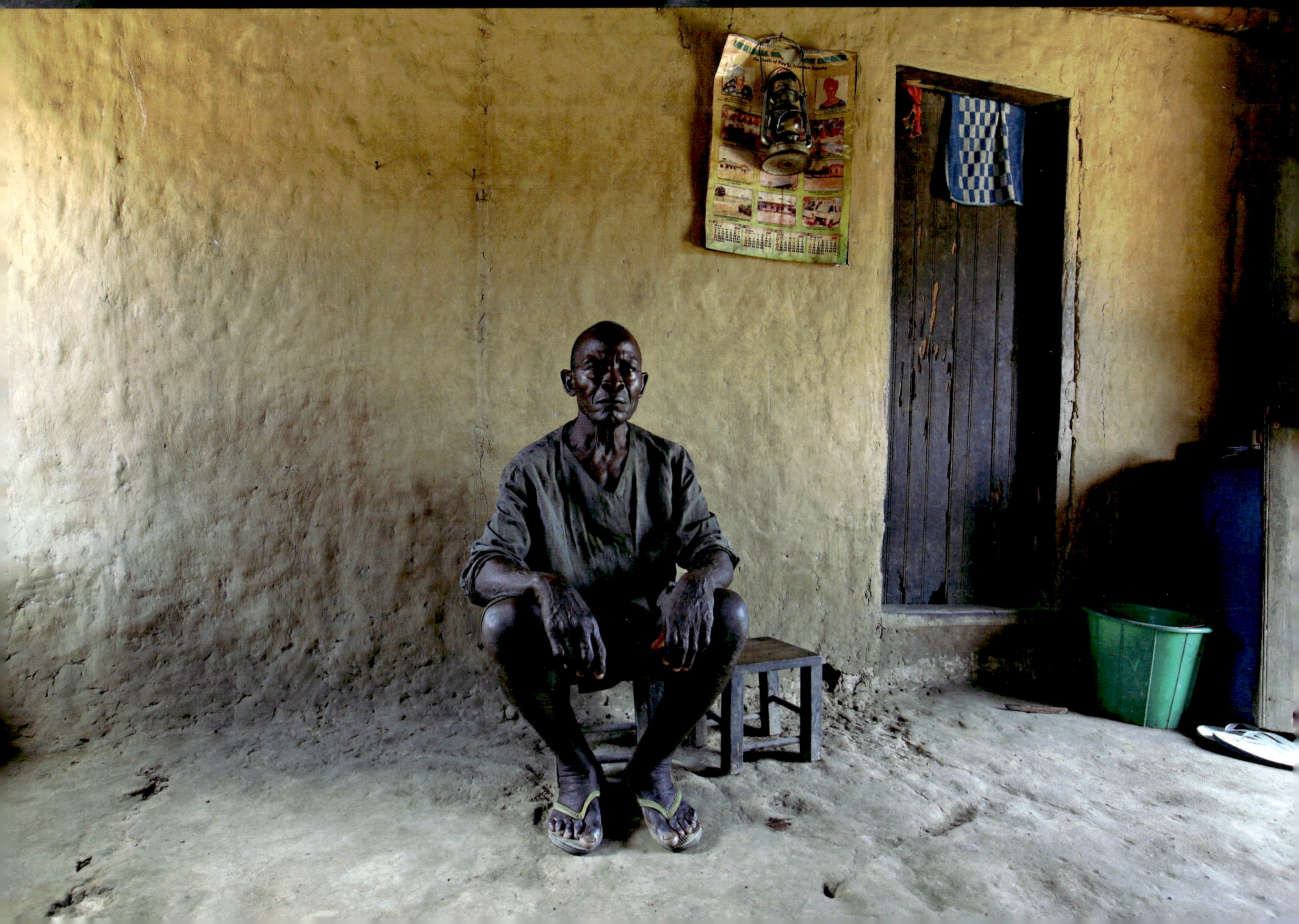

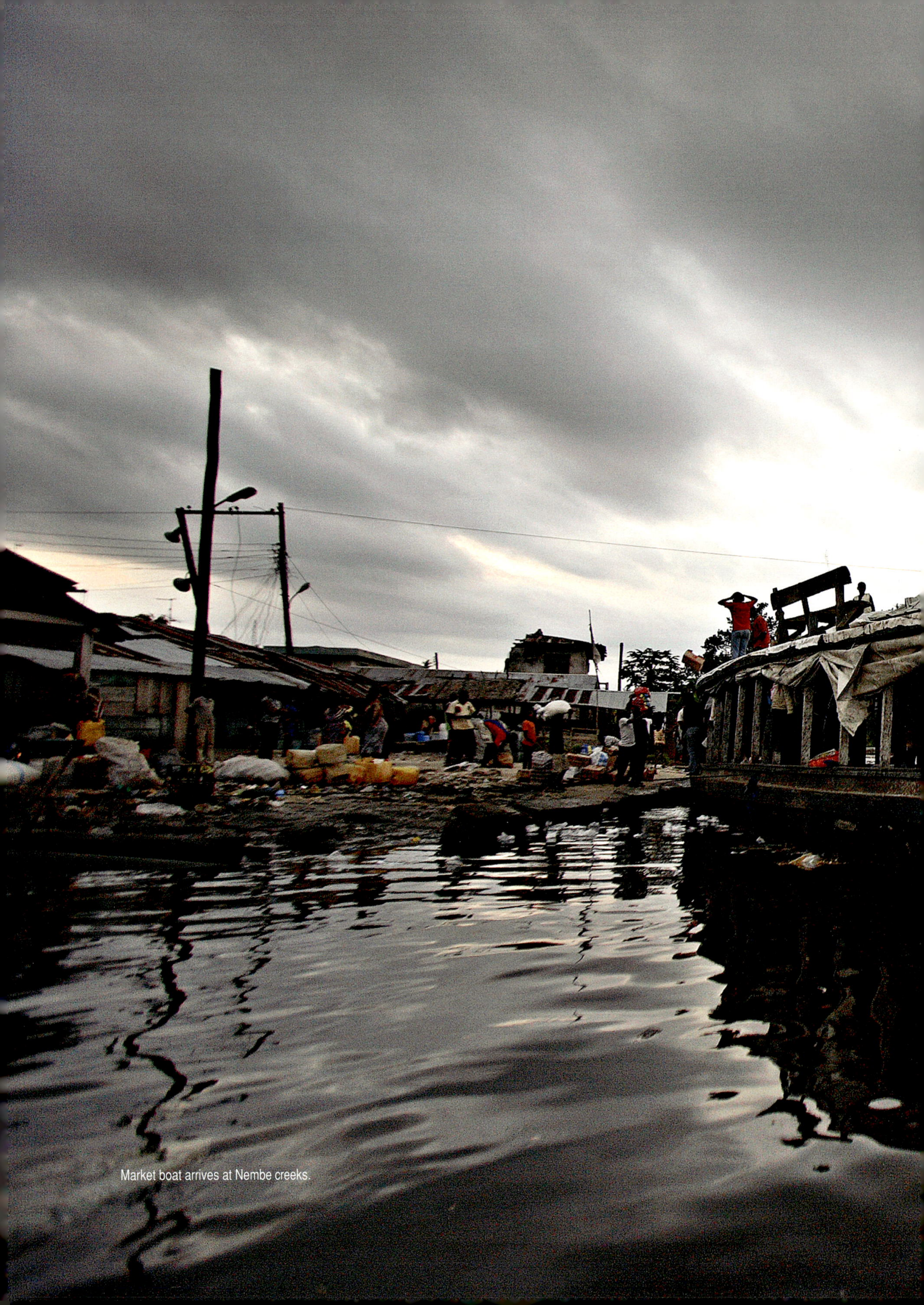

Market boat arrives at Nembe creeks.

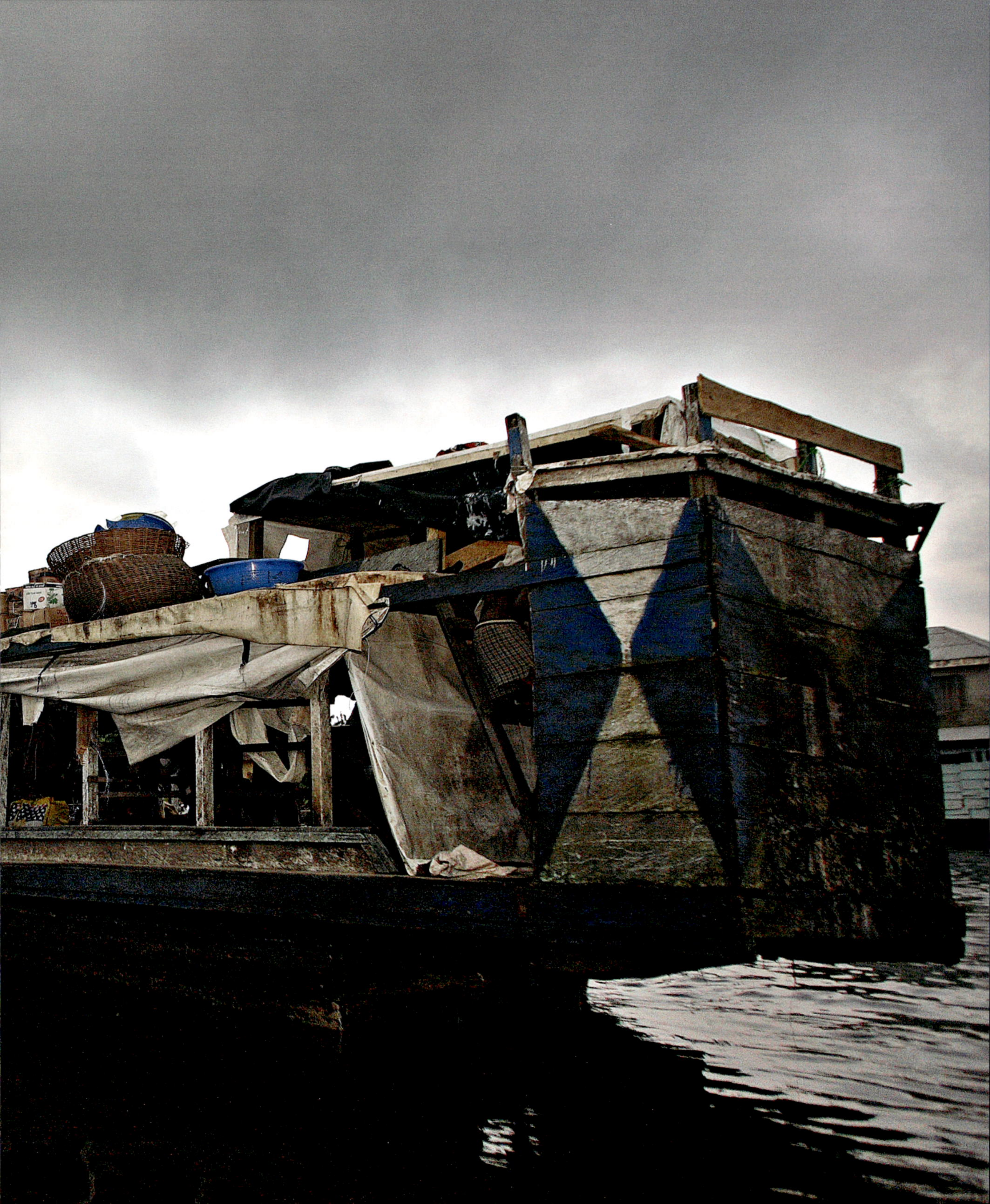

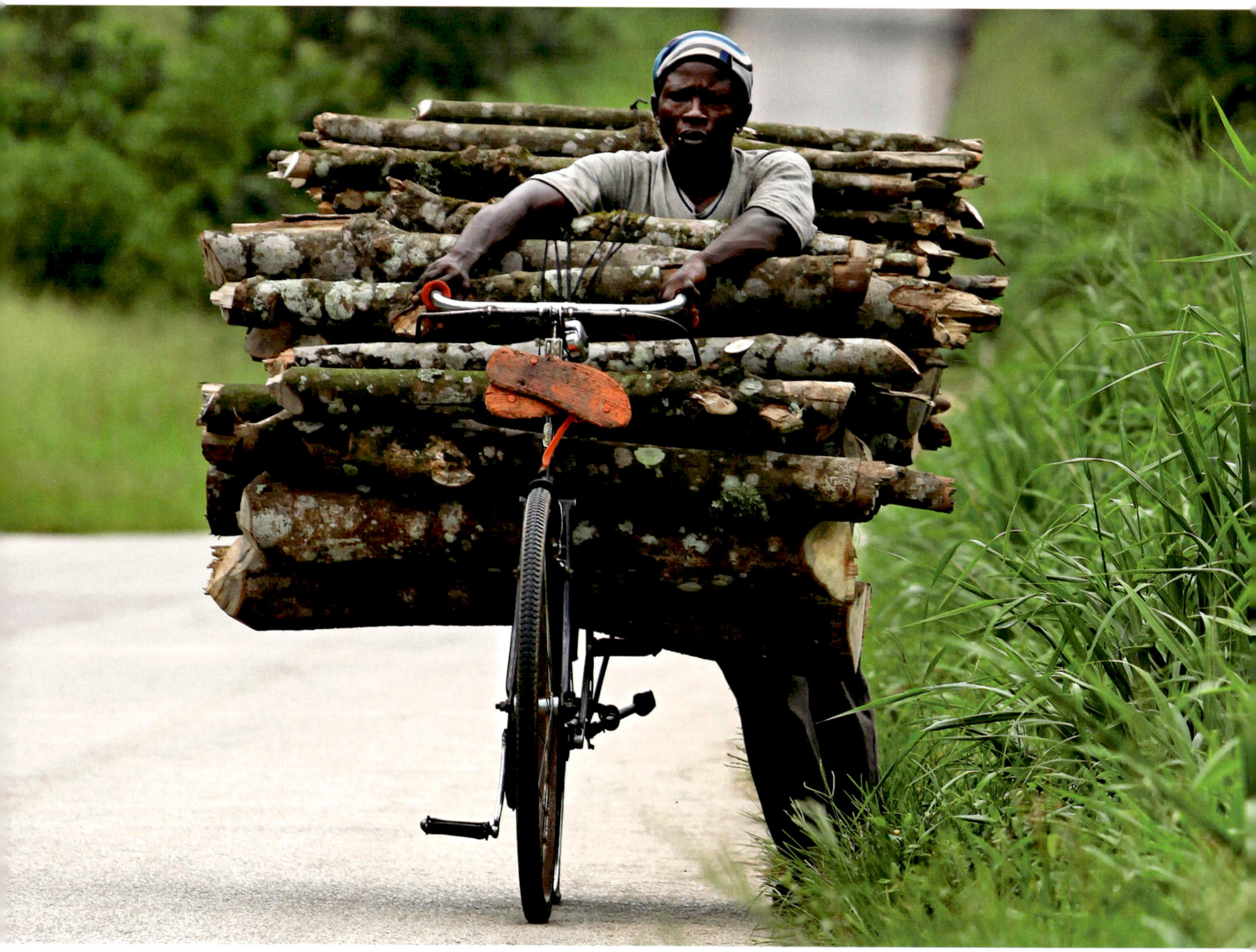

Above: A worker returns from farming with firewood, Abraka.

Right: Locals at Nun River, Odi.

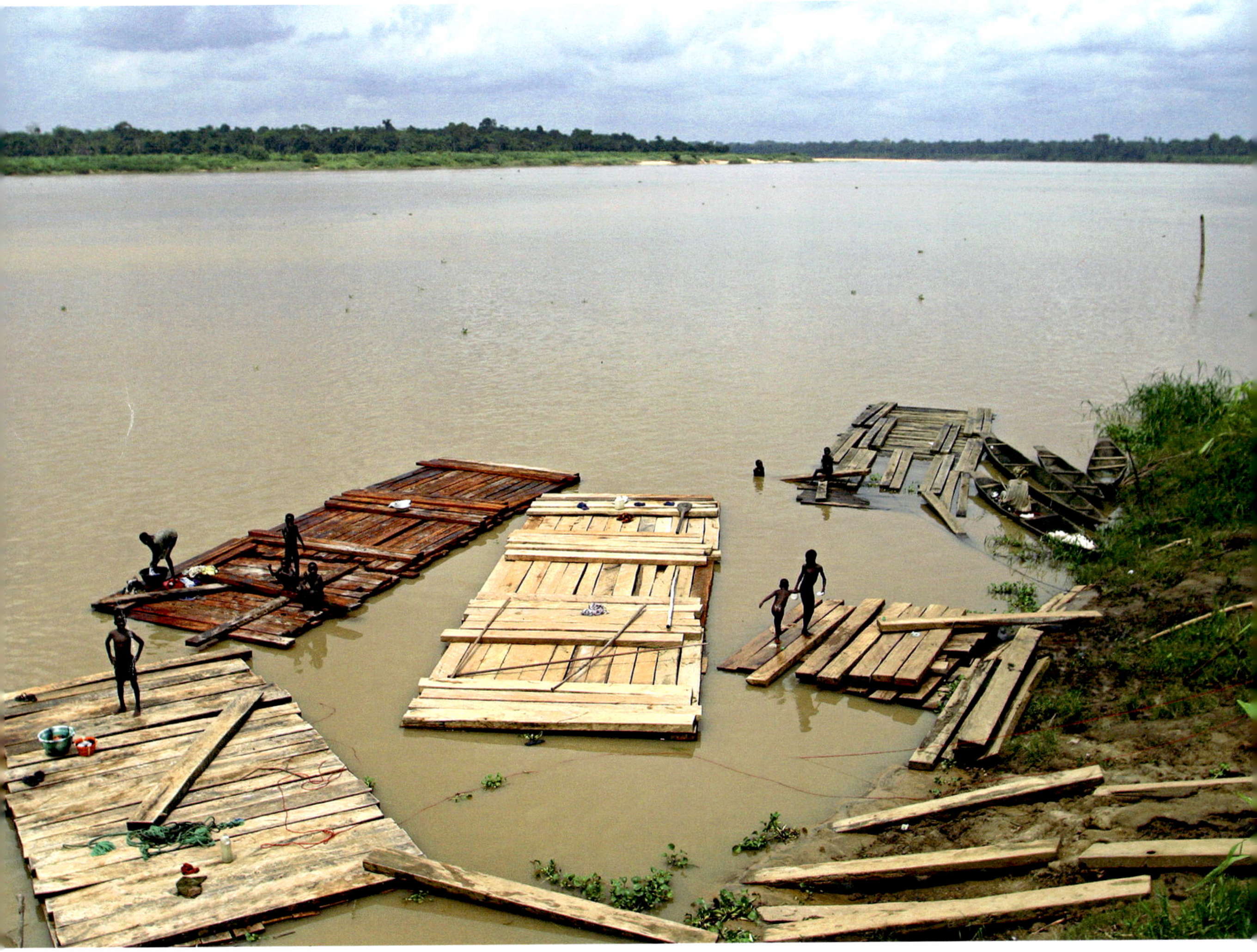

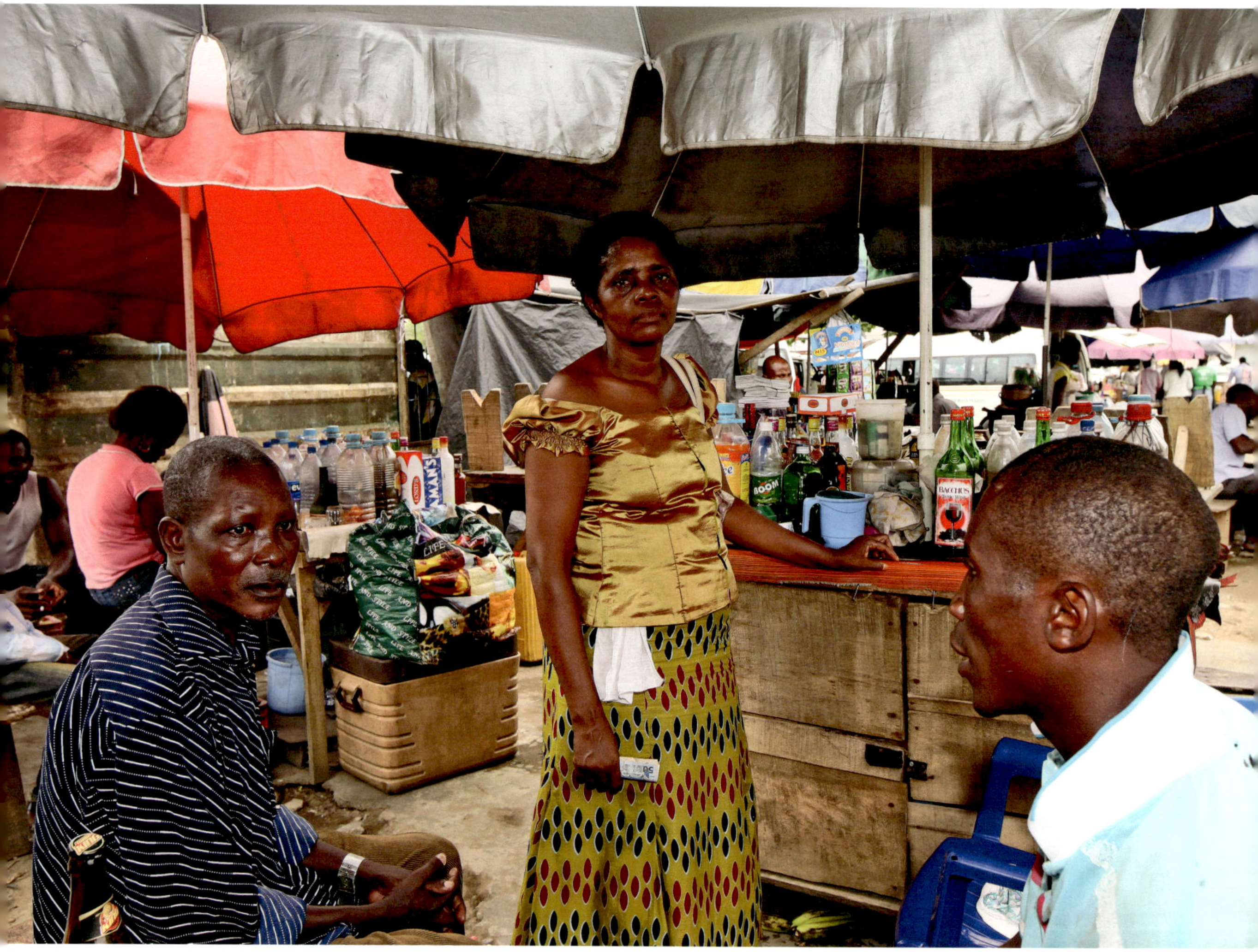

Above: Love Bassey Akpabio stands under an umbrella kiosk where she sells drinks at Mile One Market, Port Harcourt. Love is a former resident of Njemanze, an informal settlement where she lived with her husband and five children which was demolished without compensation in 2009 as part of the government's urban renewal programme for the city.

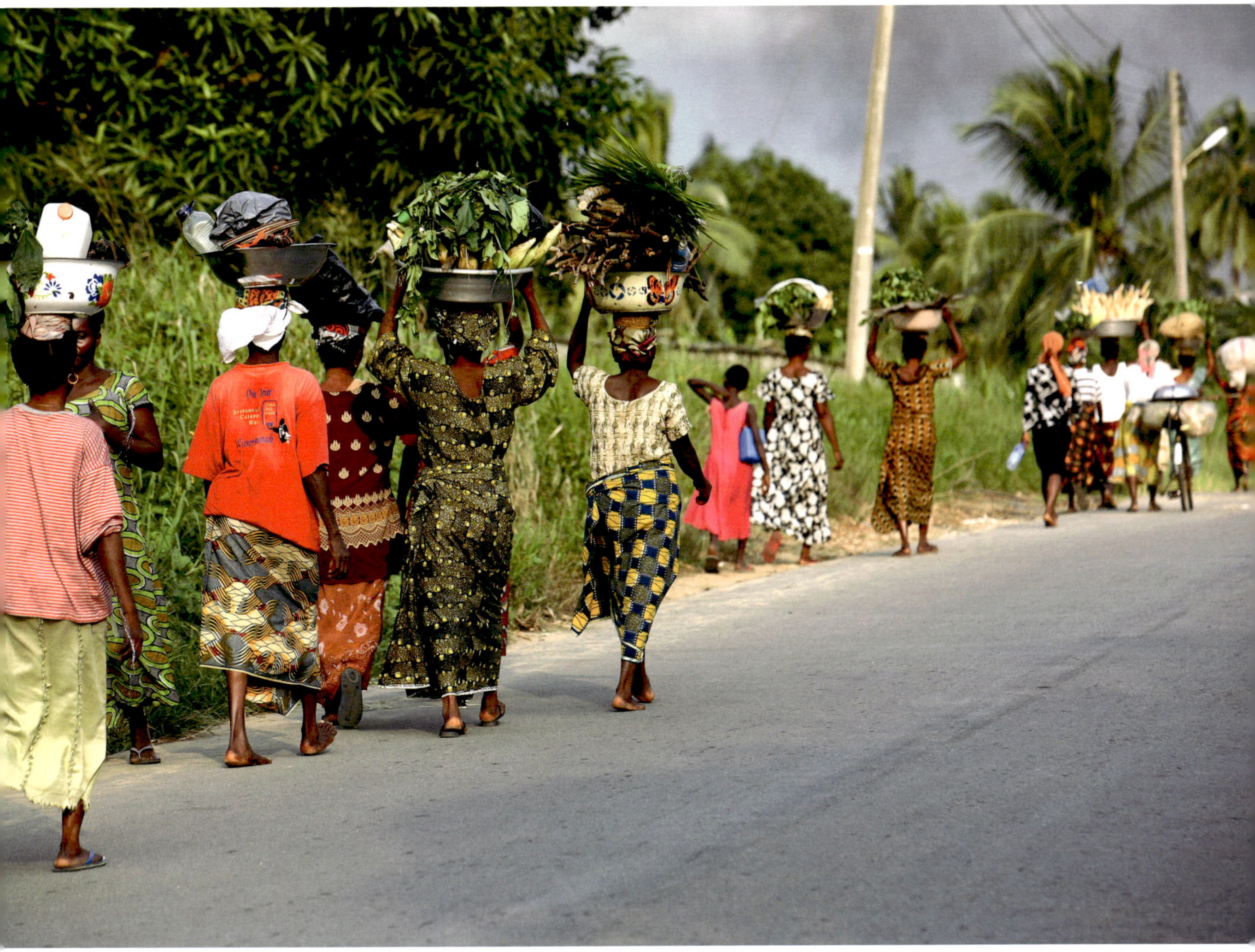

Above: Women return from farming in groups.

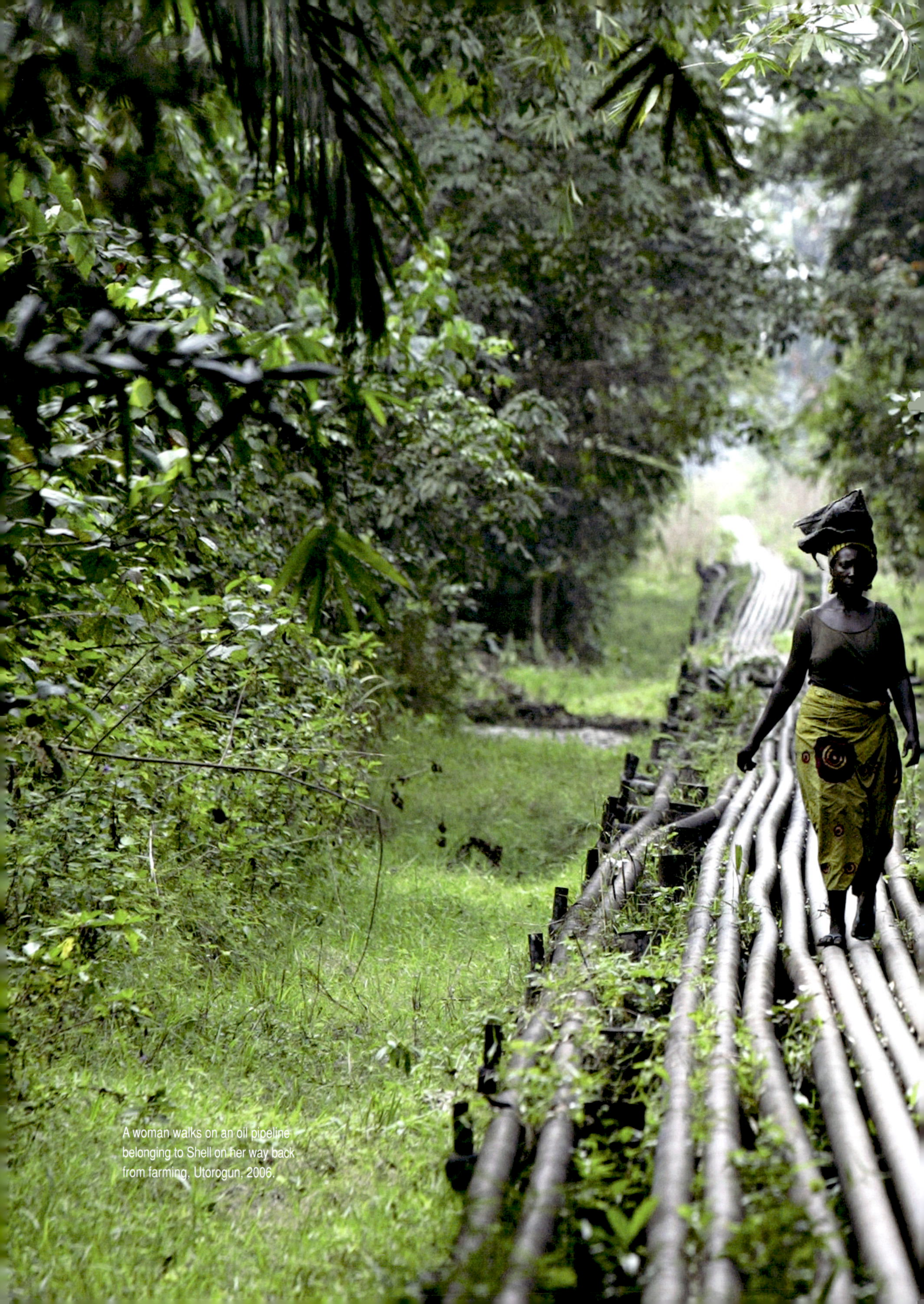

A woman walks on an oil pipeline belonging to Shell on her way back from farming, Utorogun, 2006.

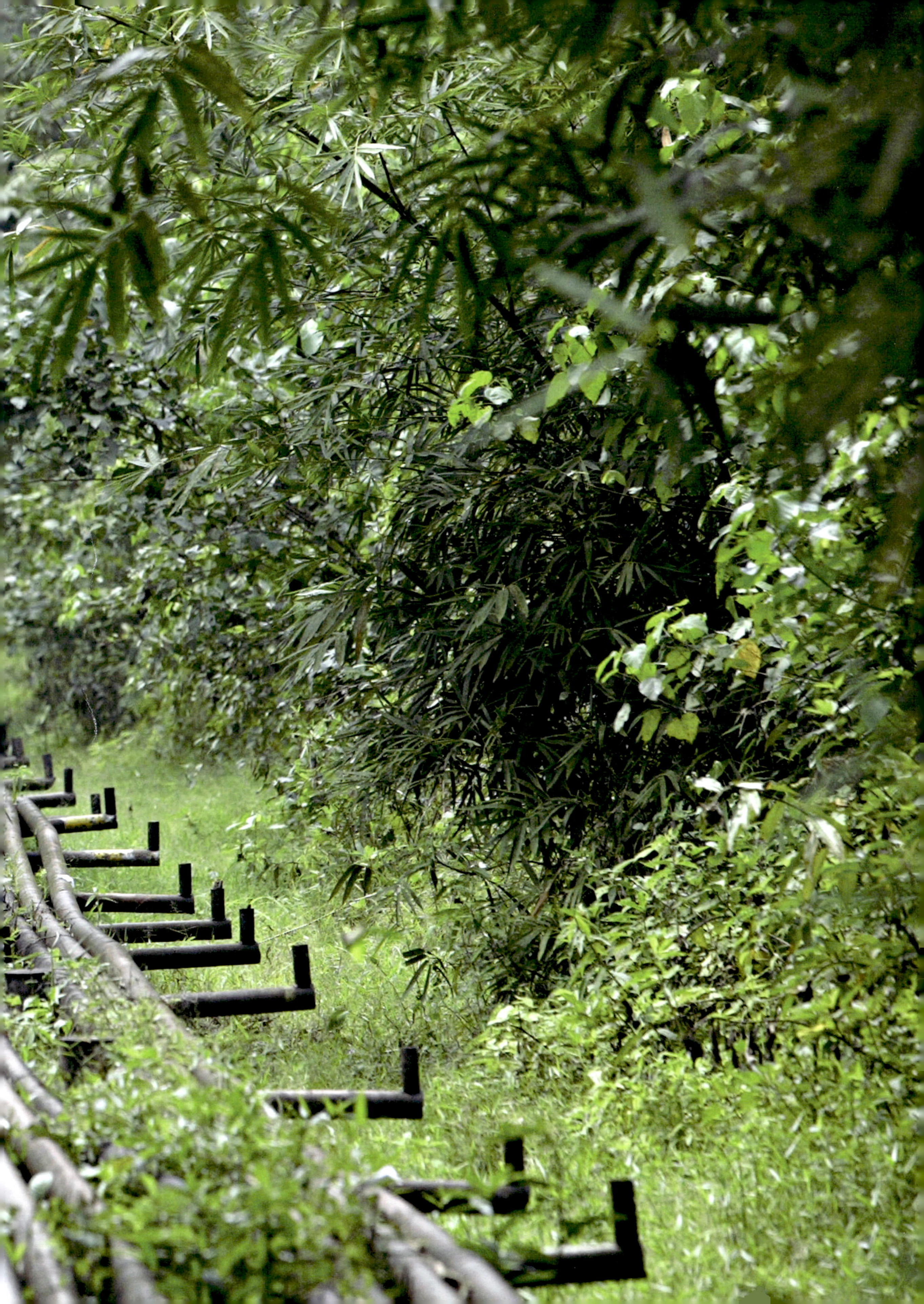

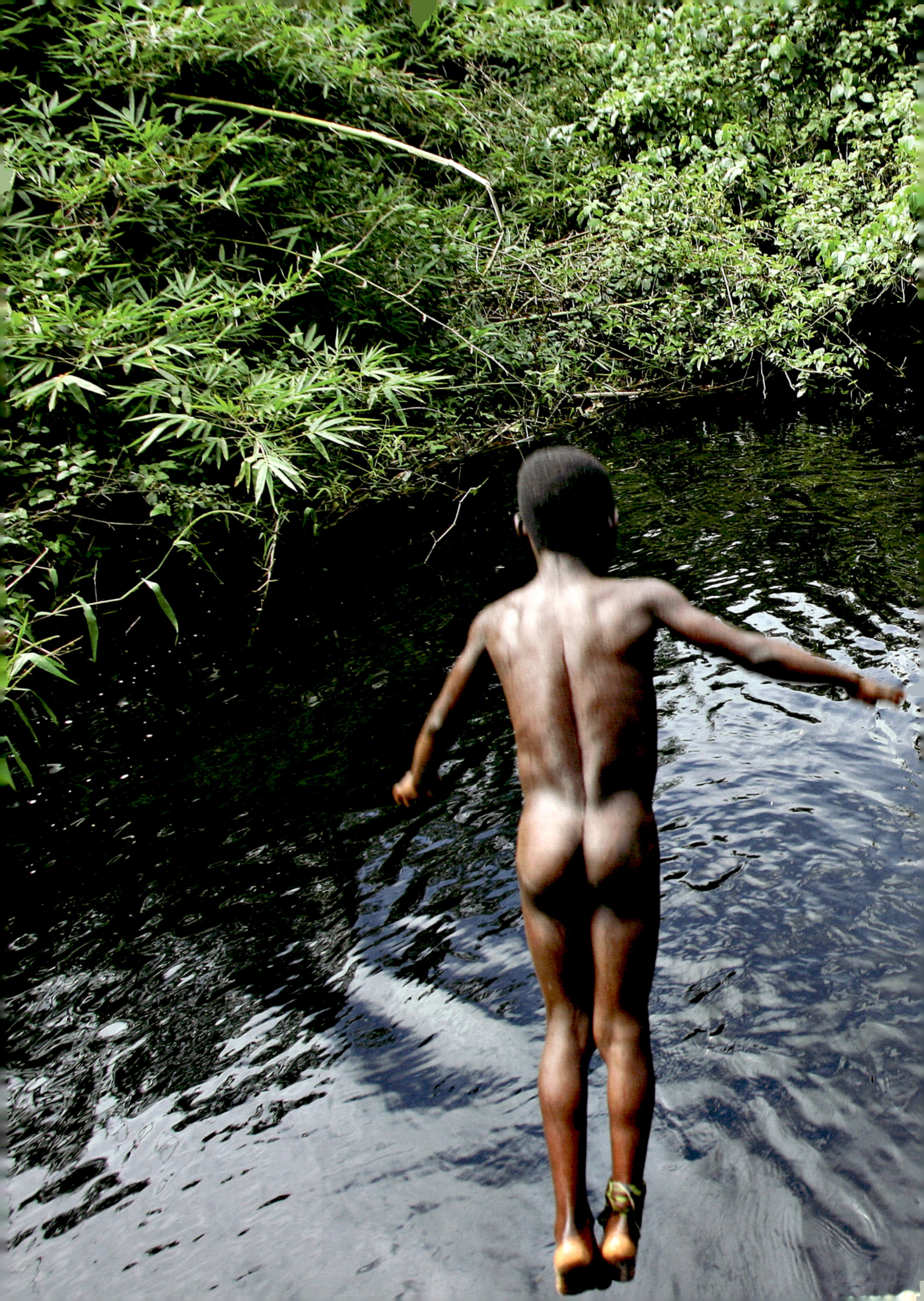

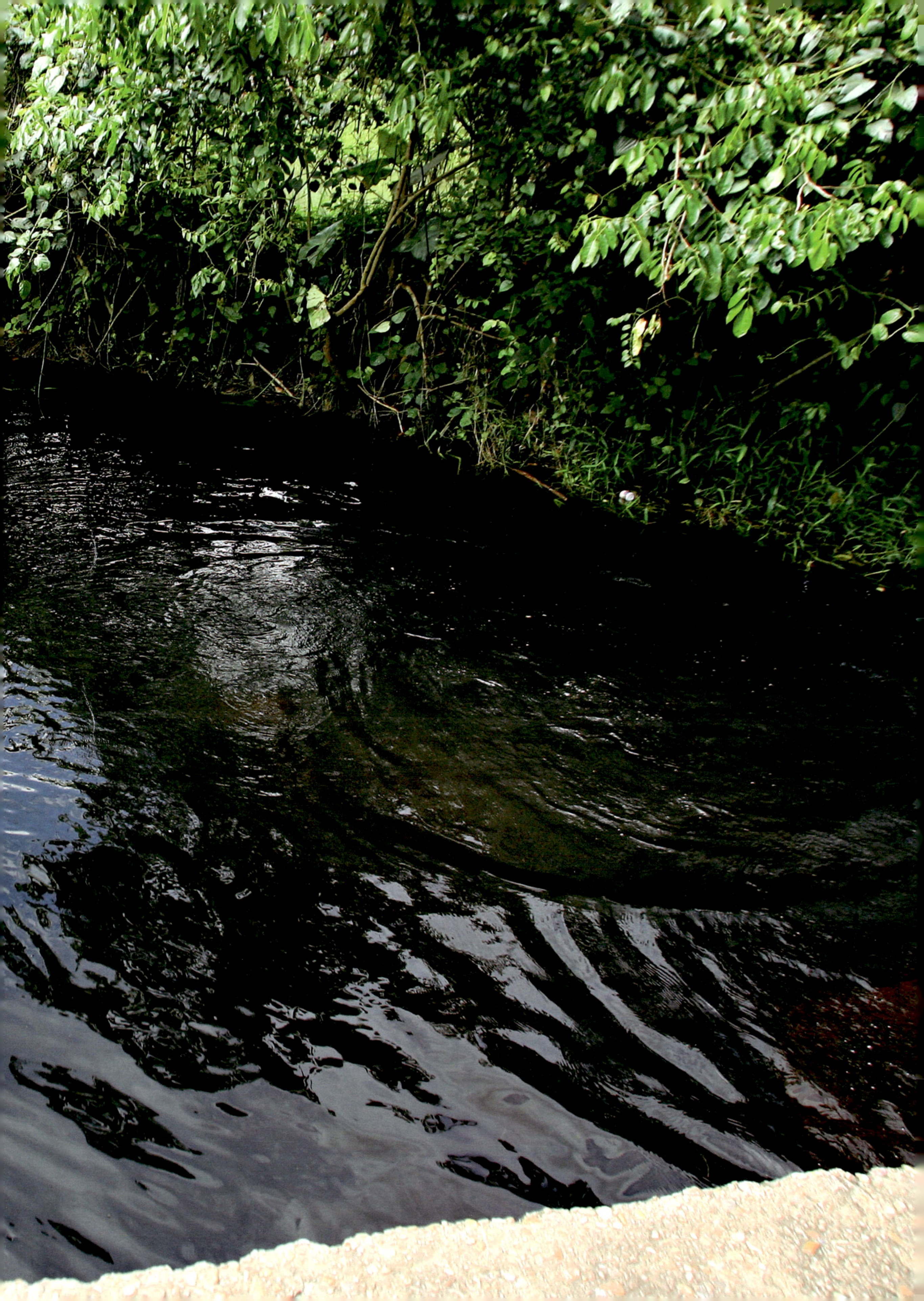

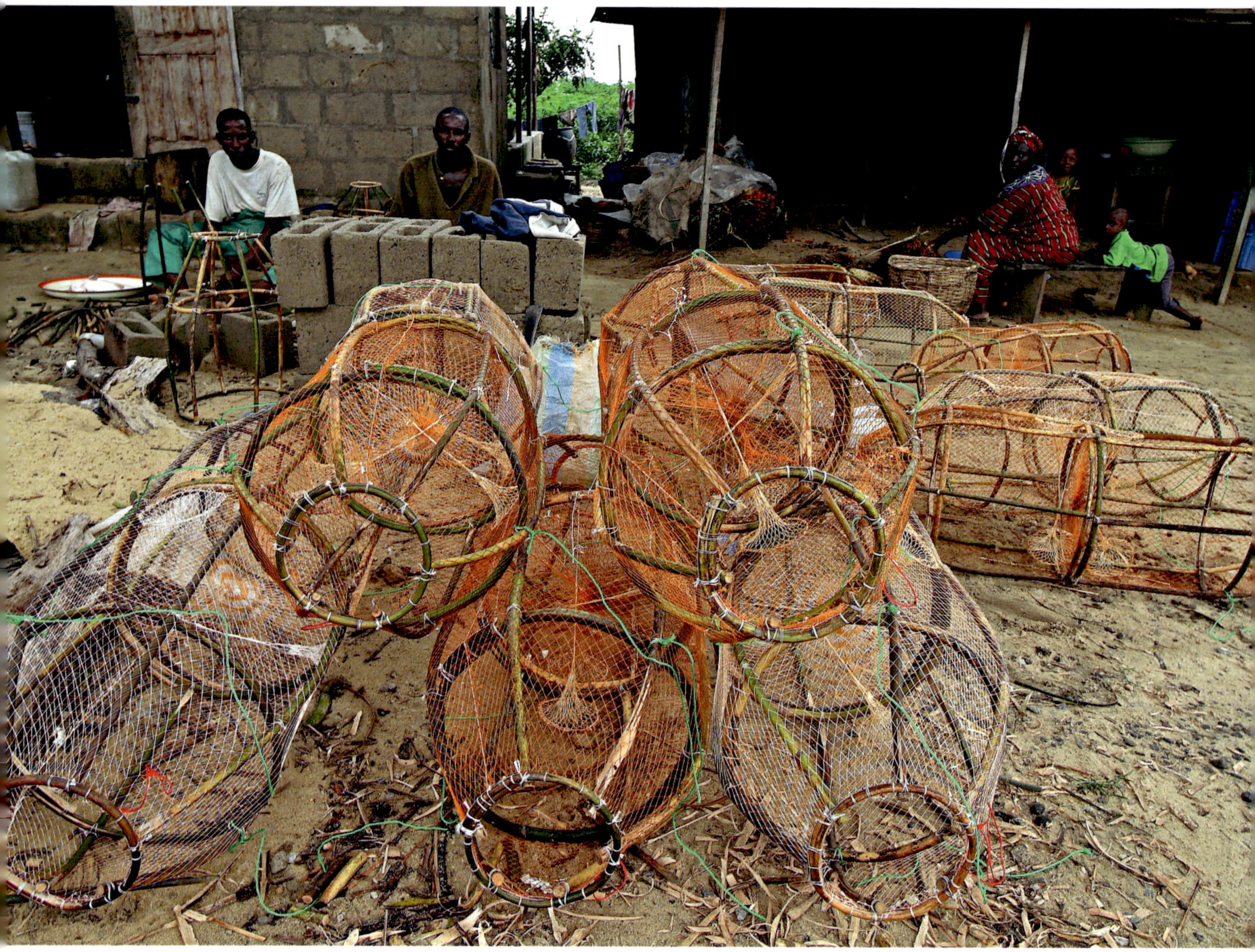

Above: Fishing traps in Kaiama, 2005.

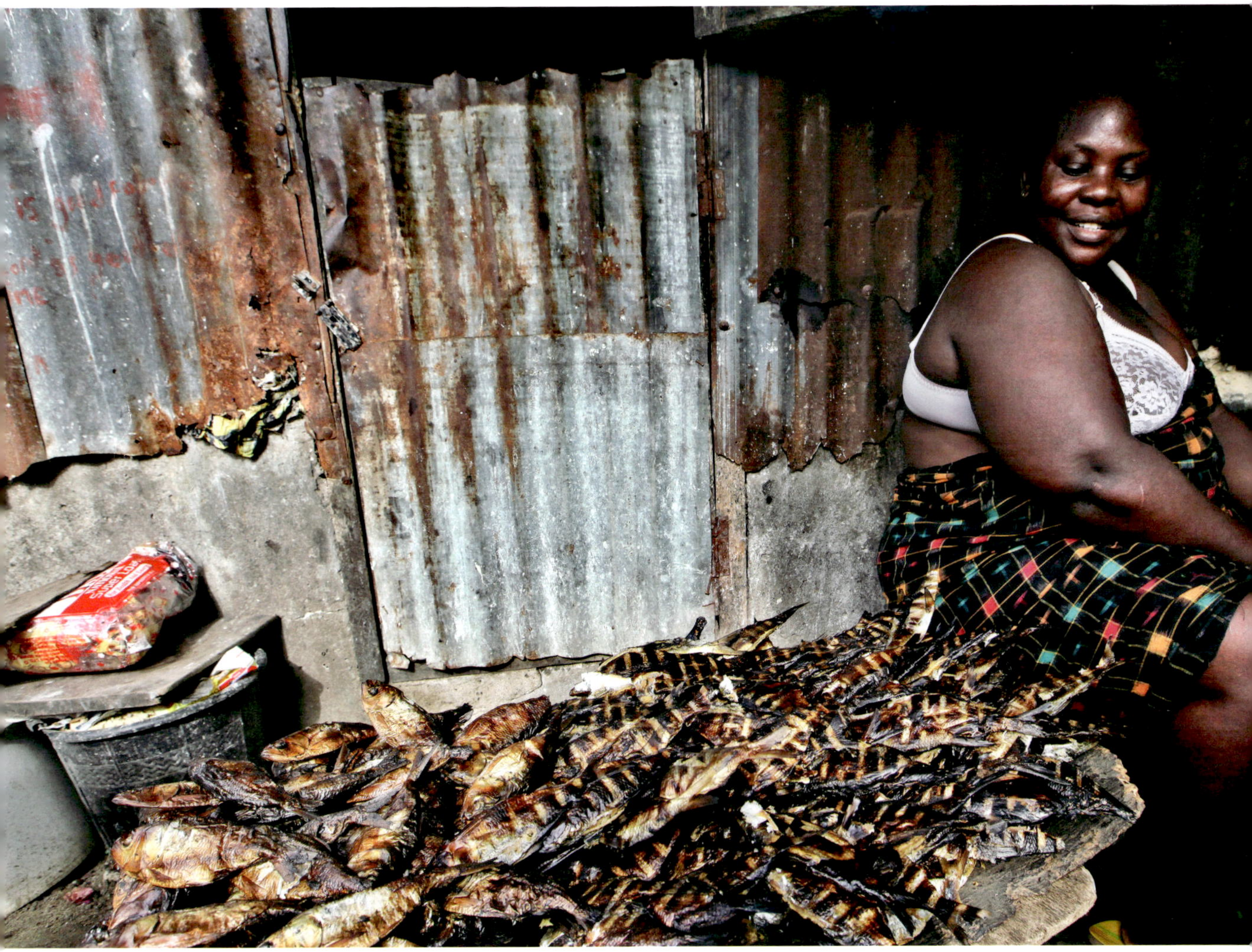

Above: Smoked fish, Port Harcourt, 2011.

Overleaf: Children at a waterside slum, Port Harcourt, 2007.

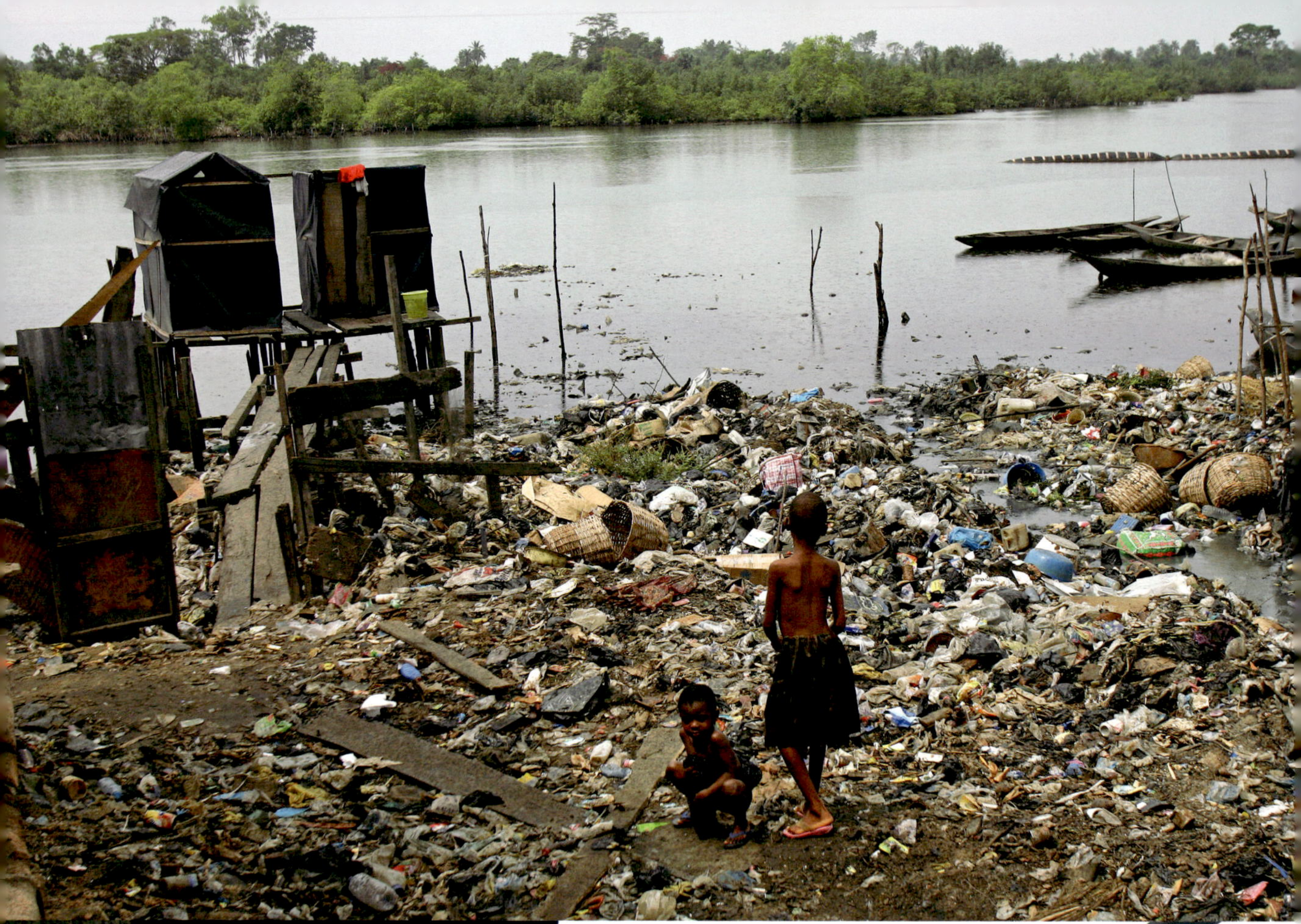
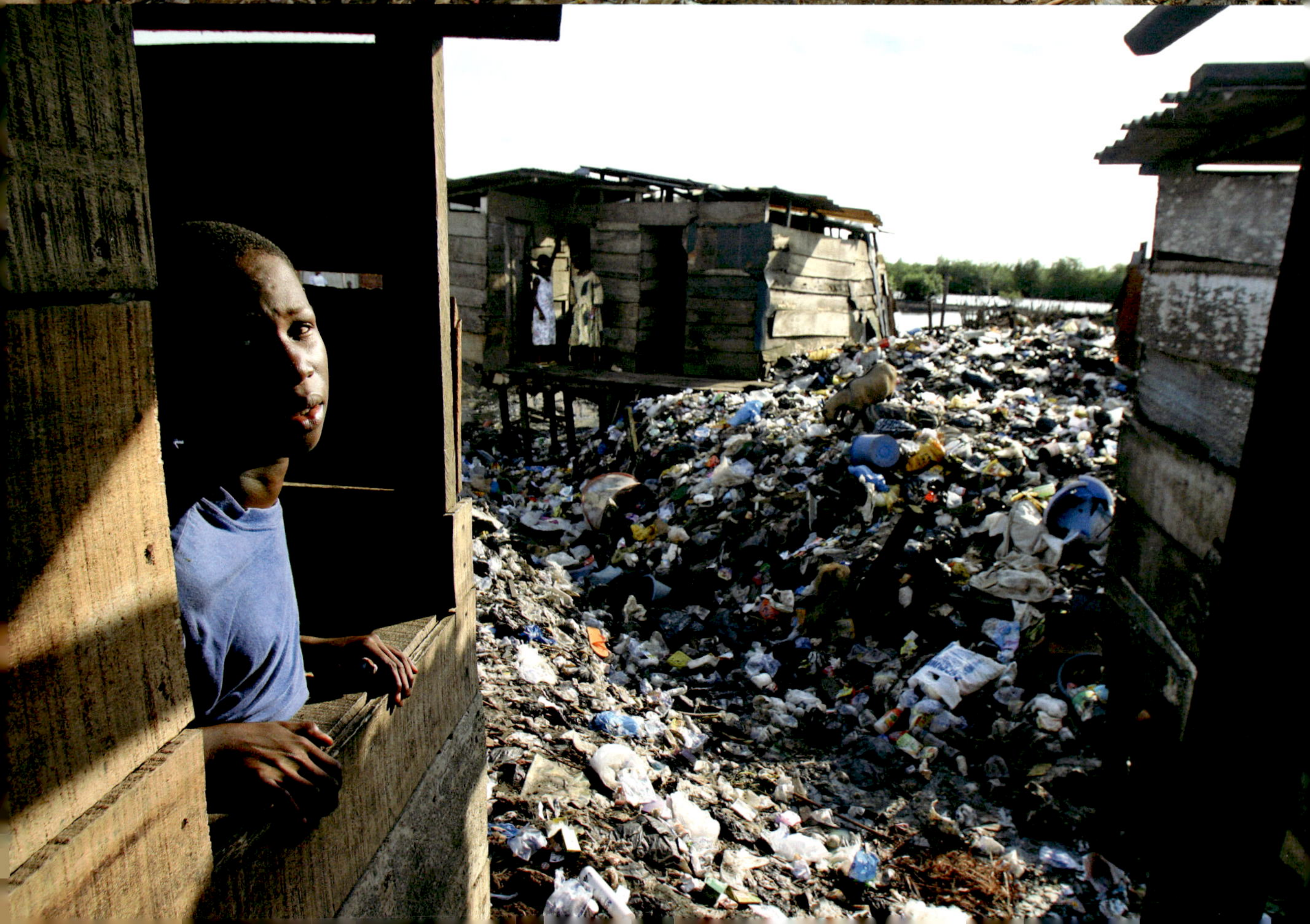

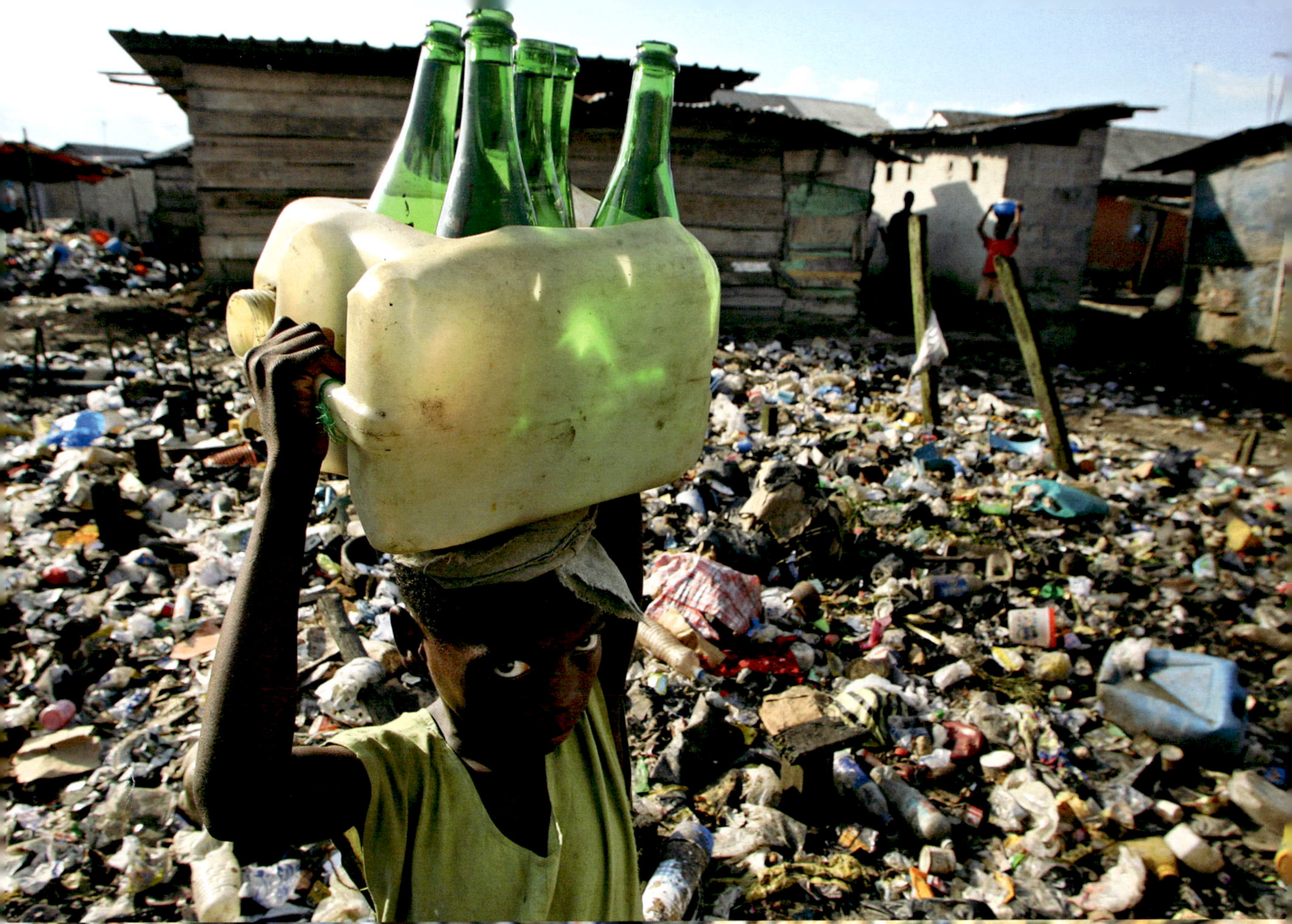
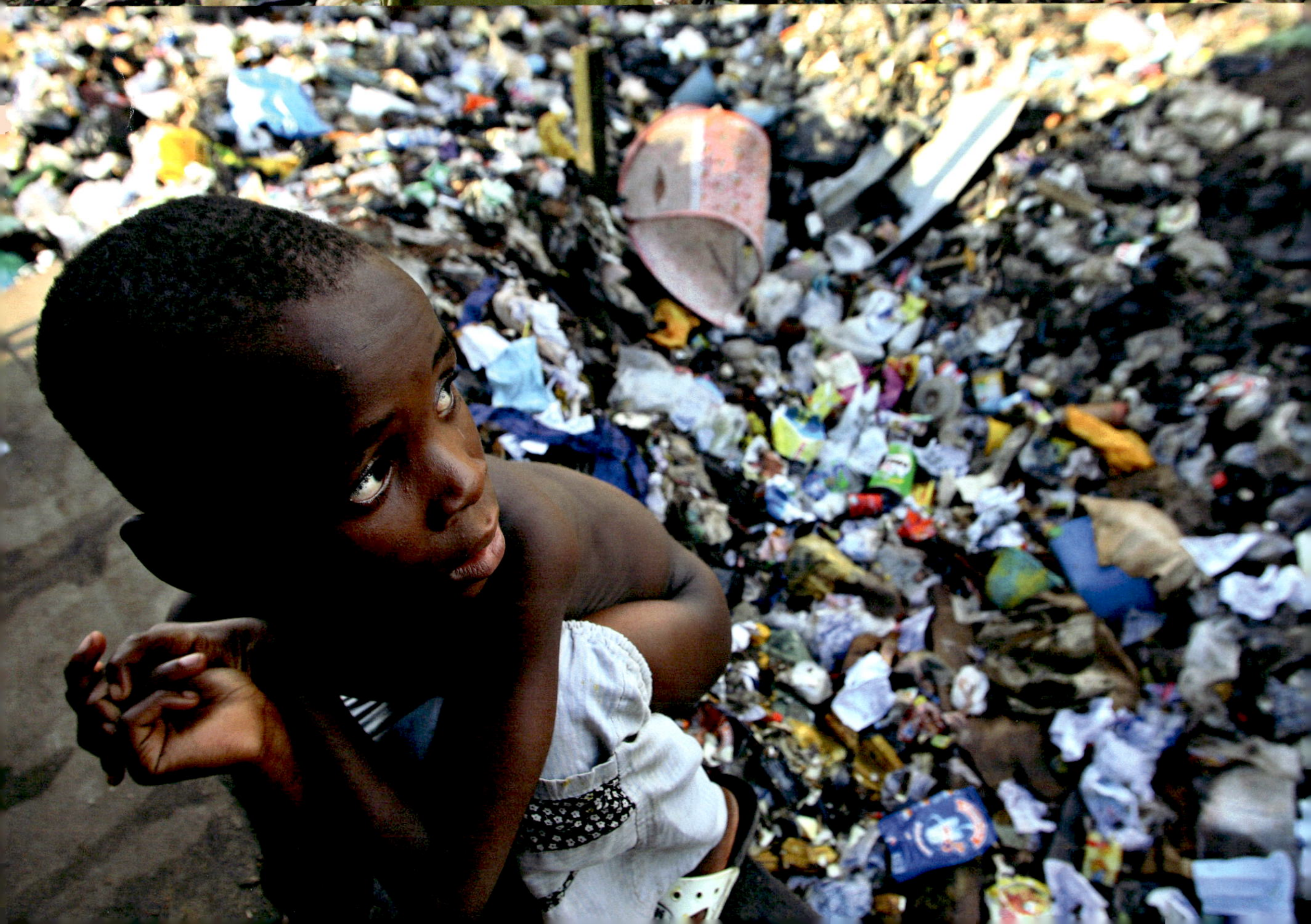

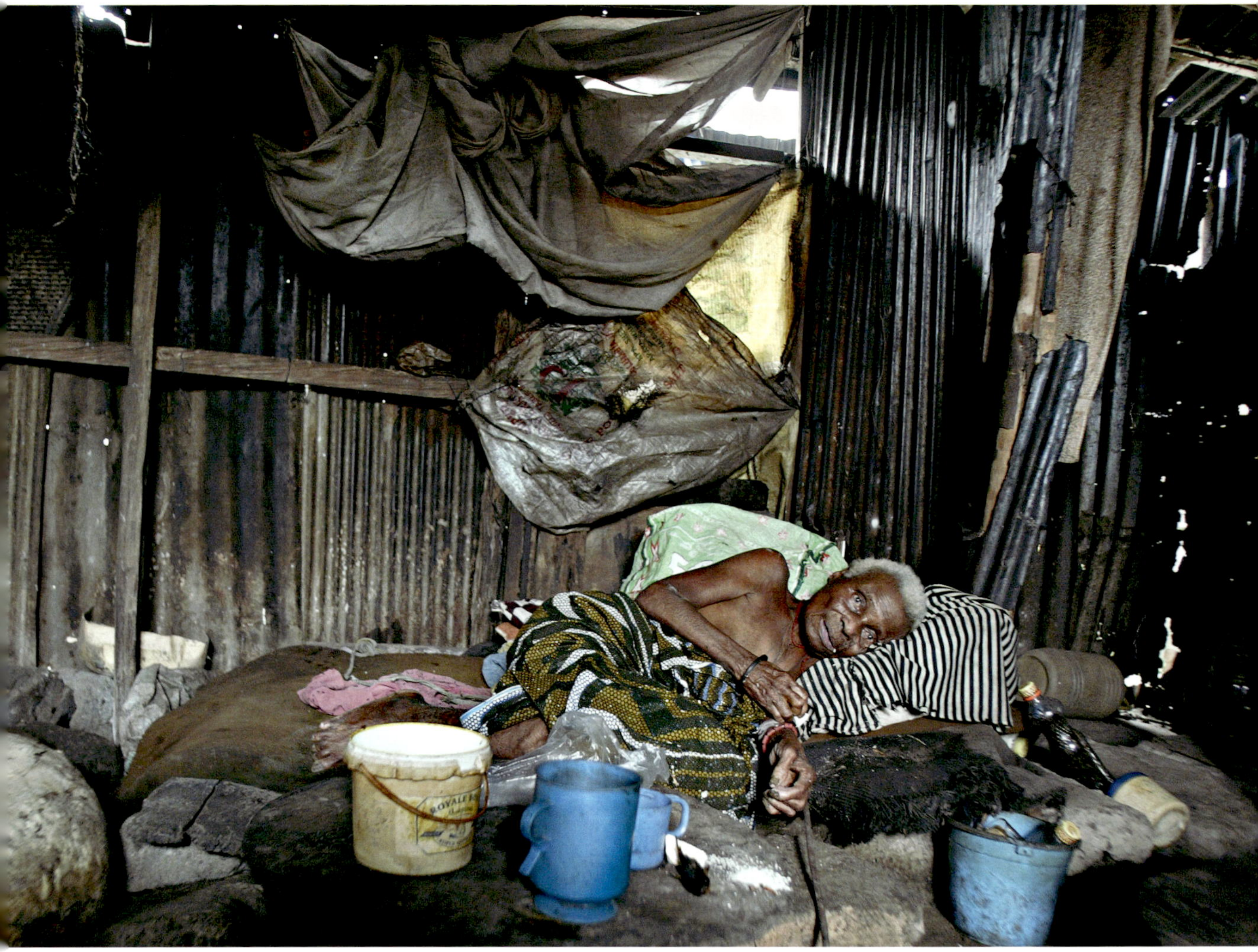

Above: An old woman lies sick at home with no money to travel to the city for medical help in Okerenkoko. Many local residents living in the villages lack access to medical facilities.

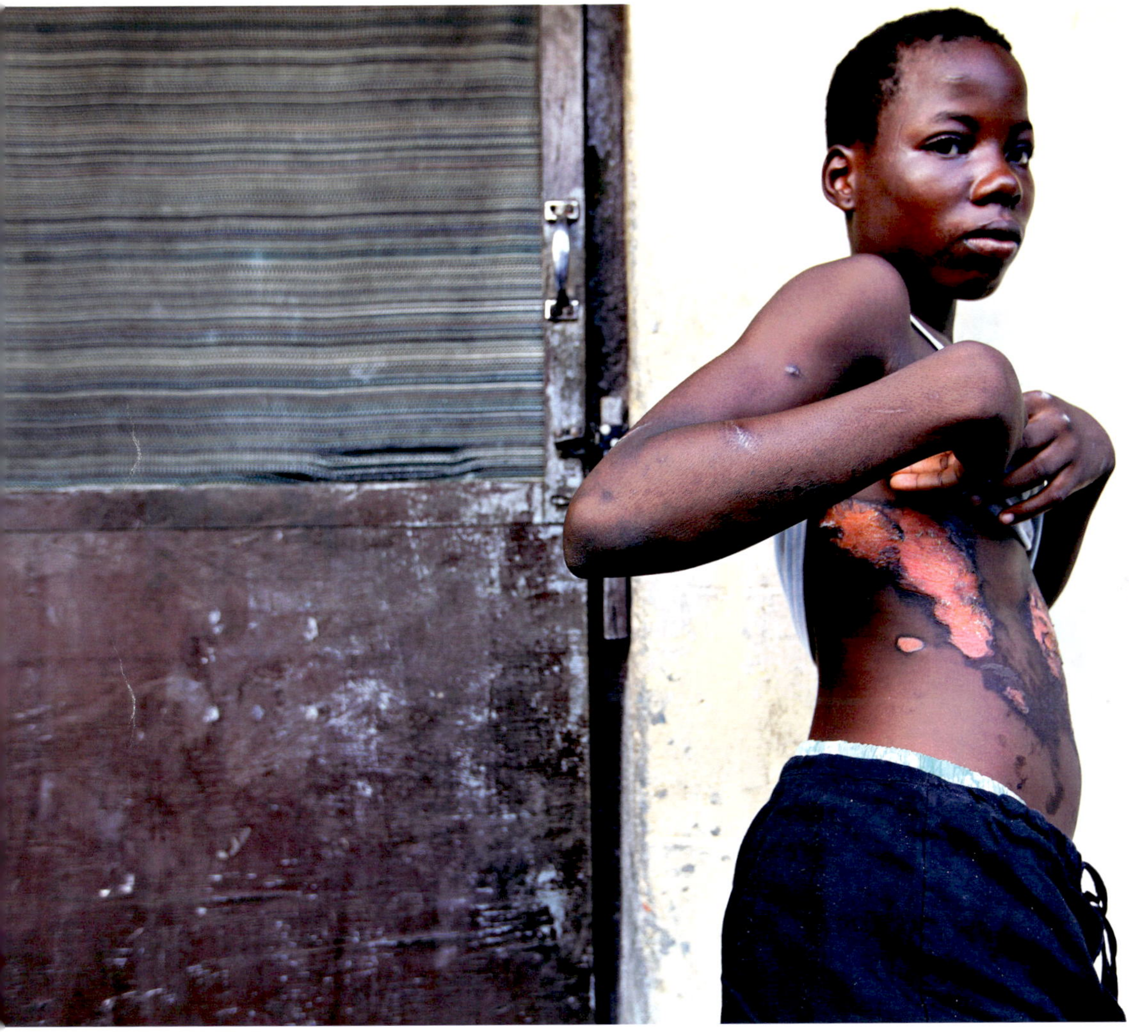

Above: Nnena Nwegewu, an orphan for 12 years, lives with her elder sister who poured hot water on her for refusing to engage in sex slave activities, Port Harcourt, 2011.

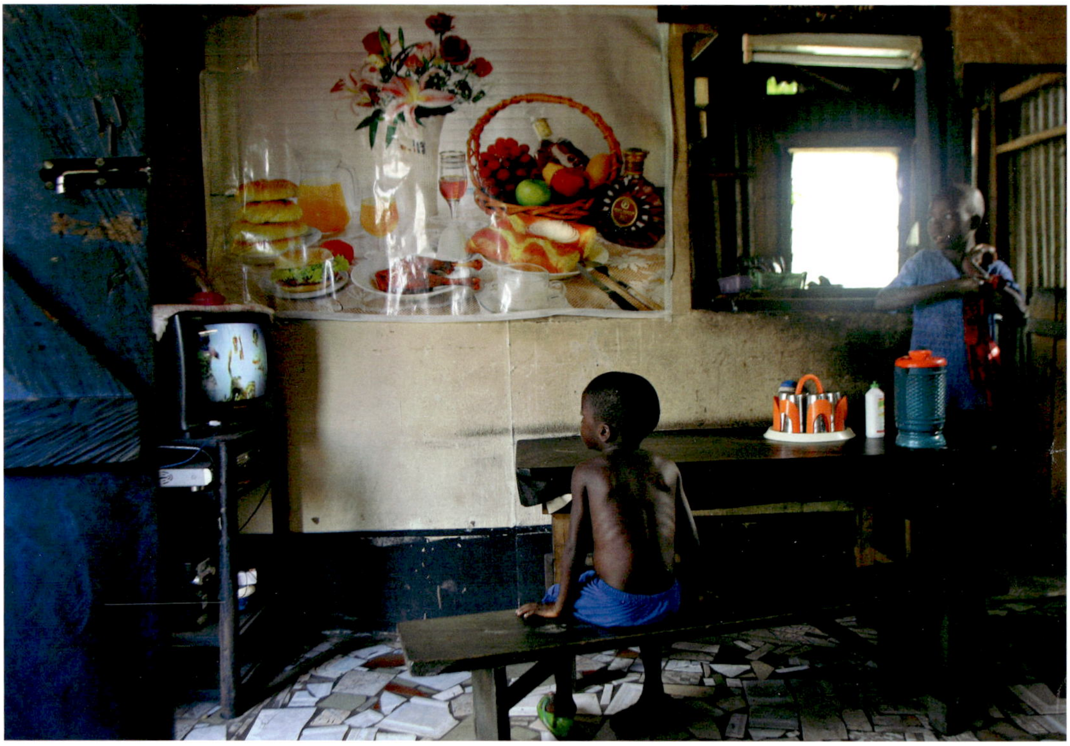

Above: A young boy sits watching television at a restaurant in Sangana.

Right: Children walk towards their house after taking their bath in Jones Creek, 2003.

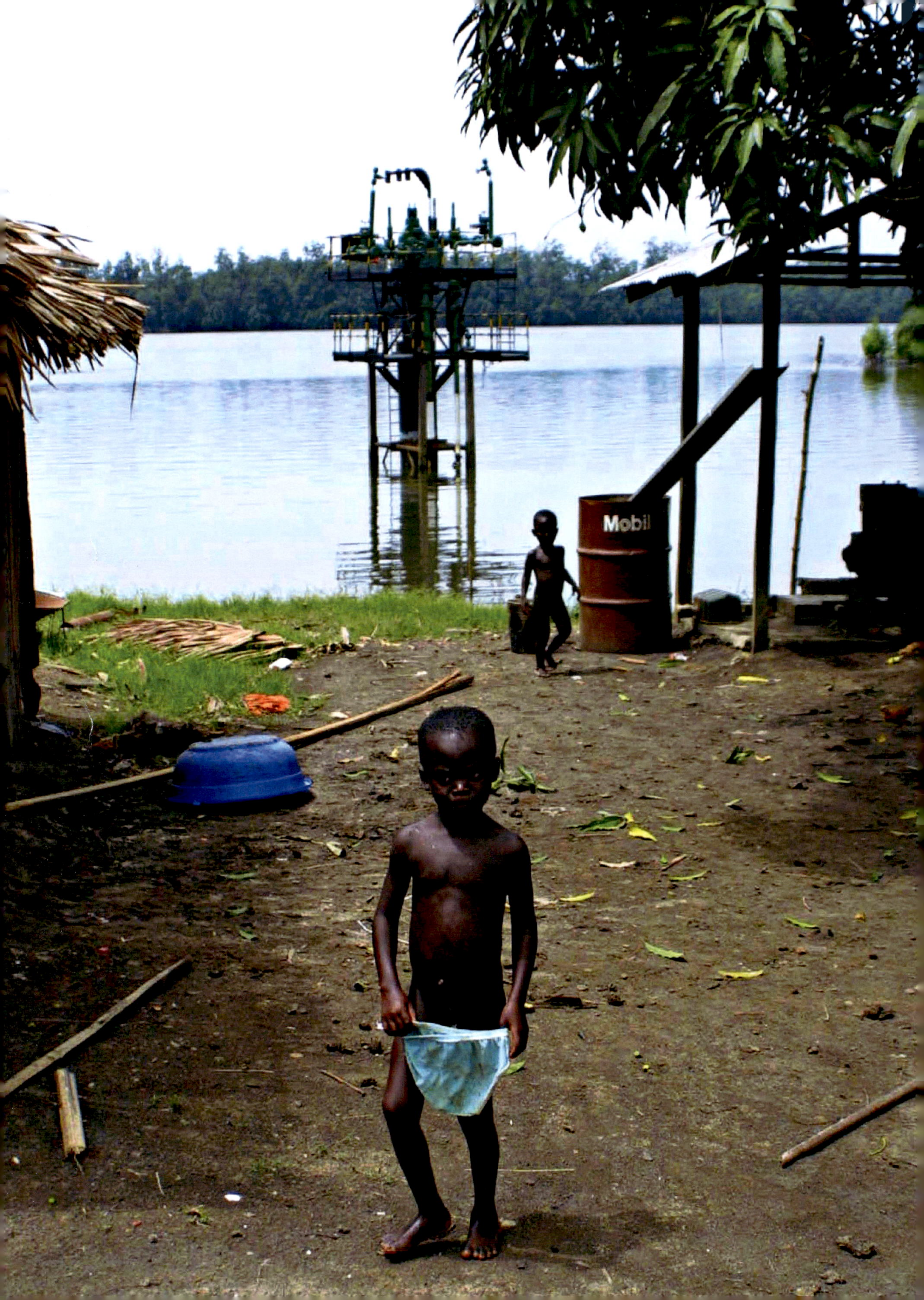
Mobil

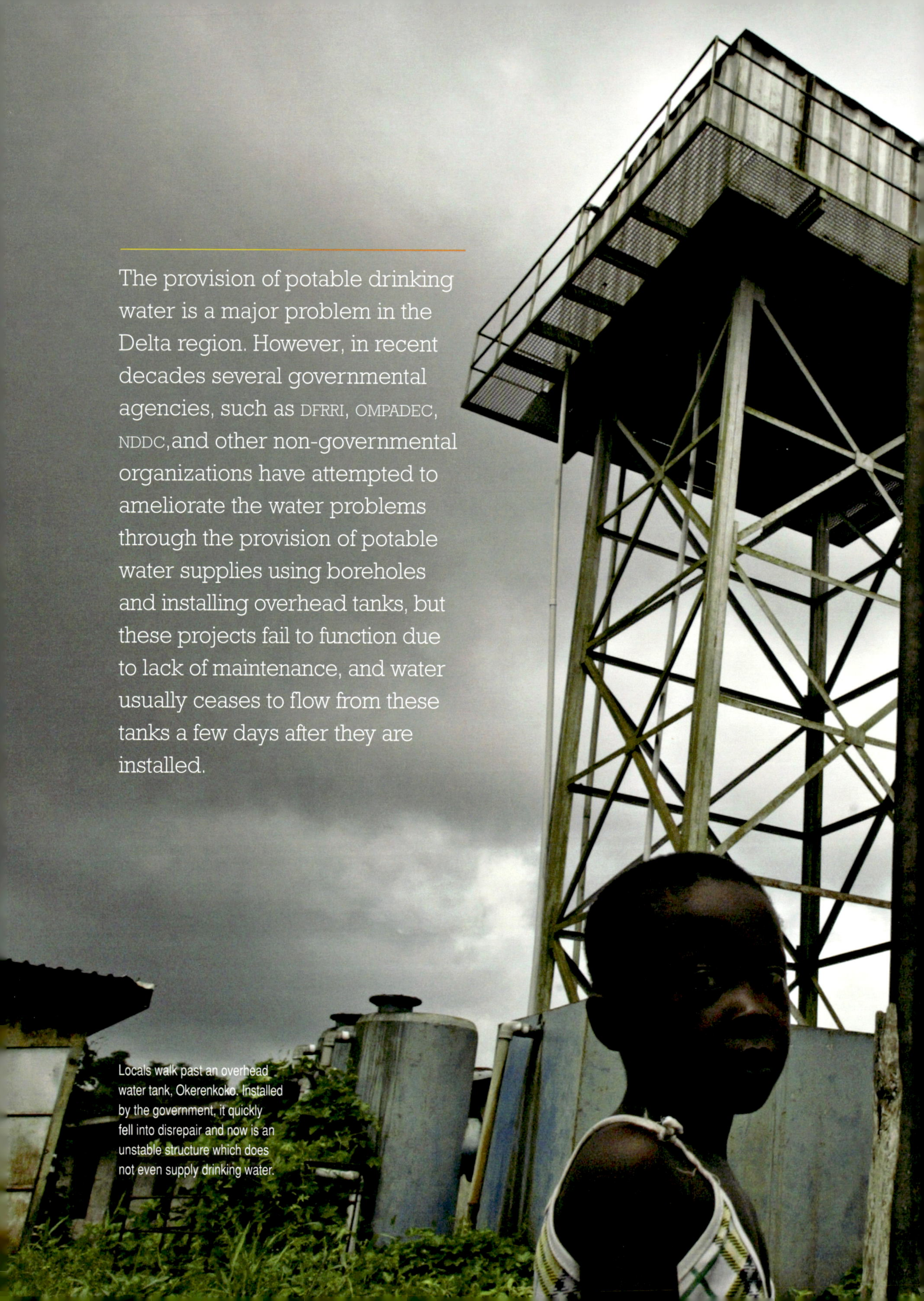

The provision of potable drinking water is a major problem in the Delta region. However, in recent decades several governmental agencies, such as DFRRI, OMPADEC, NDDC, and other non-governmental organizations have attempted to ameliorate the water problems through the provision of potable water supplies using boreholes and installing overhead tanks, but these projects fail to function due to lack of maintenance, and water usually ceases to flow from these tanks a few days after they are installed.

Locals walk past an overhead water tank, Okerenkoko. Installed by the government, it quickly fell into disrepair and now is an unstable structure which does not even supply drinking water.

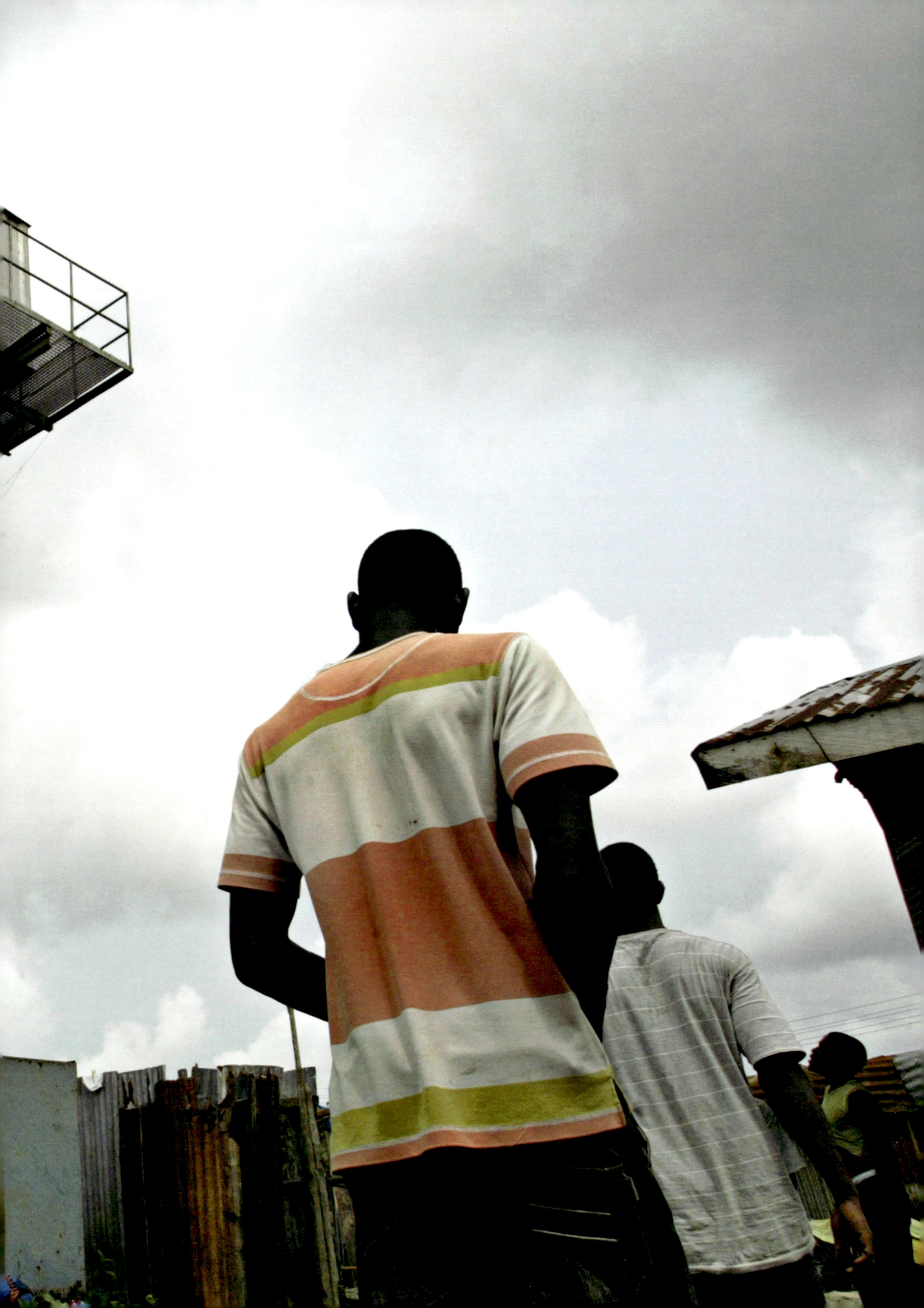

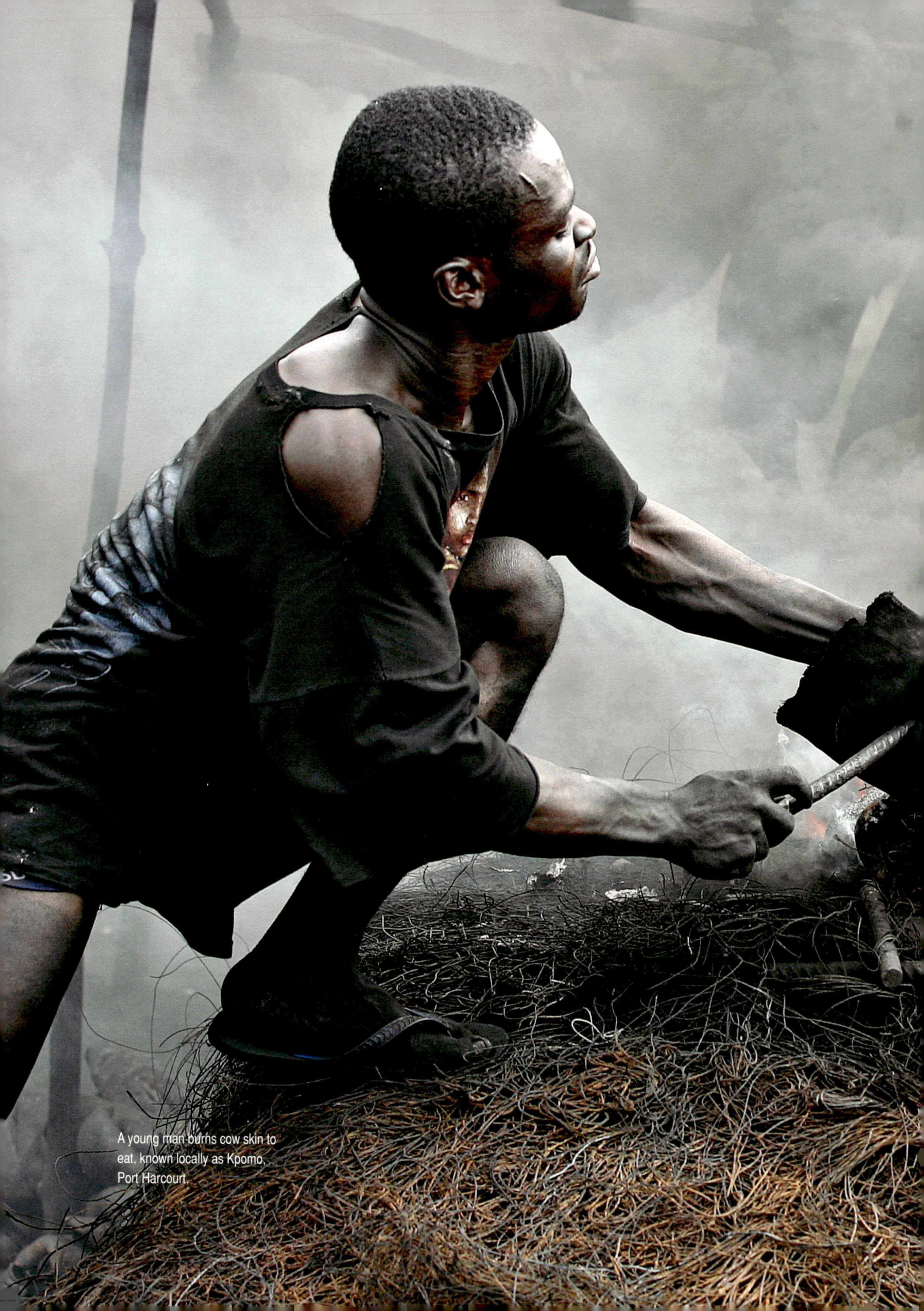

A young man burns cow skin to eat, known locally as Kpomo, Port Harcourt.

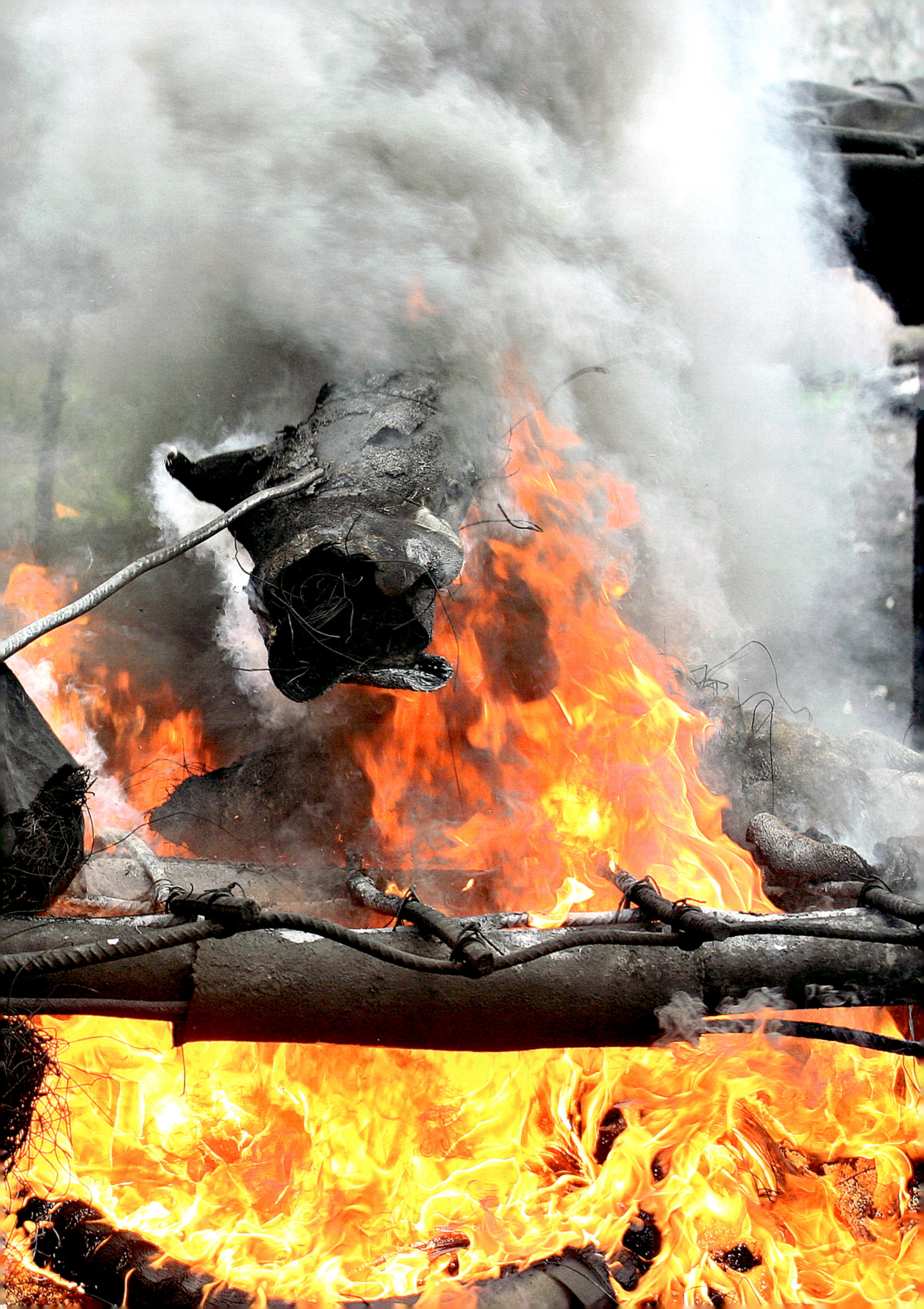

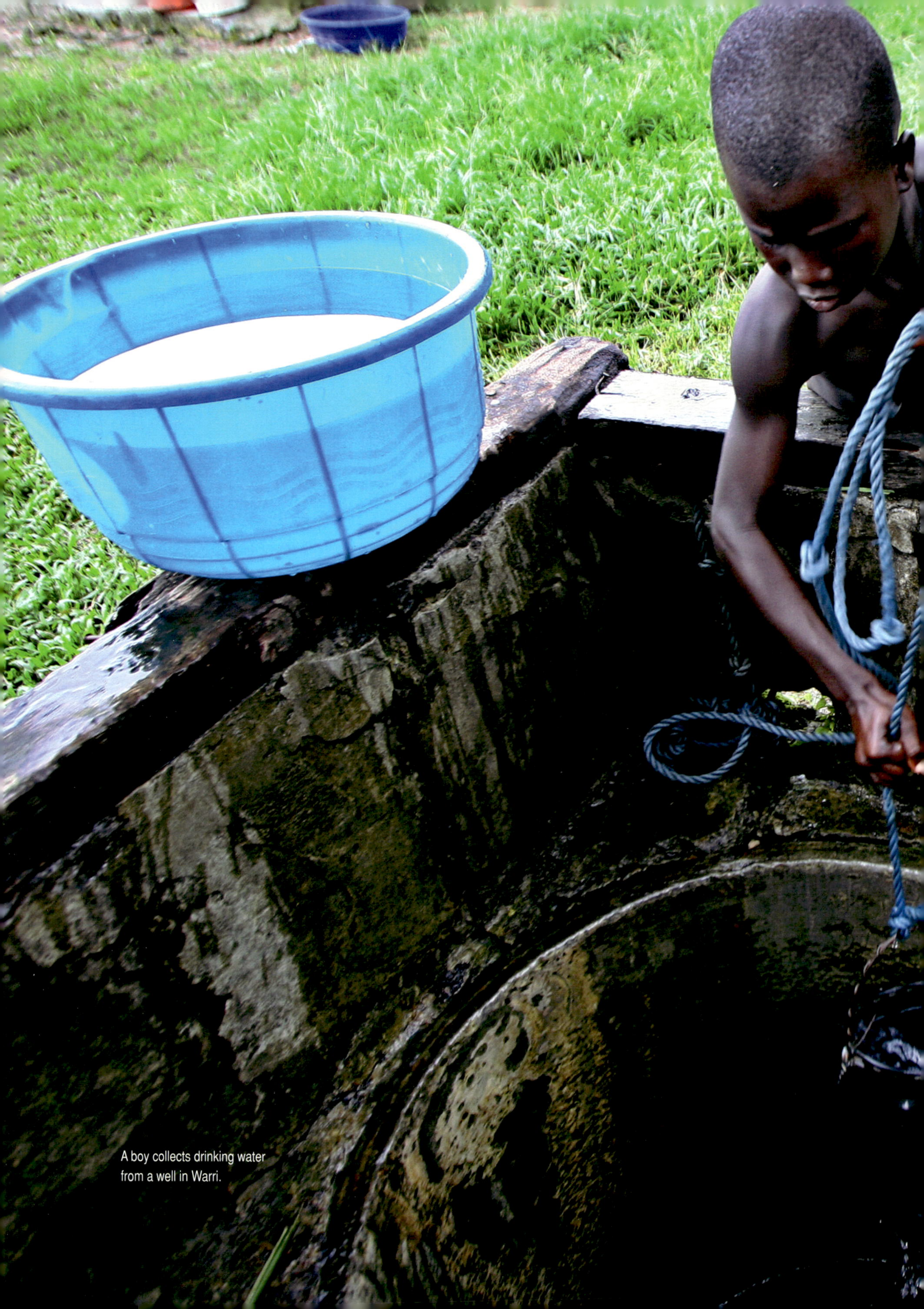

A boy collects drinking water
from a well in Warri.

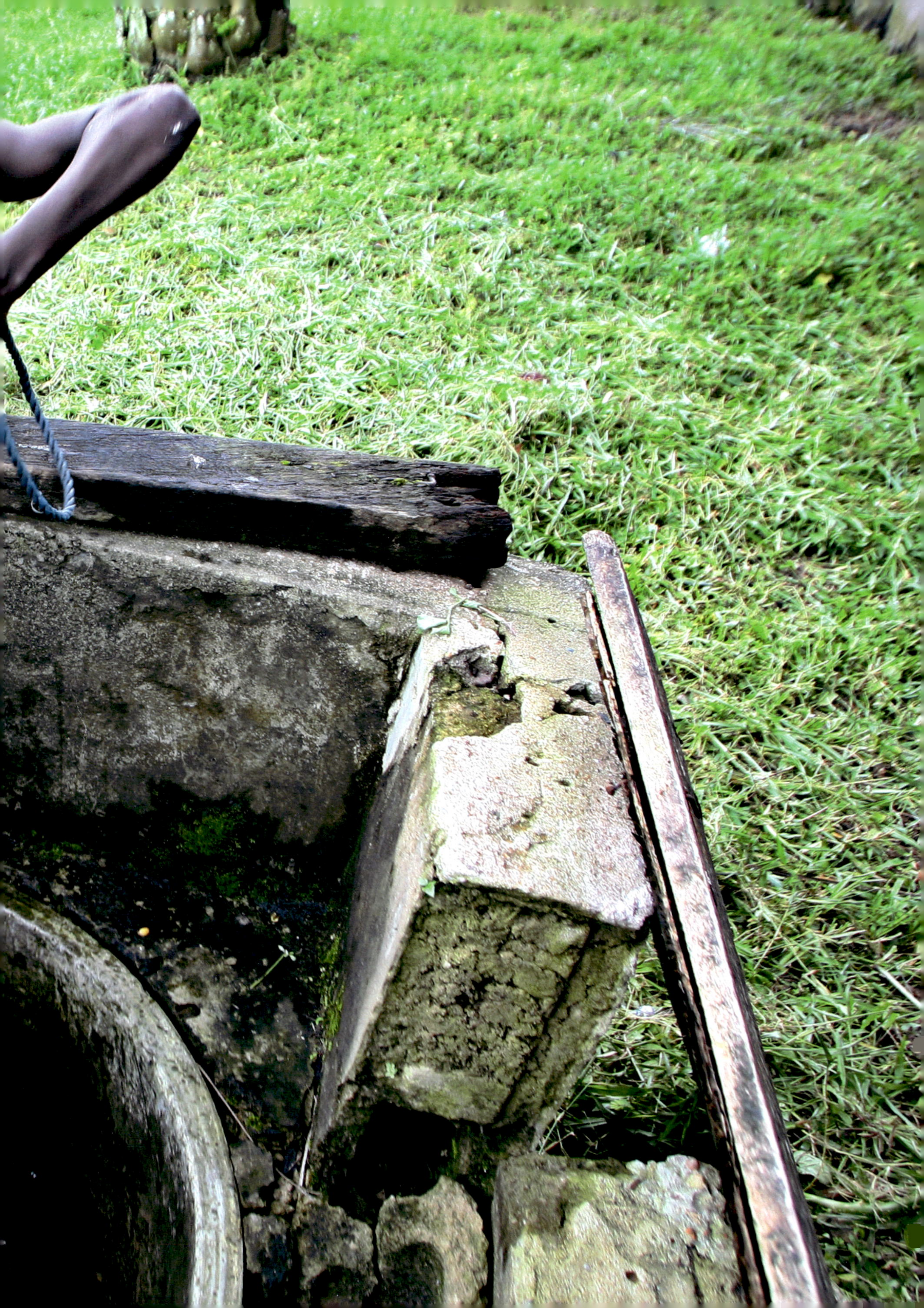

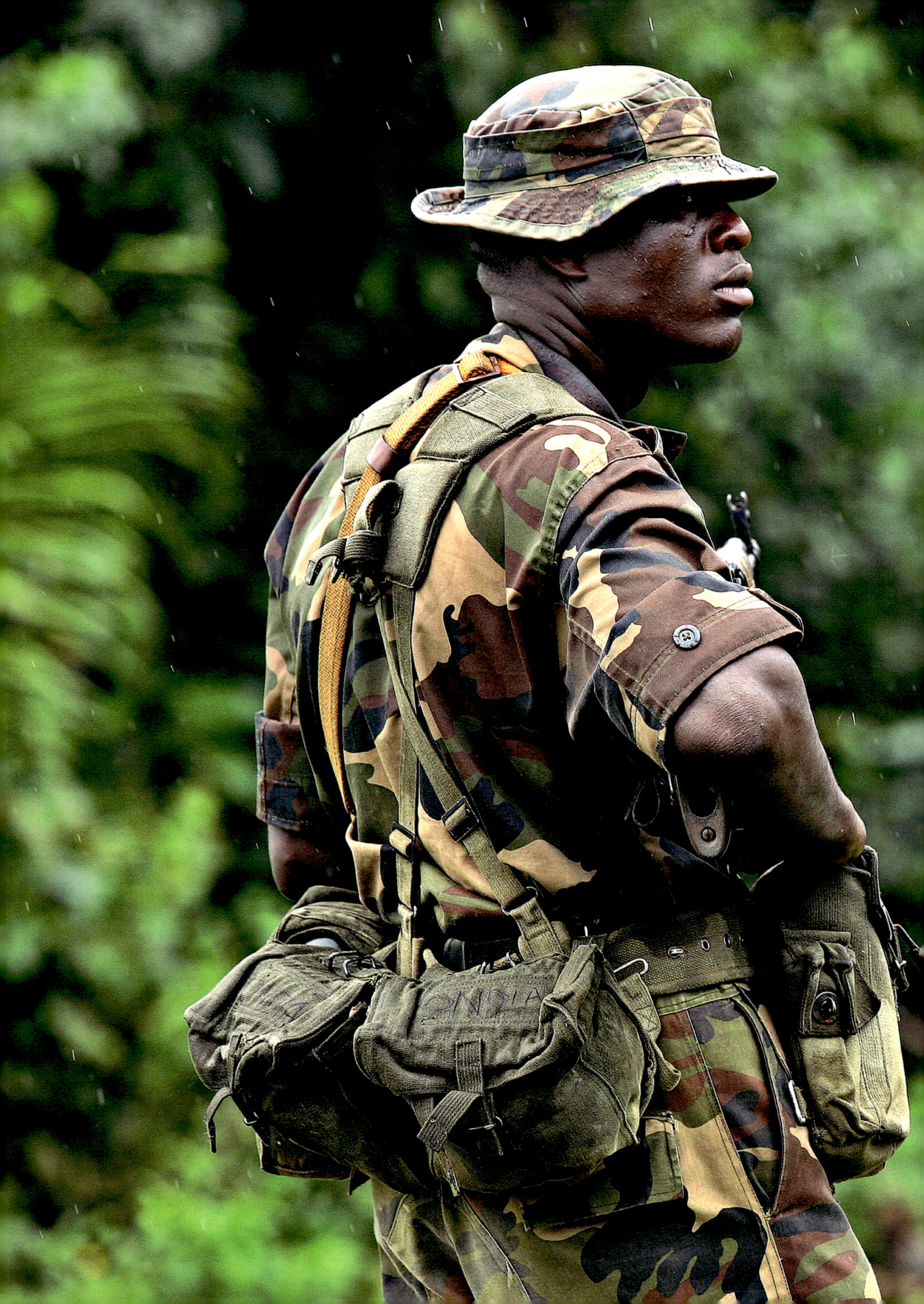

A government soldier stands guard outside international oil installation in the Delta region.

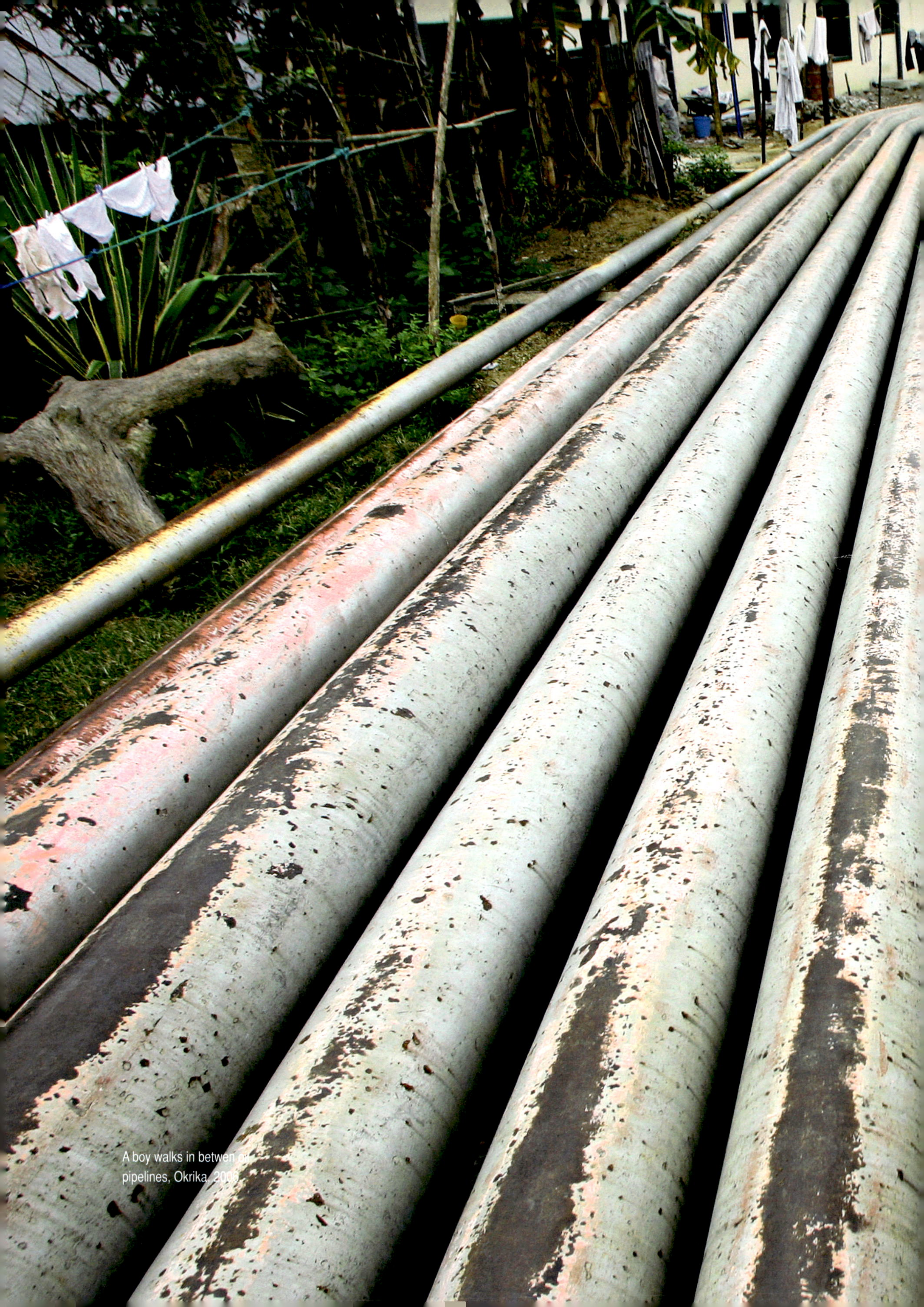

A boy walks in betwen oil pipelines, Okrika, 2006.

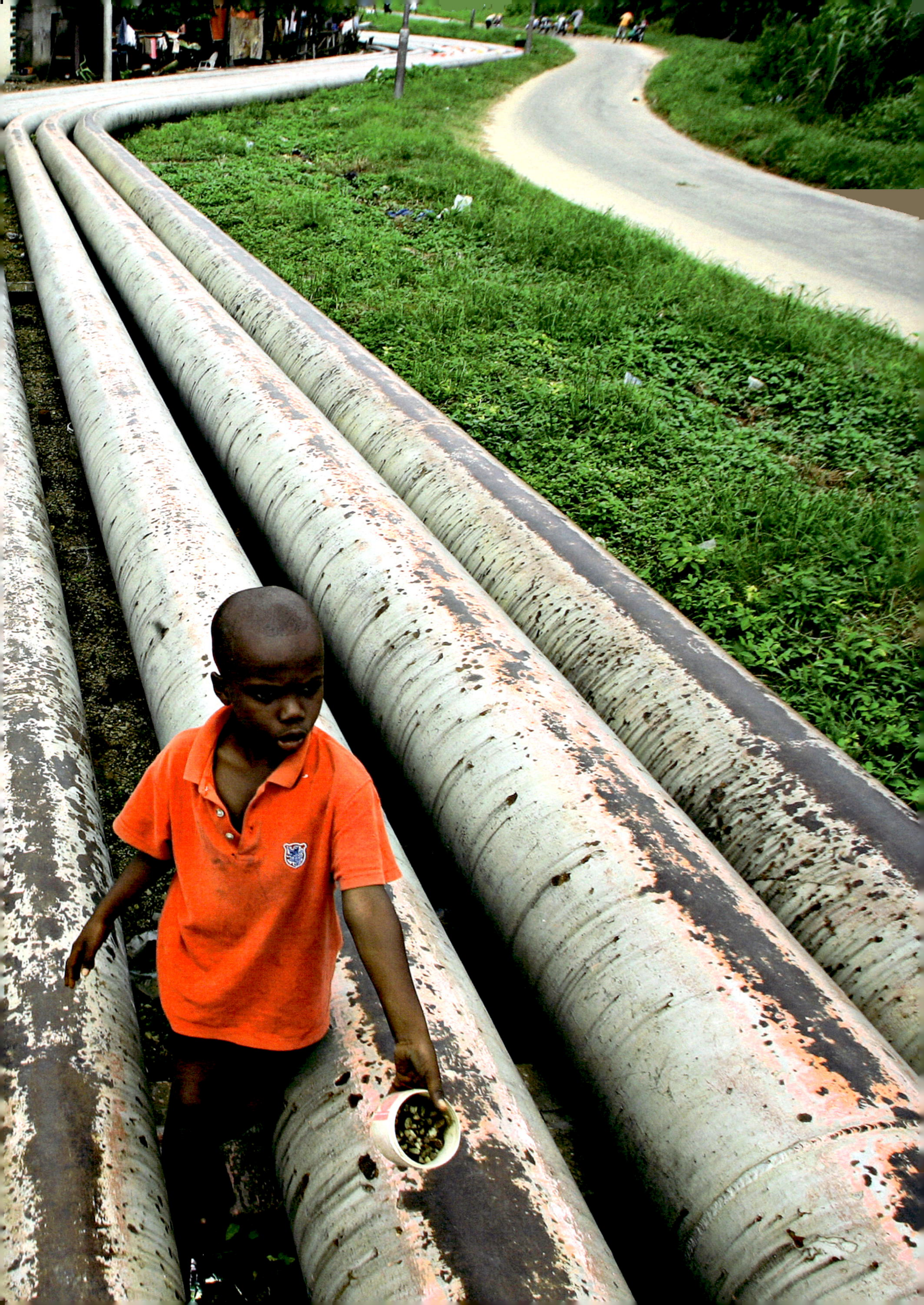

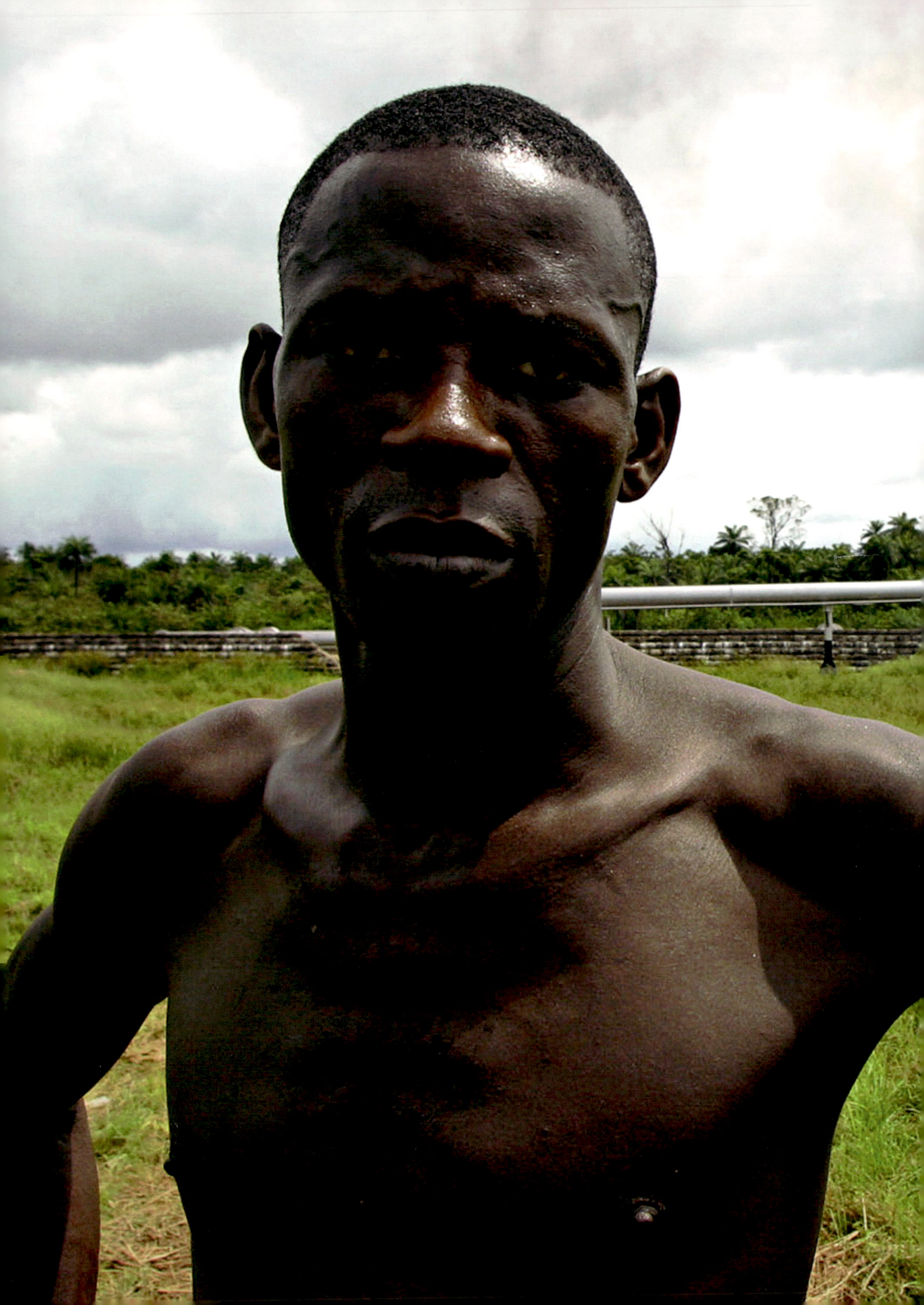

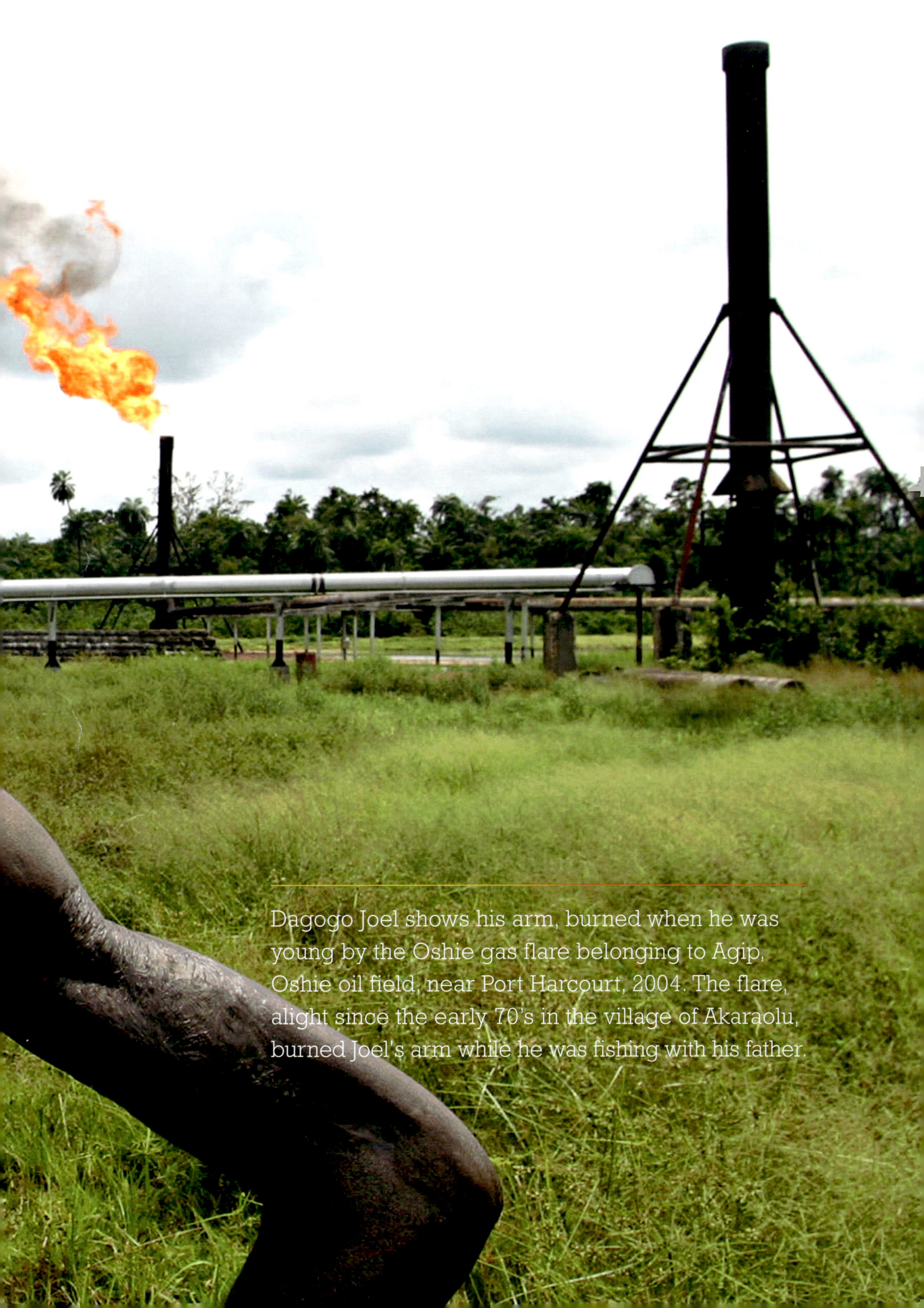

Dagogo Joel shows his arm, burned when he was young by the Oshie gas flare belonging to Agip, Oshie oil field, near Port Harcourt, 2004. The flare, alight since the early 70's in the village of Akaraolu, burned Joel's arm while he was fishing with his father.

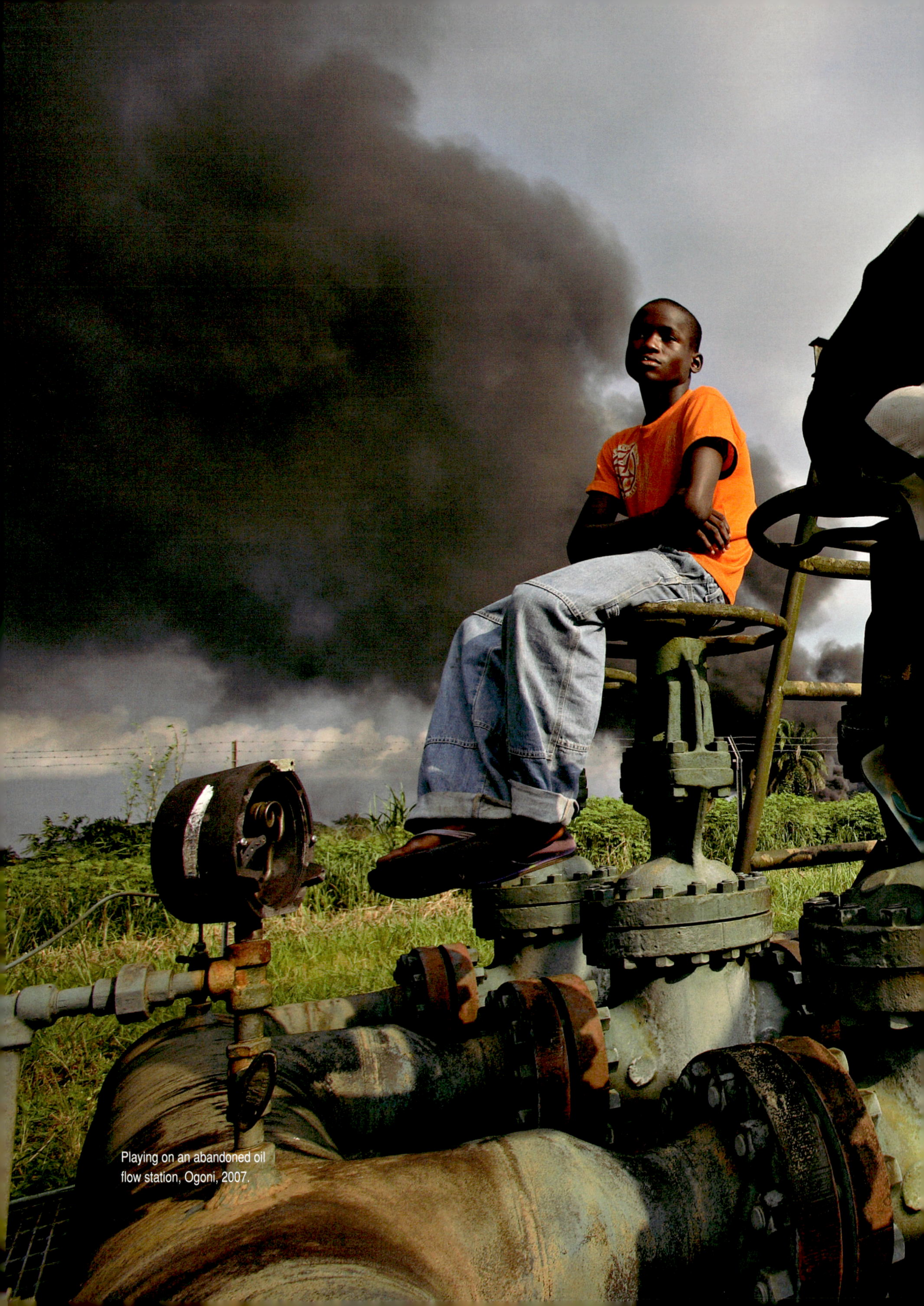

Playing on an abandoned oil flow station, Ogoni, 2007.

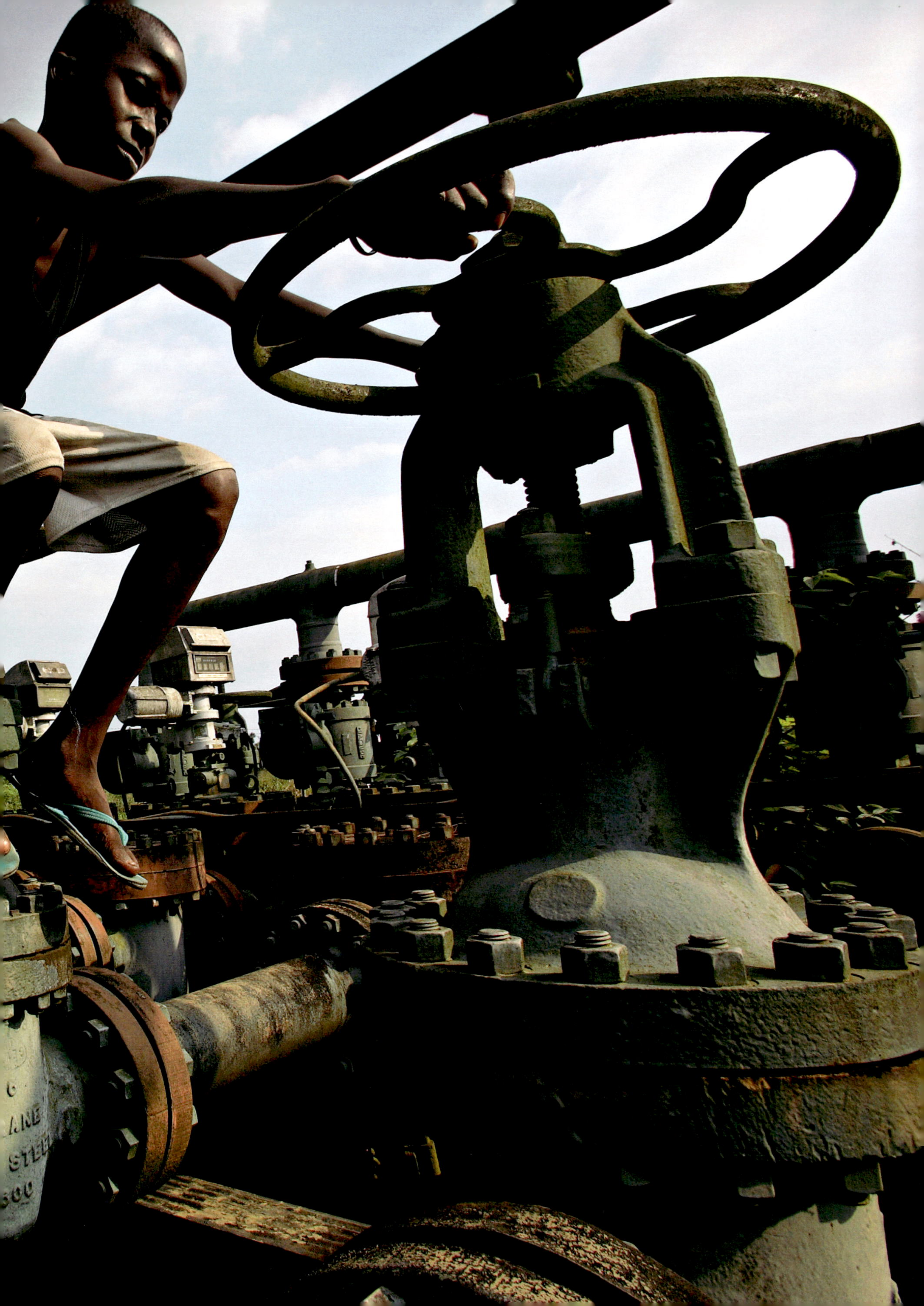

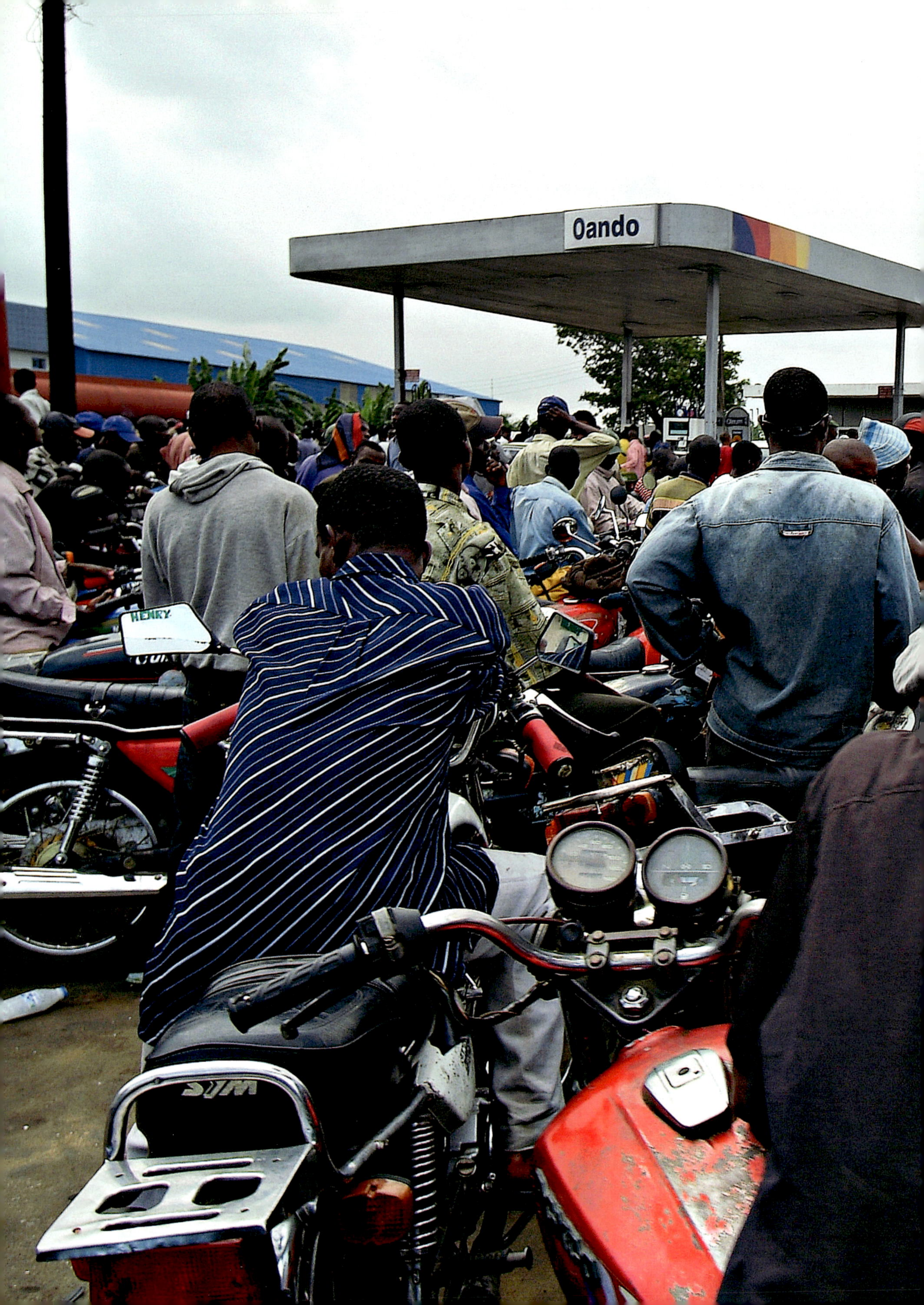
Oando
HENRY

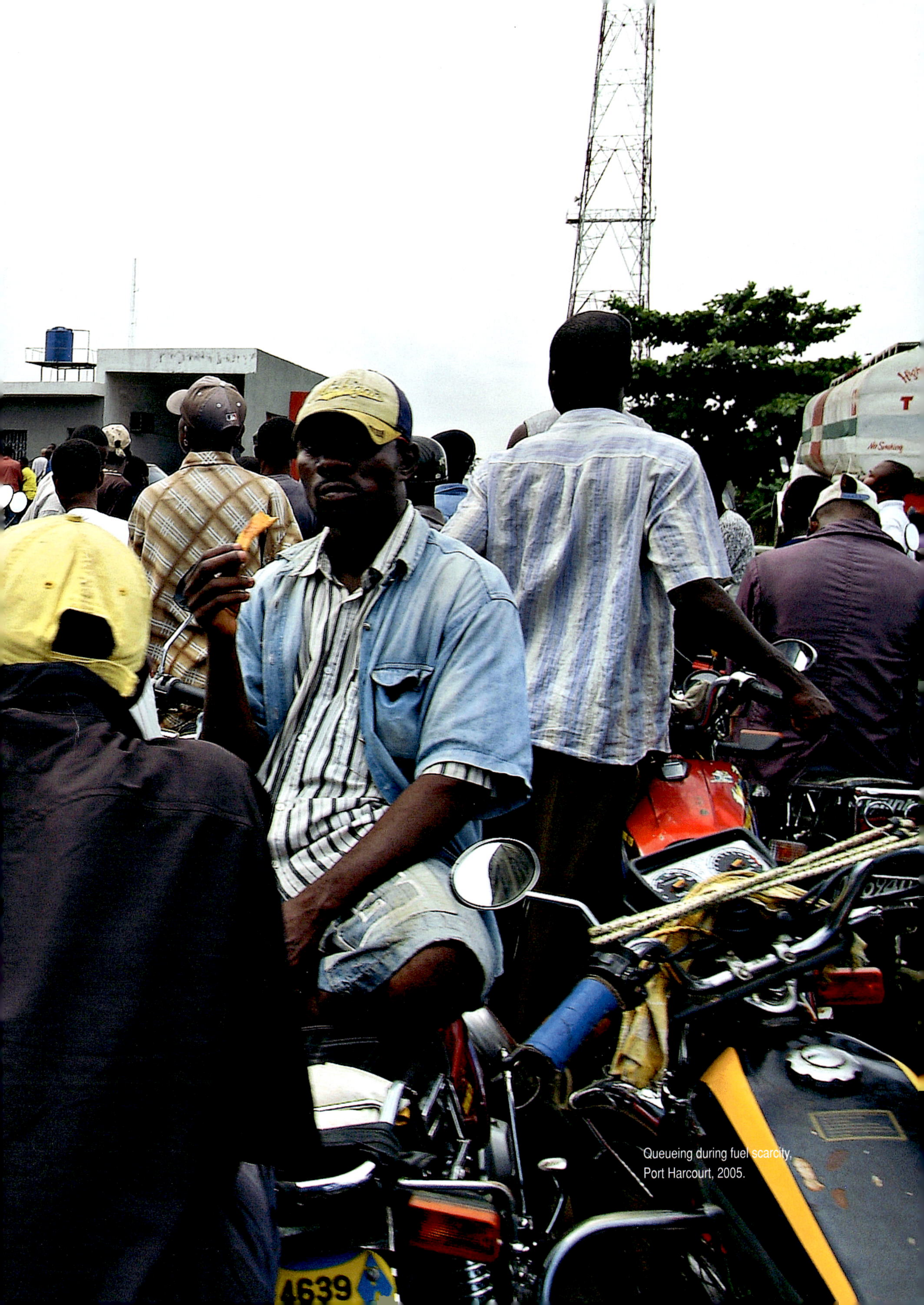

Queueing during fuel scarcity,
Port Harcourt, 2005.

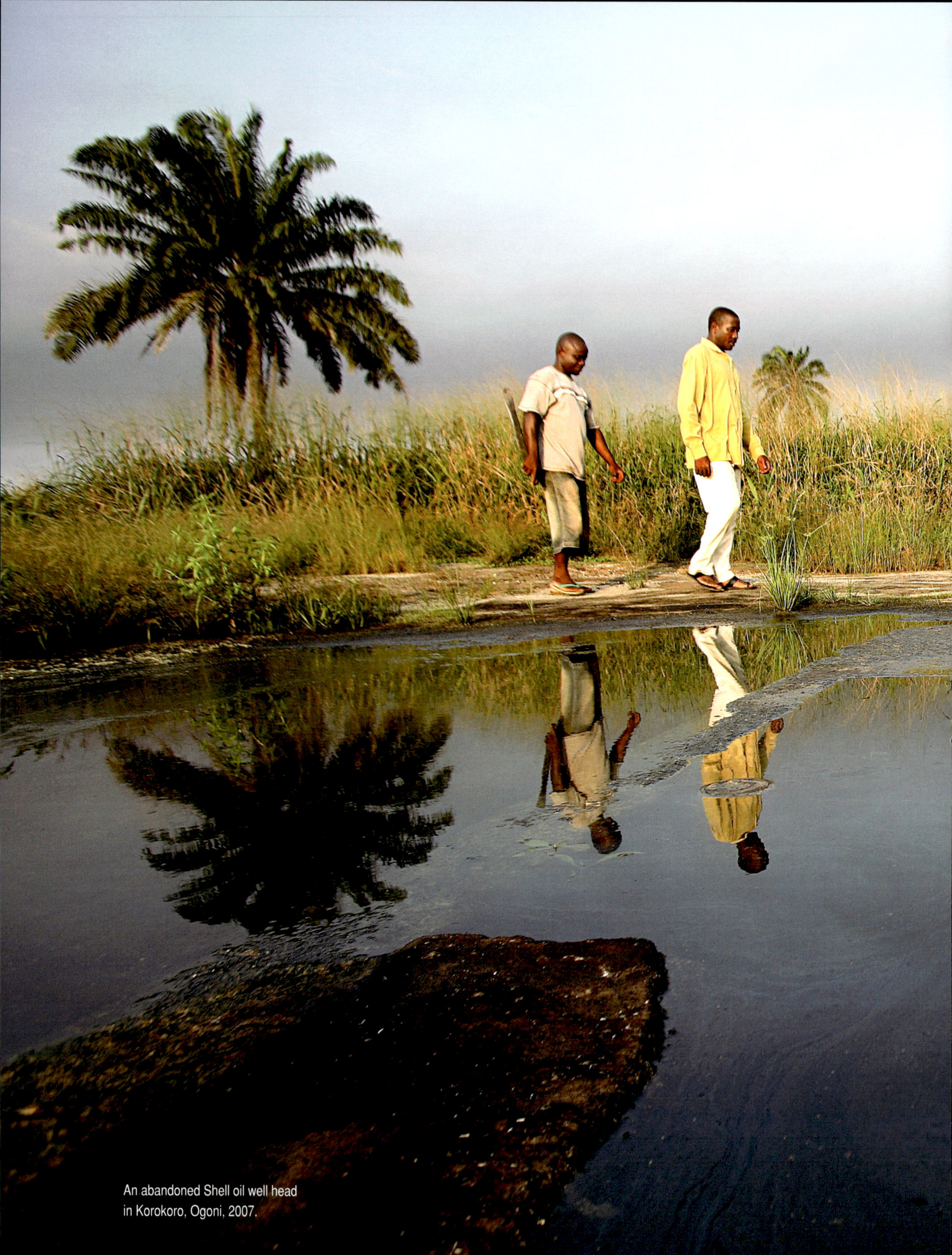

An abandoned Shell oil well head
in Korokoro, Ogoni, 2007.

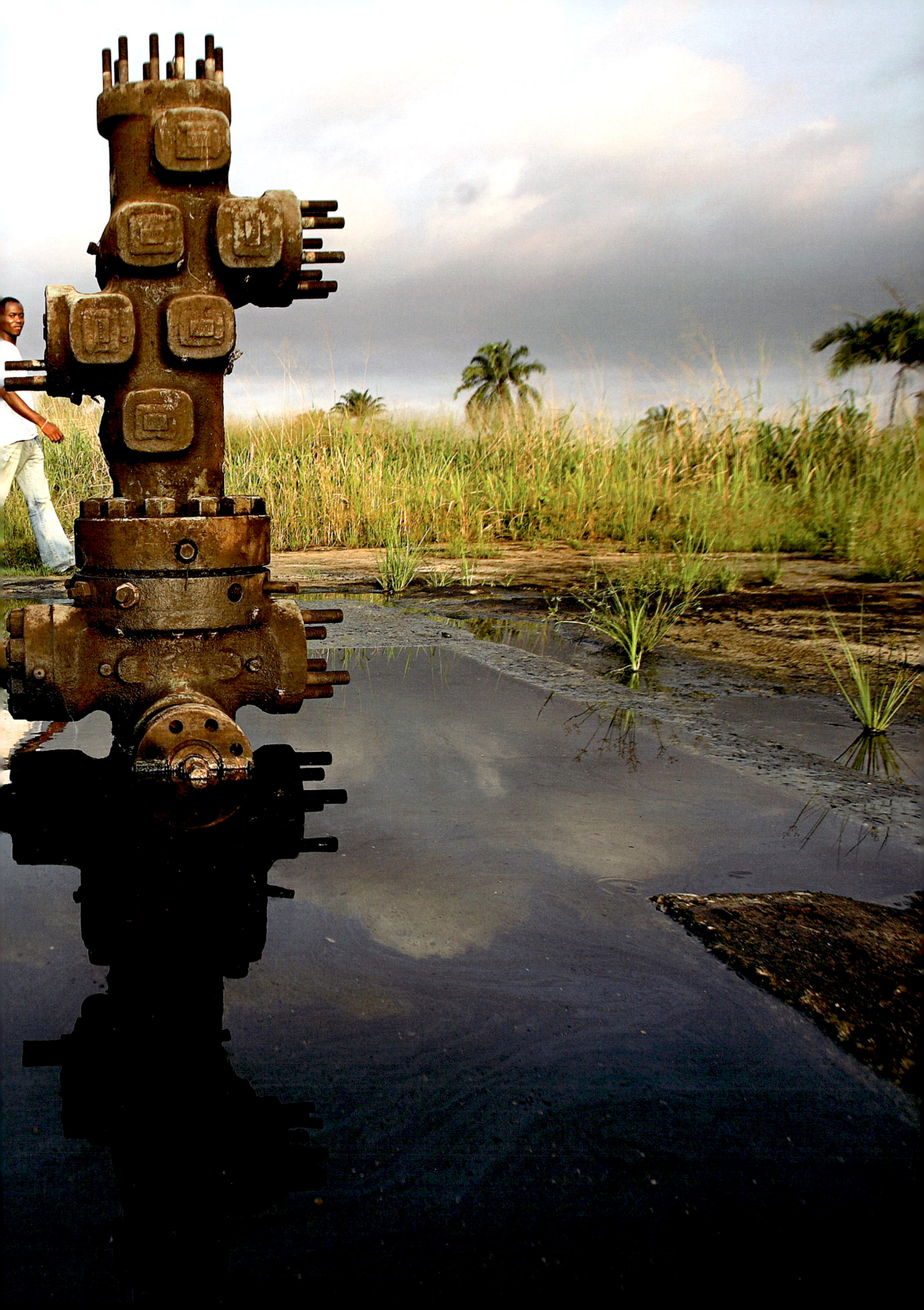

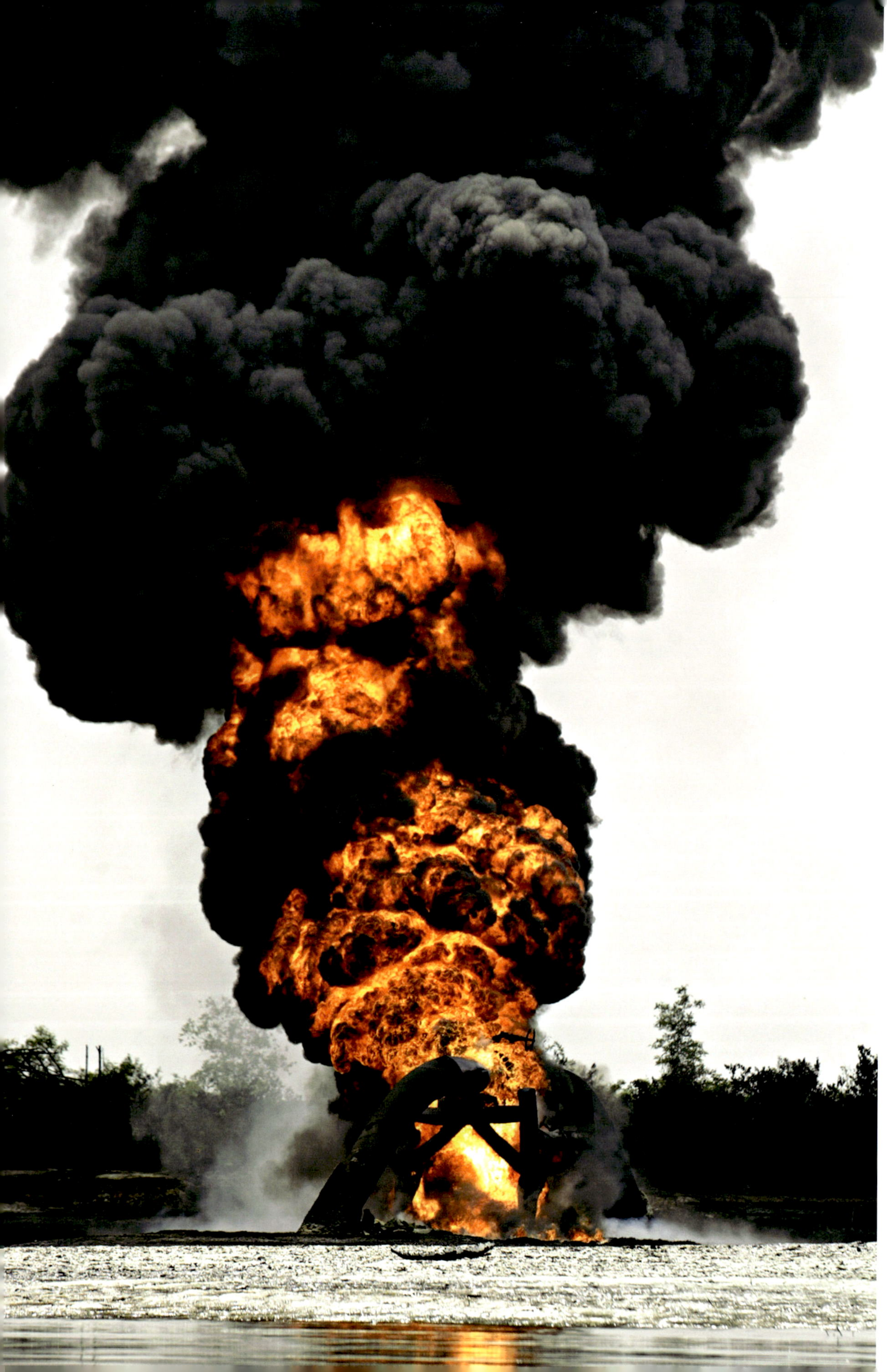

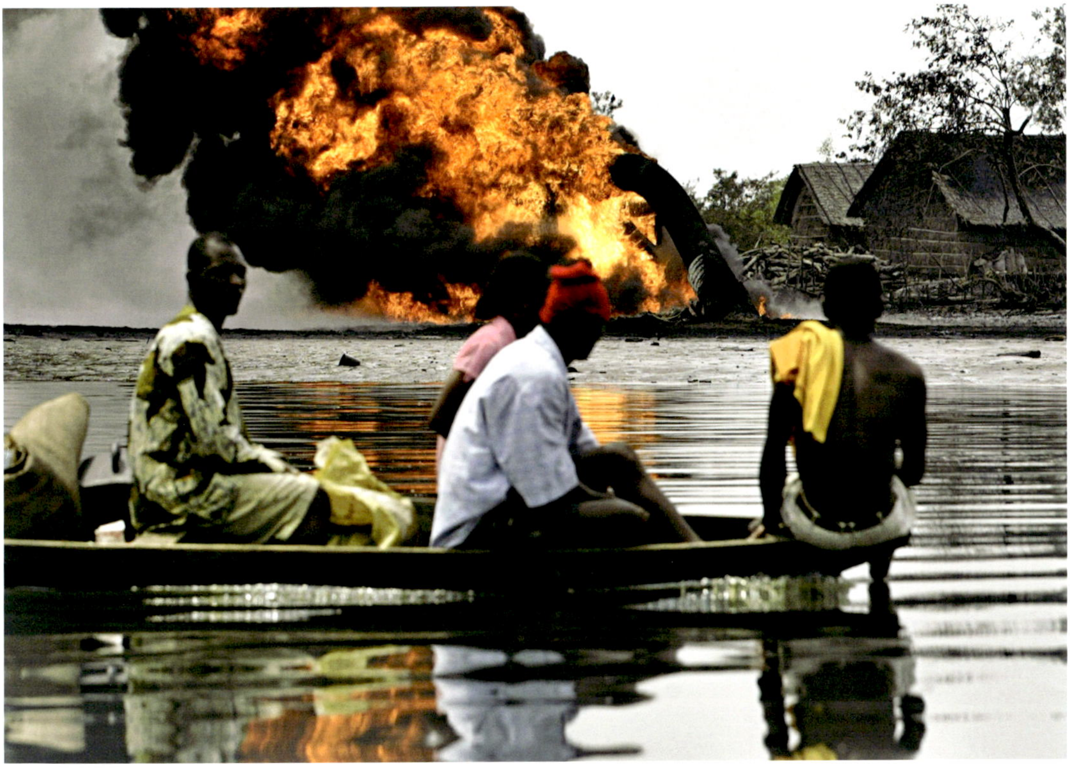

Smoke and flames billow from a burning oil pipeline blown out by MEND militants (Movement for the Emancipation of the Niger Delta) across the Opobo Channel, south-west Port Harcourt, 2005. Violence has been escalating in the region as the government military intensifies operations to flush out these rebels who are battling for a larger share of the country's oil revenues. The federal government announced a loss of about 1,101,488 barrels of oil per day (bopd) during the first quarter of 2009 alone, mainly due to the threat of these militants in the Delta region.

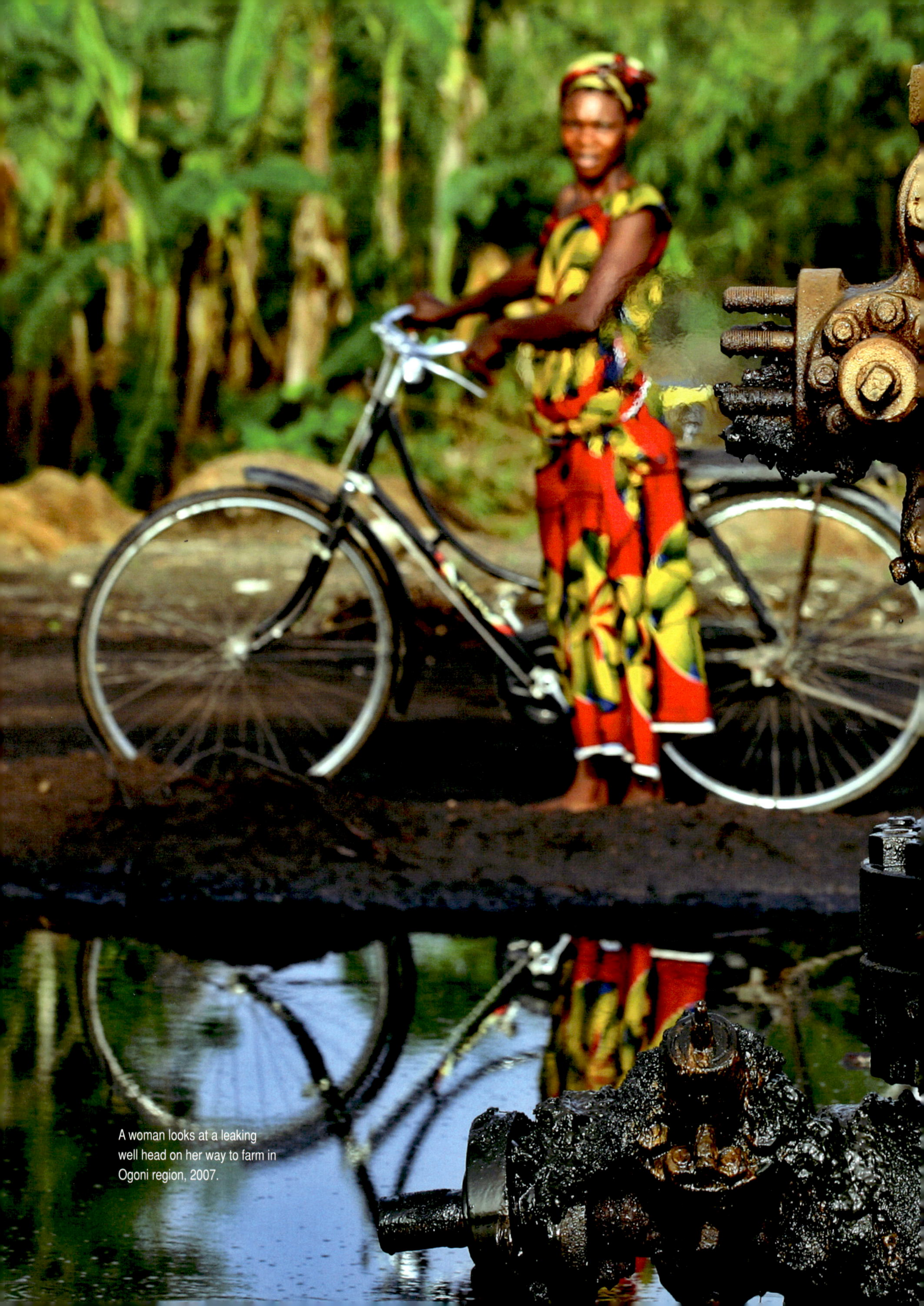

A woman looks at a leaking well head on her way to farm in Ogoni region, 2007.

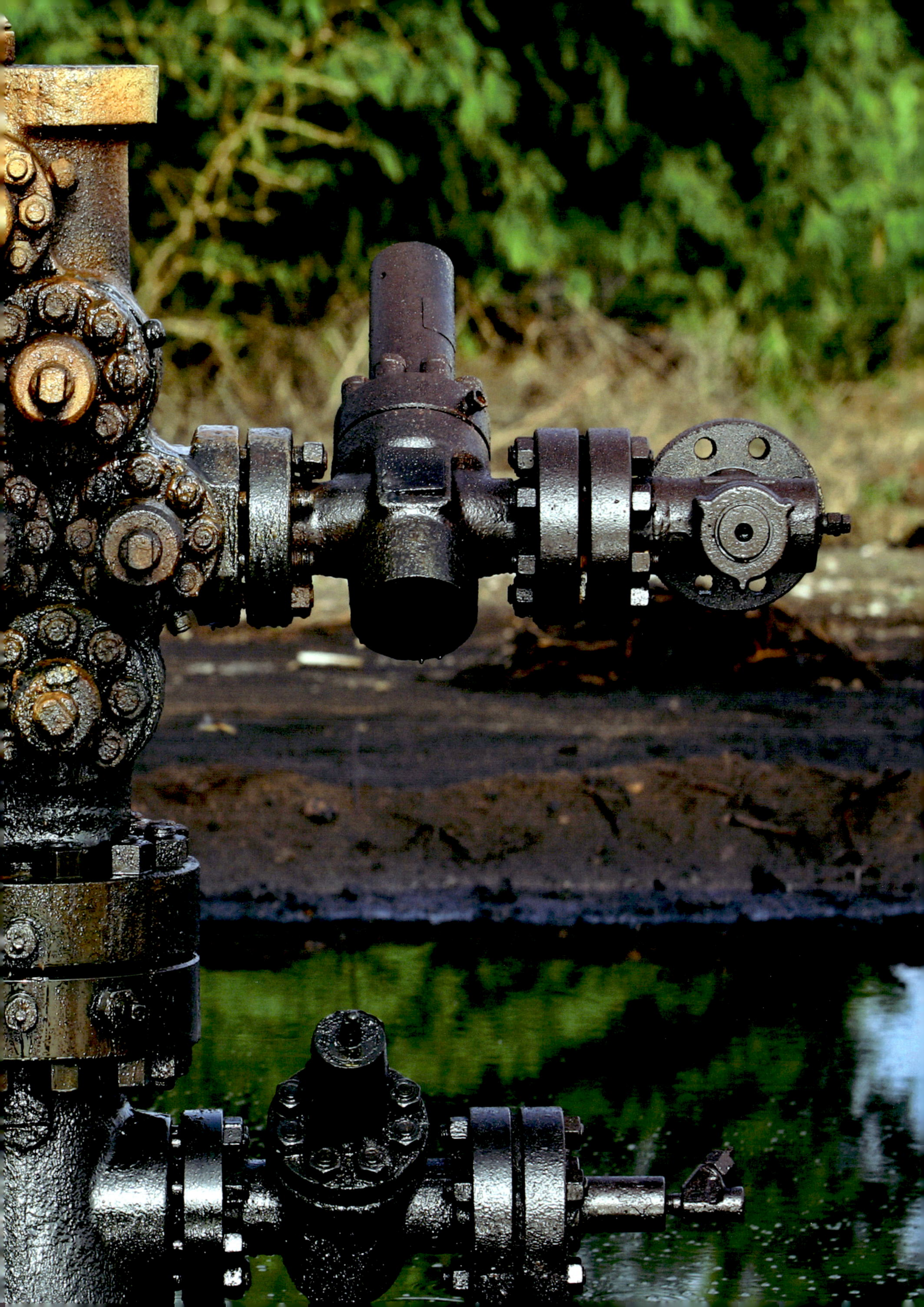

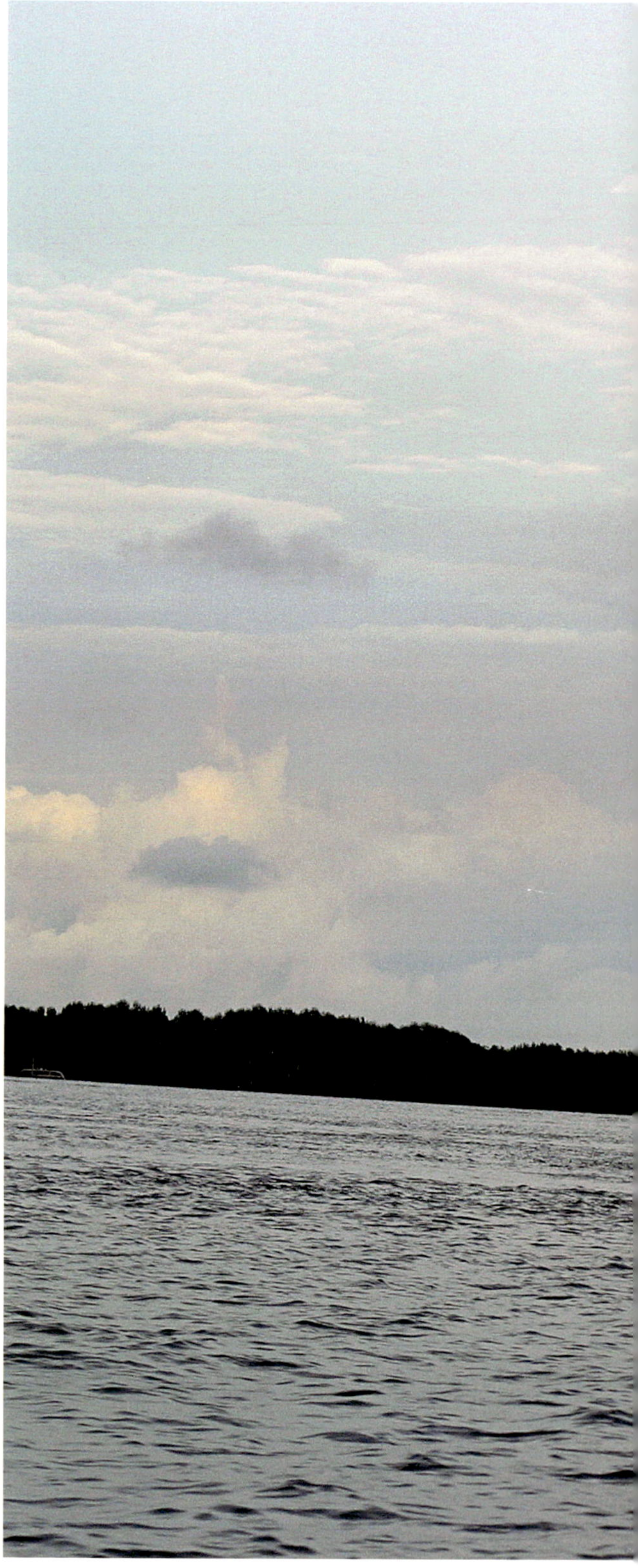

Abandoned oil vessel in Delta creek, 2003.

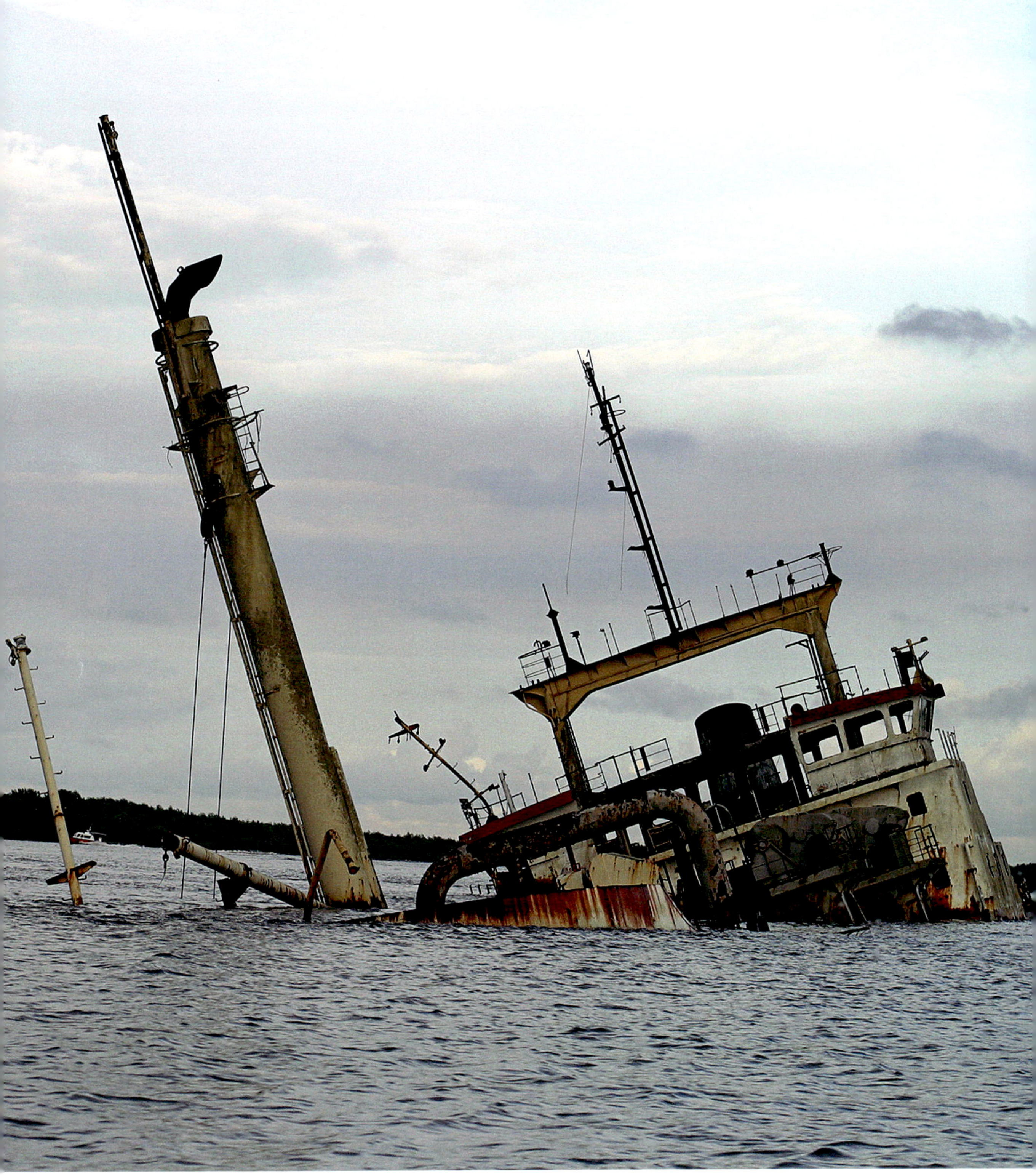

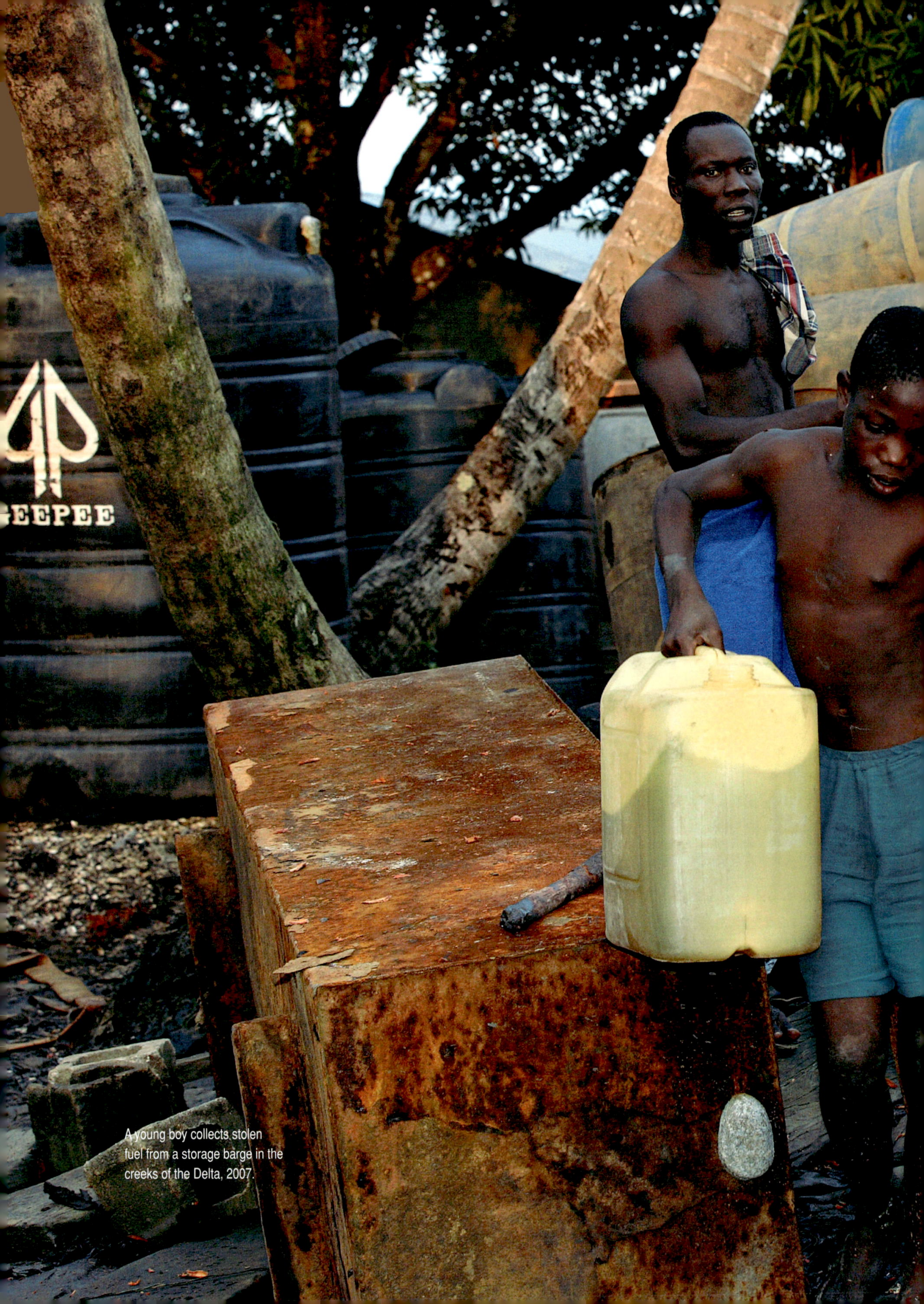

A young boy collects stolen fuel from a storage barge in the creeks of the Delta, 2007.

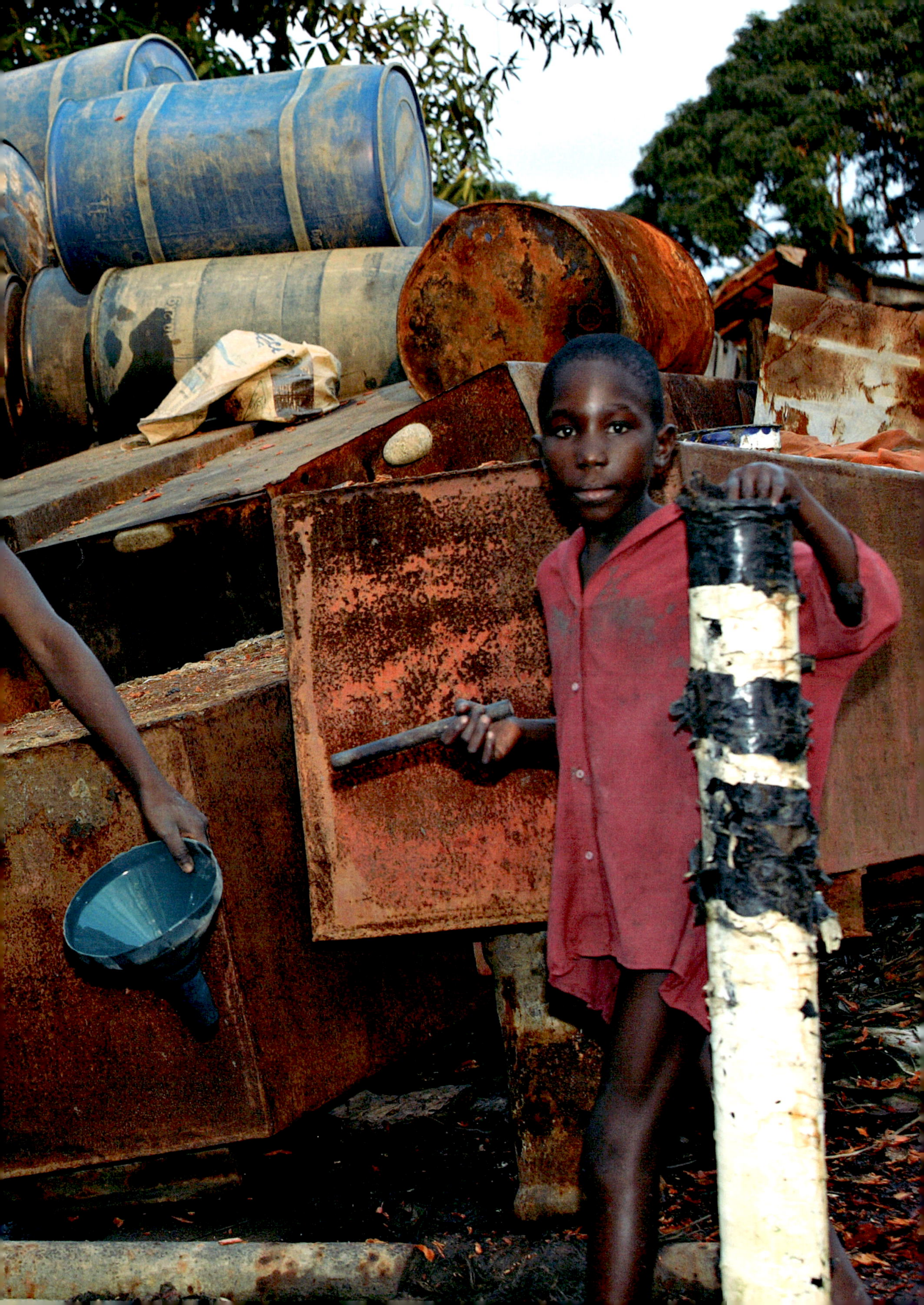

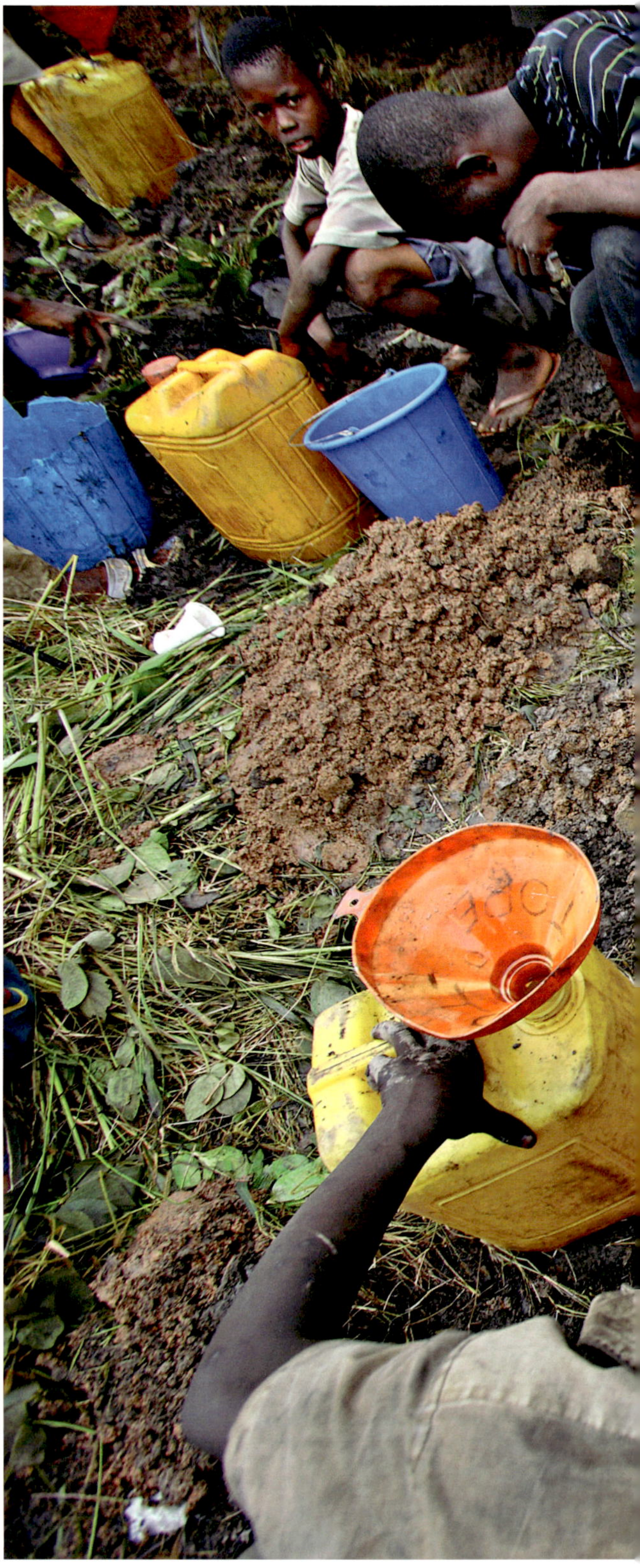

Local youths scoop fuel in Ondo 2008. Poverty, neglect and the quest for survival drives youths to break into pipelines for fuel theft, without care for the danger of this act to their health and the economy.

Overleaf left: Charred remains of a body, Onicha Amiyi-Uhu, Abia state, 2003. The fuel gushing from a vandalised pipeline exploded at the time, killing at least 105 villagers in rural eastern Nigeria as they scavenged the gasoline. Pipeline vandalism or "scooping" is common in Nigeria despite the risk of fire, prosecution or being shot on sight by security forces. Scores of people die regularly from pipeline explosions nationwide.

Overleaf Right: Oil pool in Kigbara-Dere, an Ogoni farming/fishing village, Gohana District, 60 km from Port Harcourt, 2007.

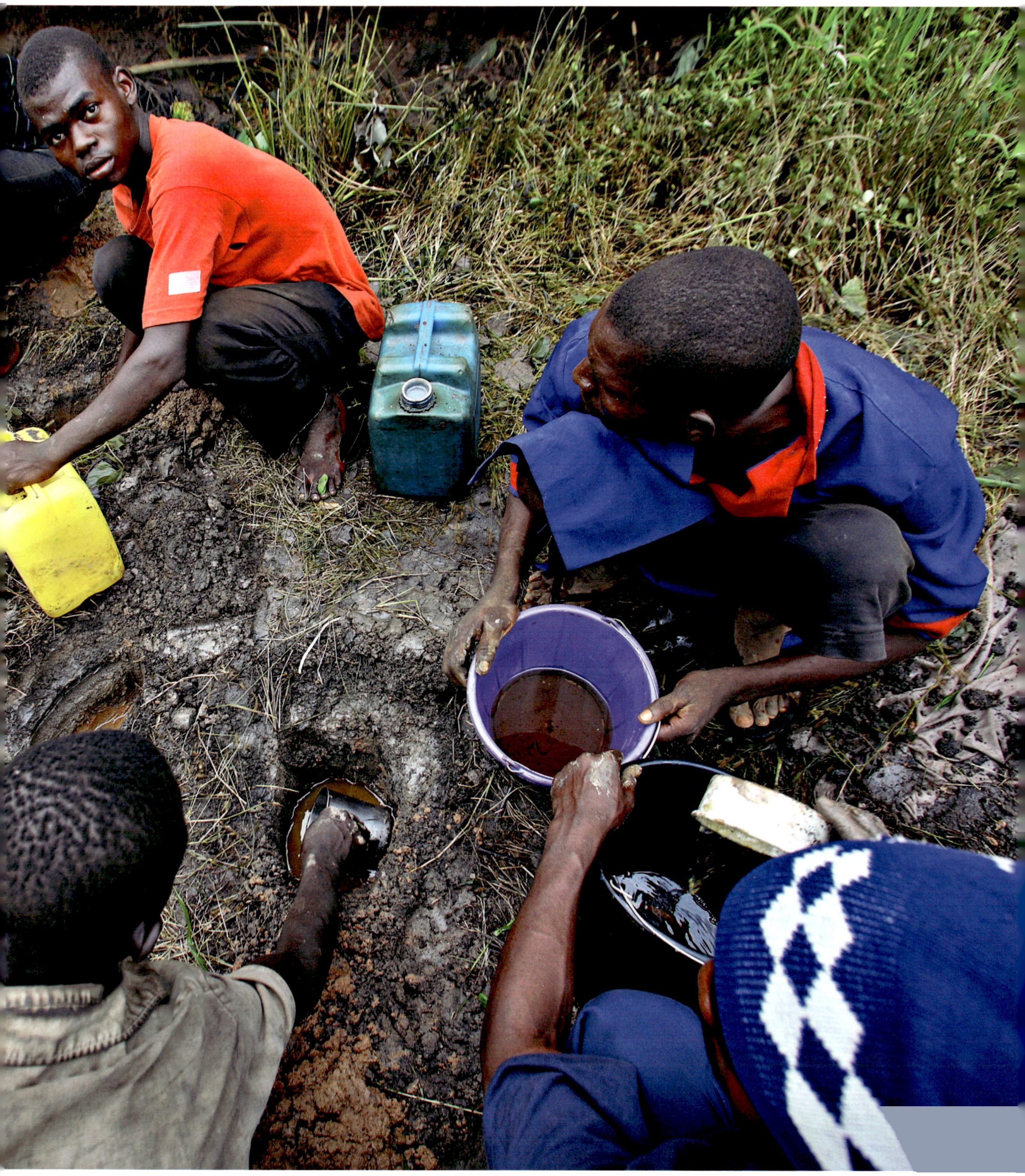

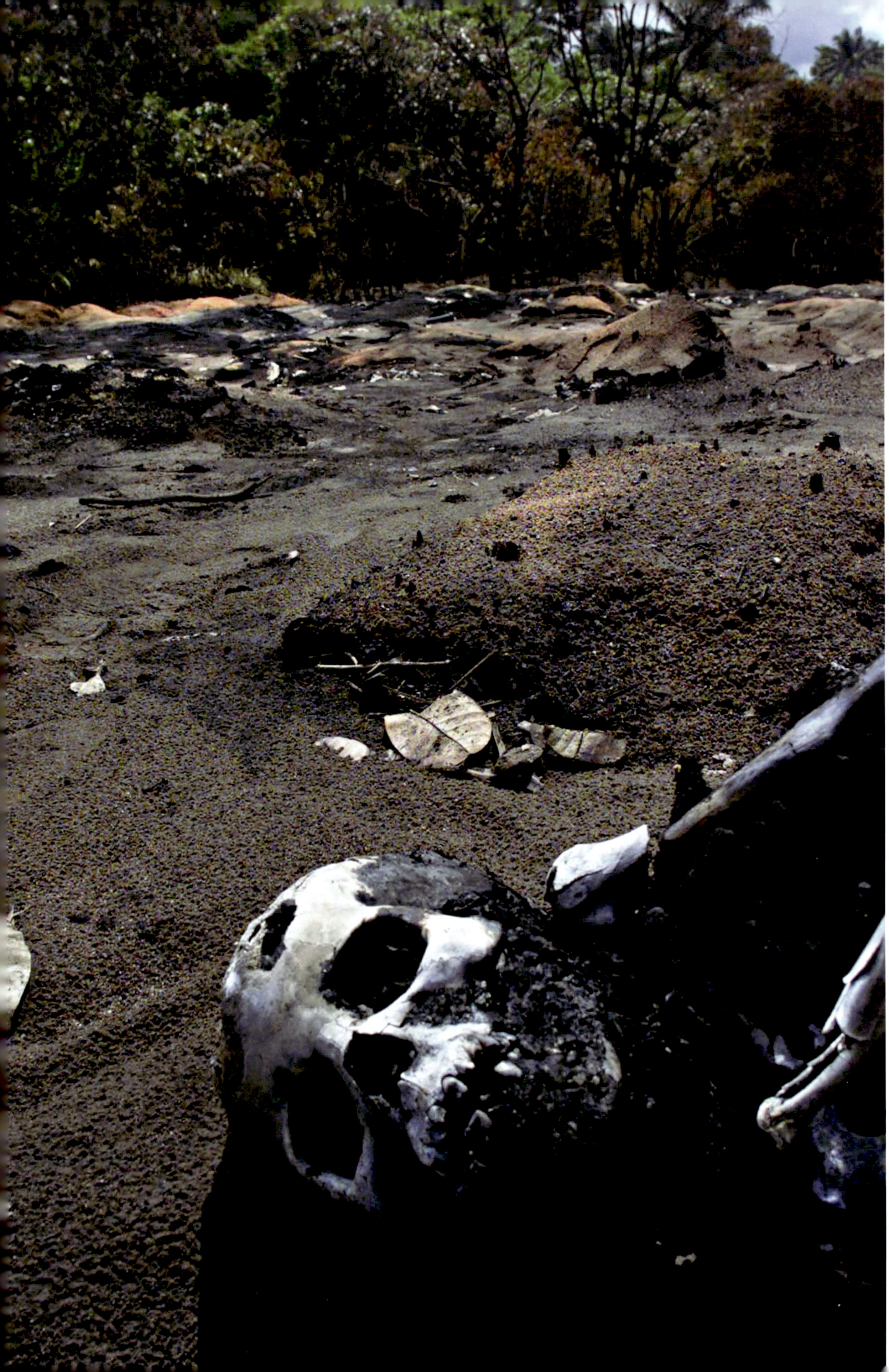

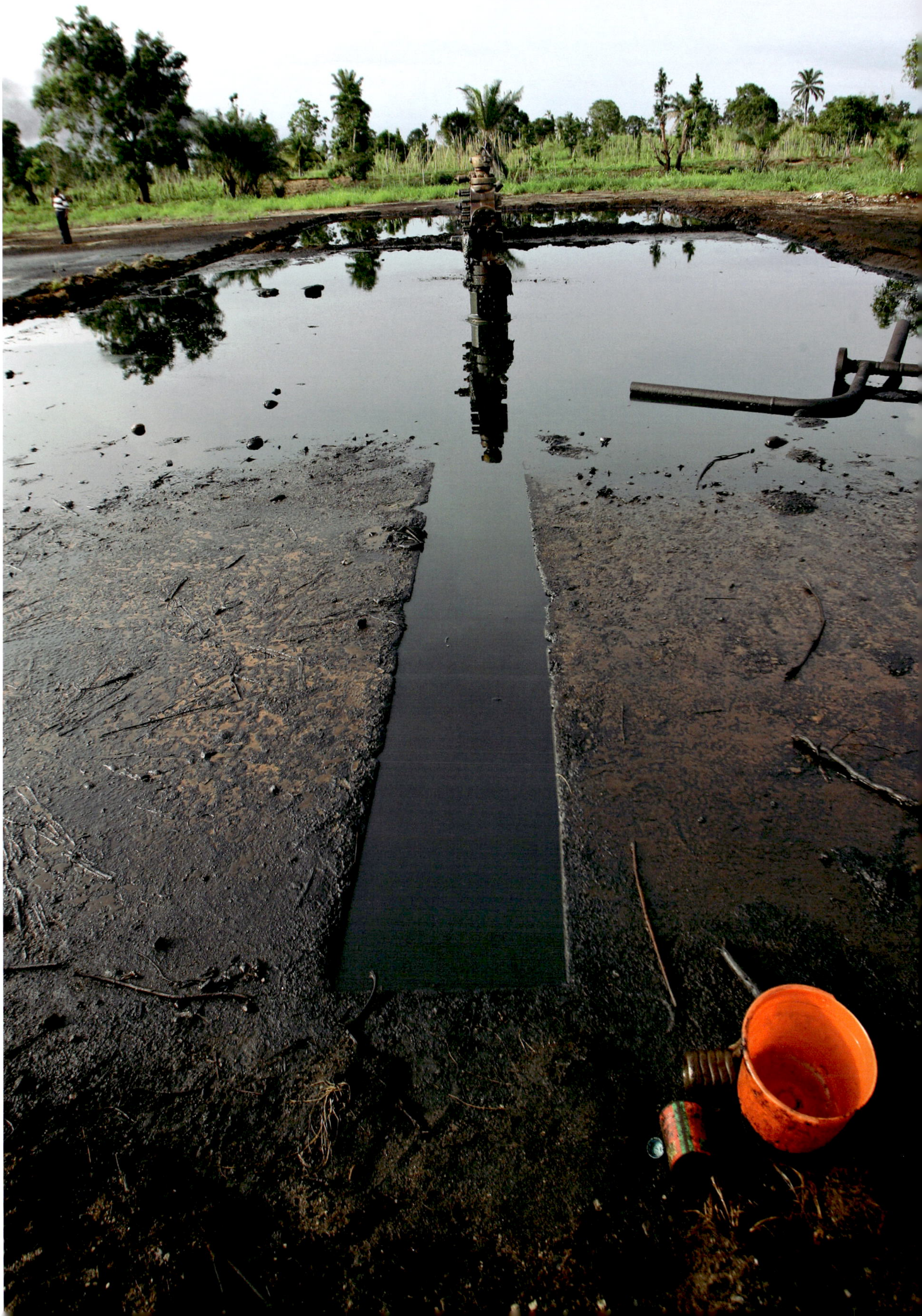

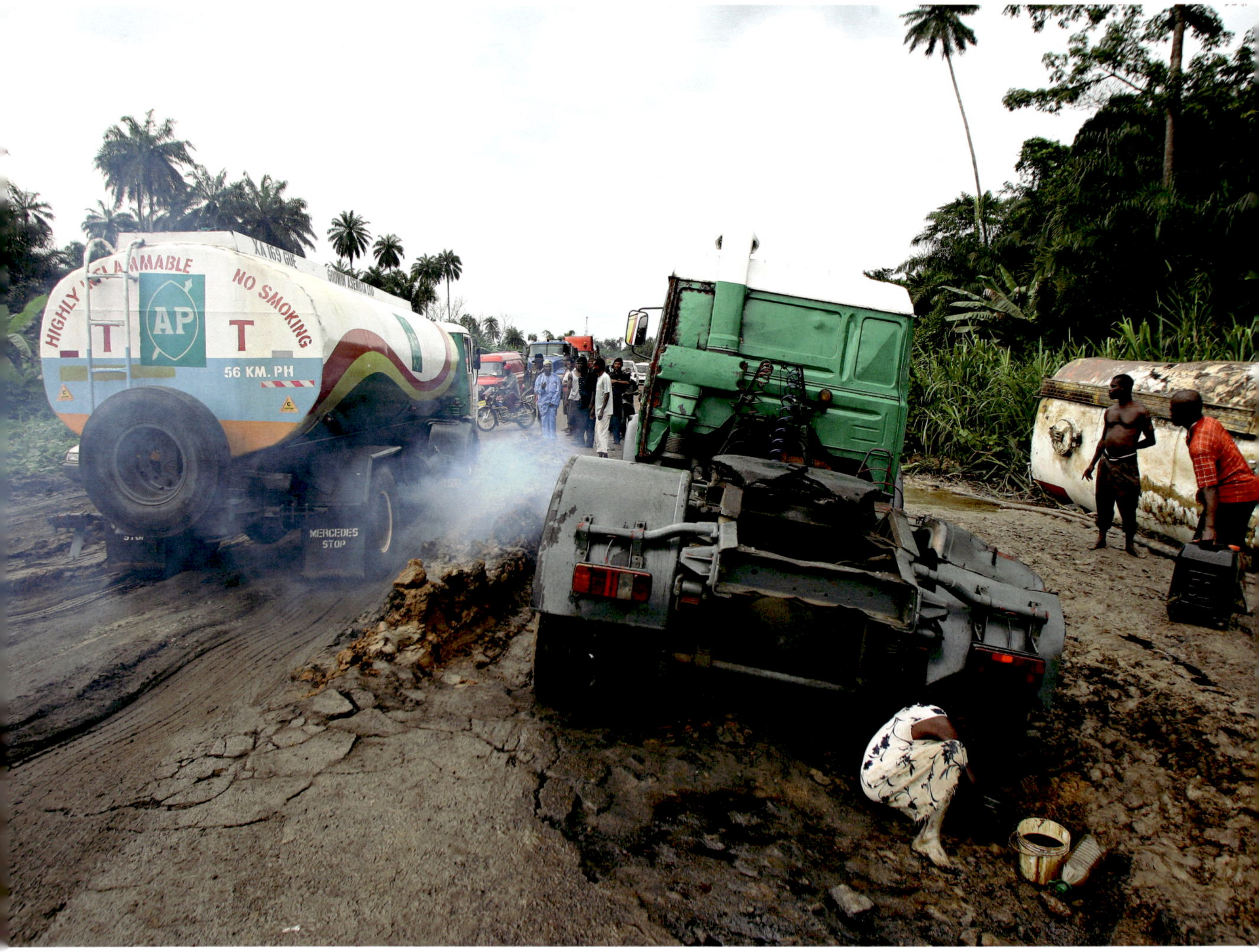

Above and right: Locals collect diesel from an overturned tanker on the pot-holed Warri-Port Harcourt road, Oshie, 2006.

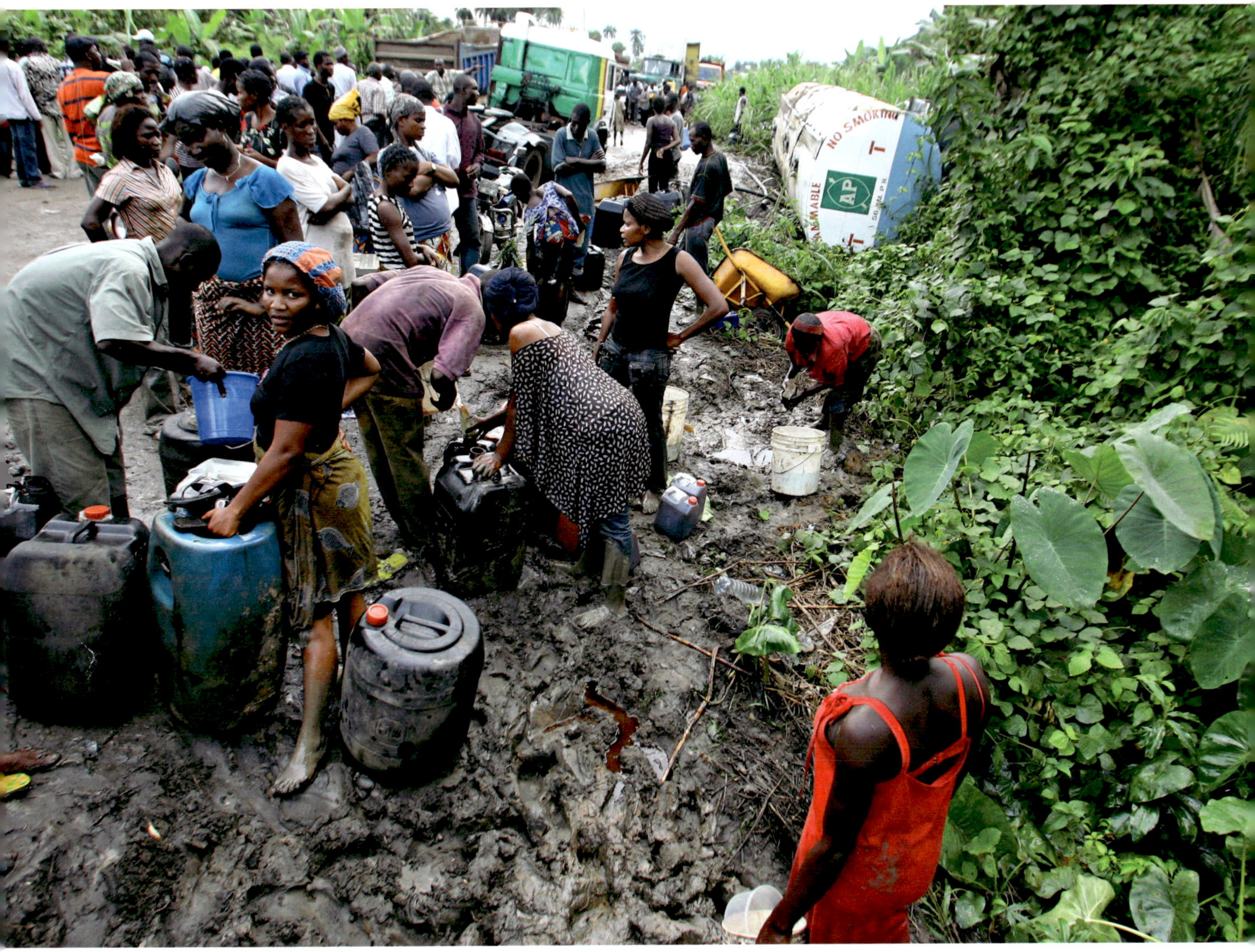
AP

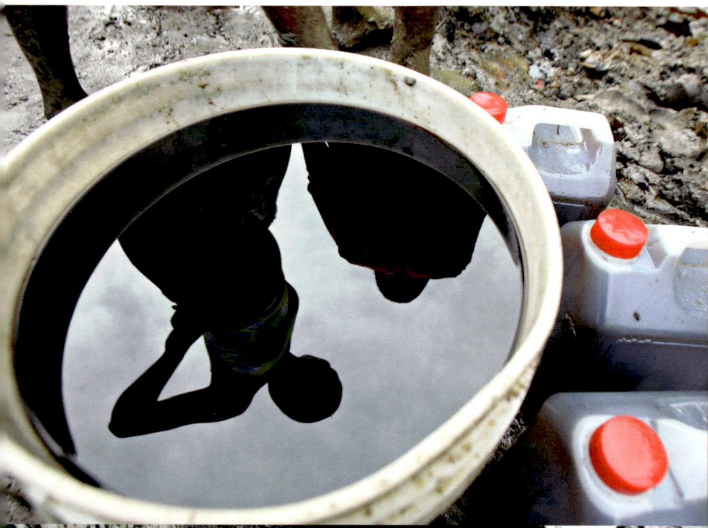
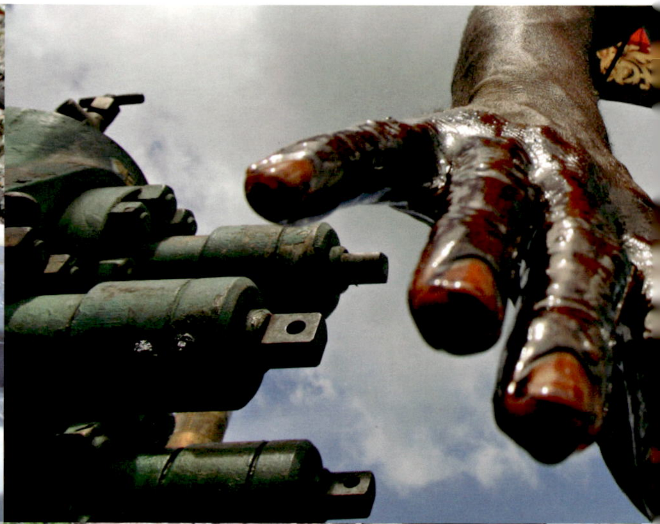
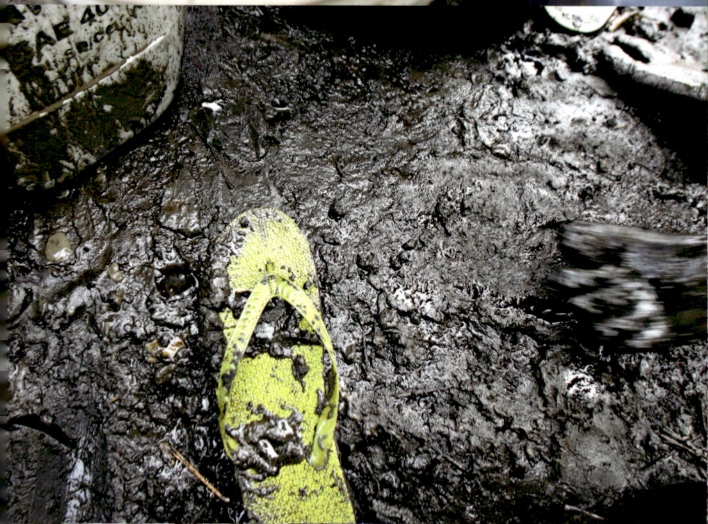
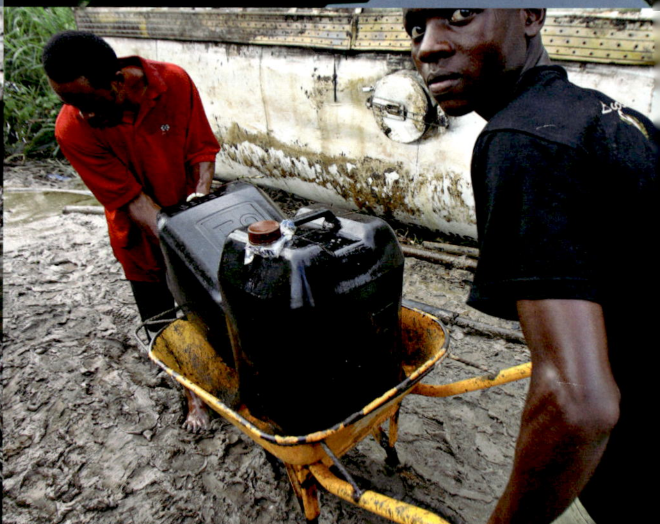
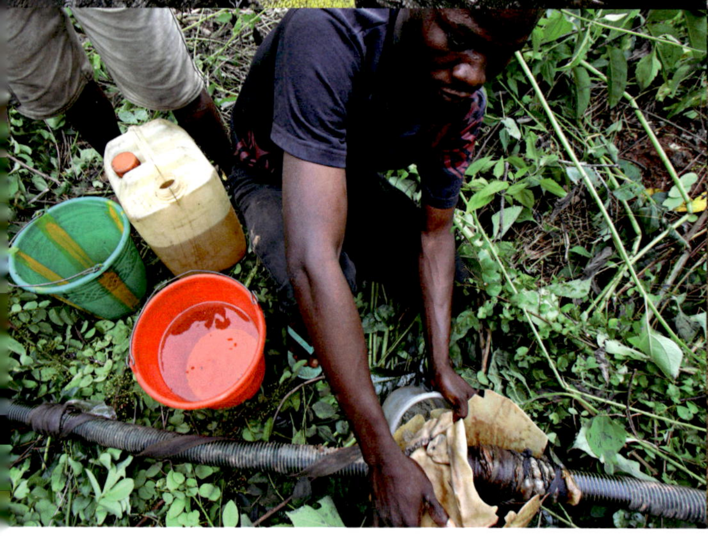
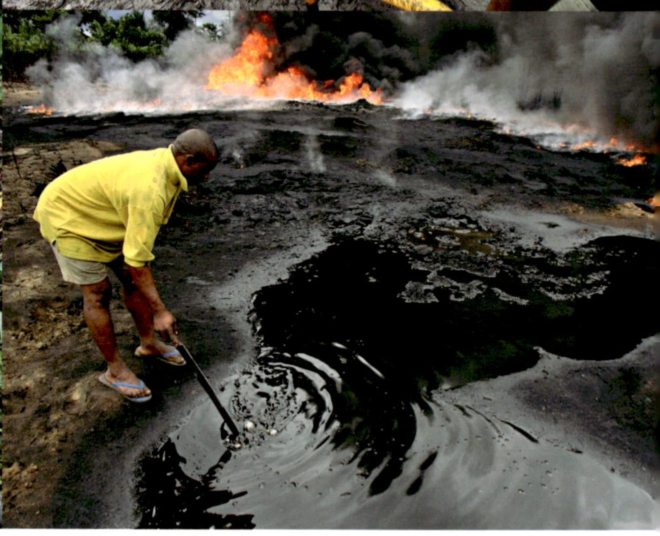

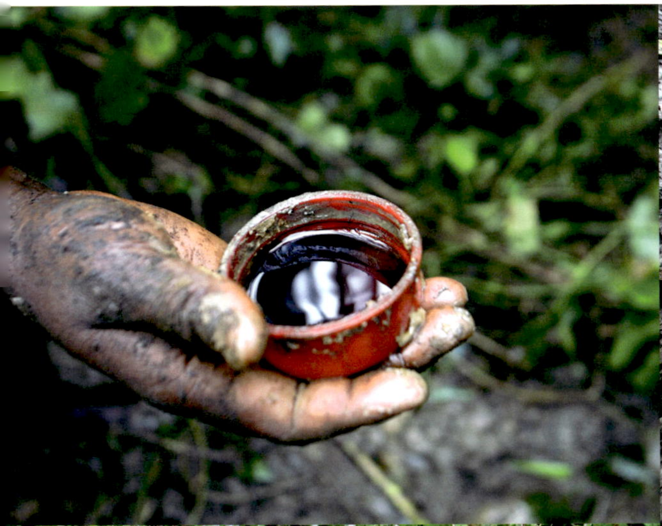
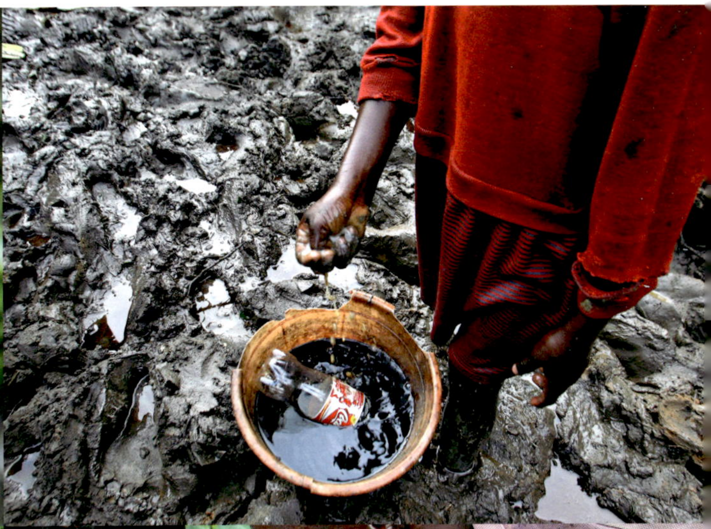
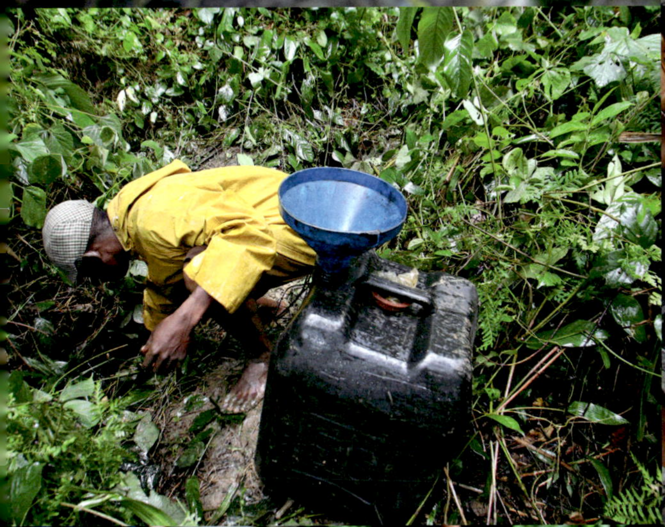
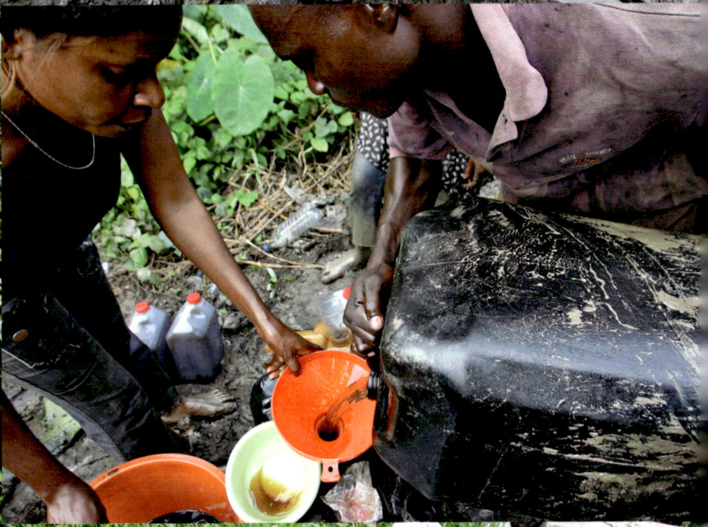
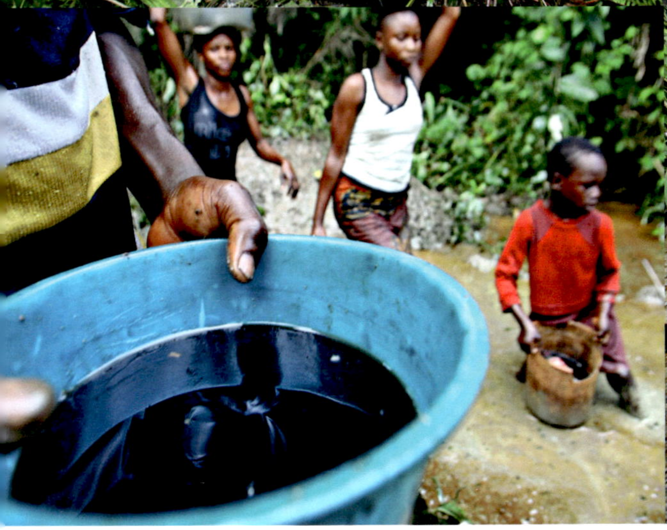
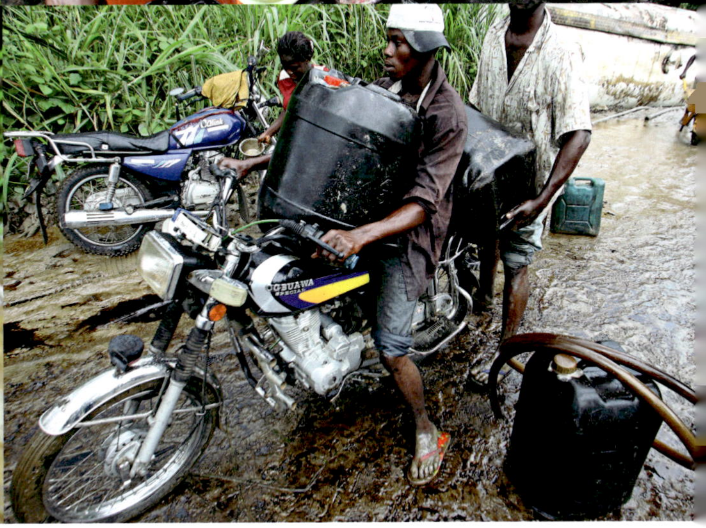

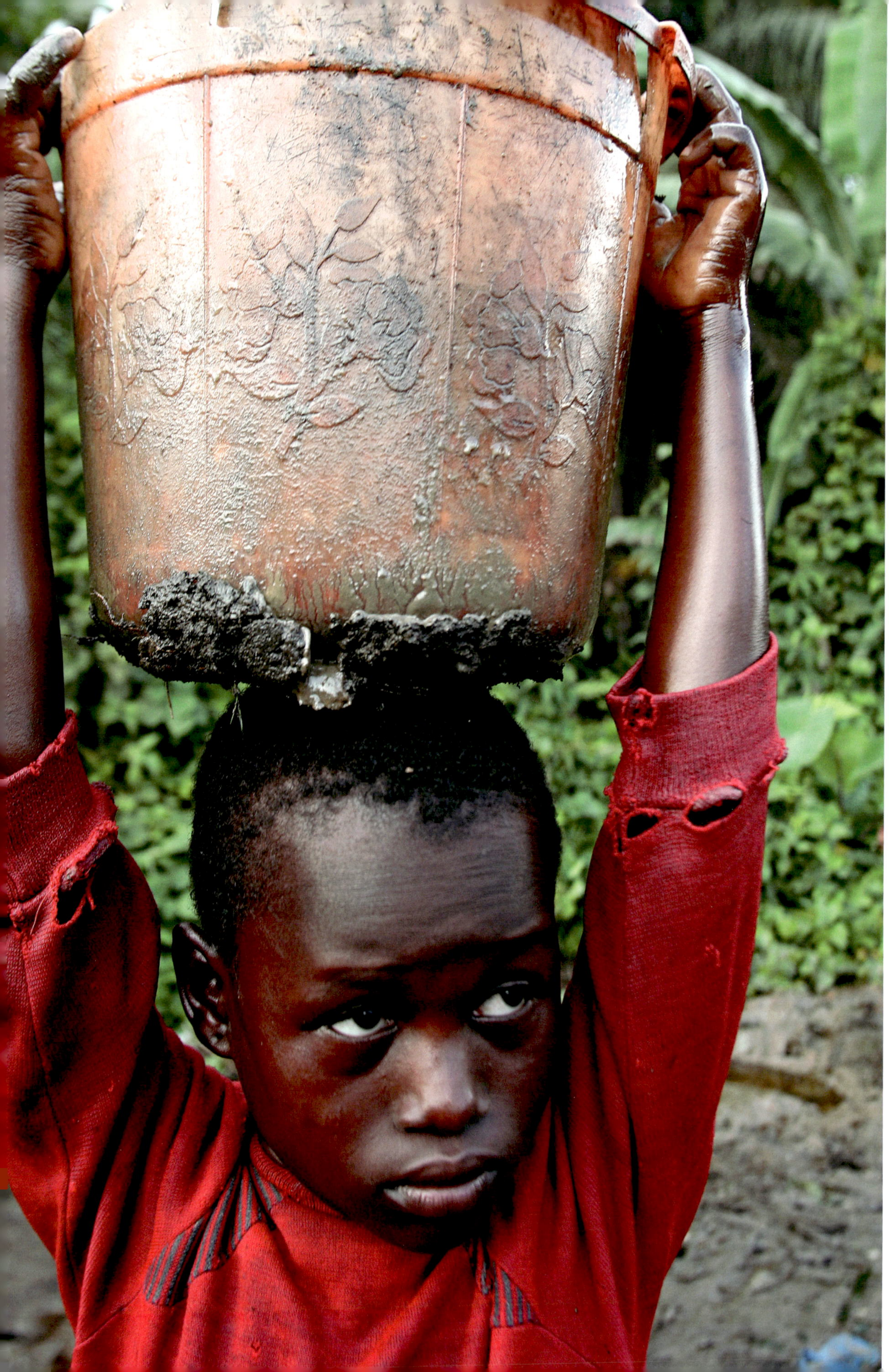

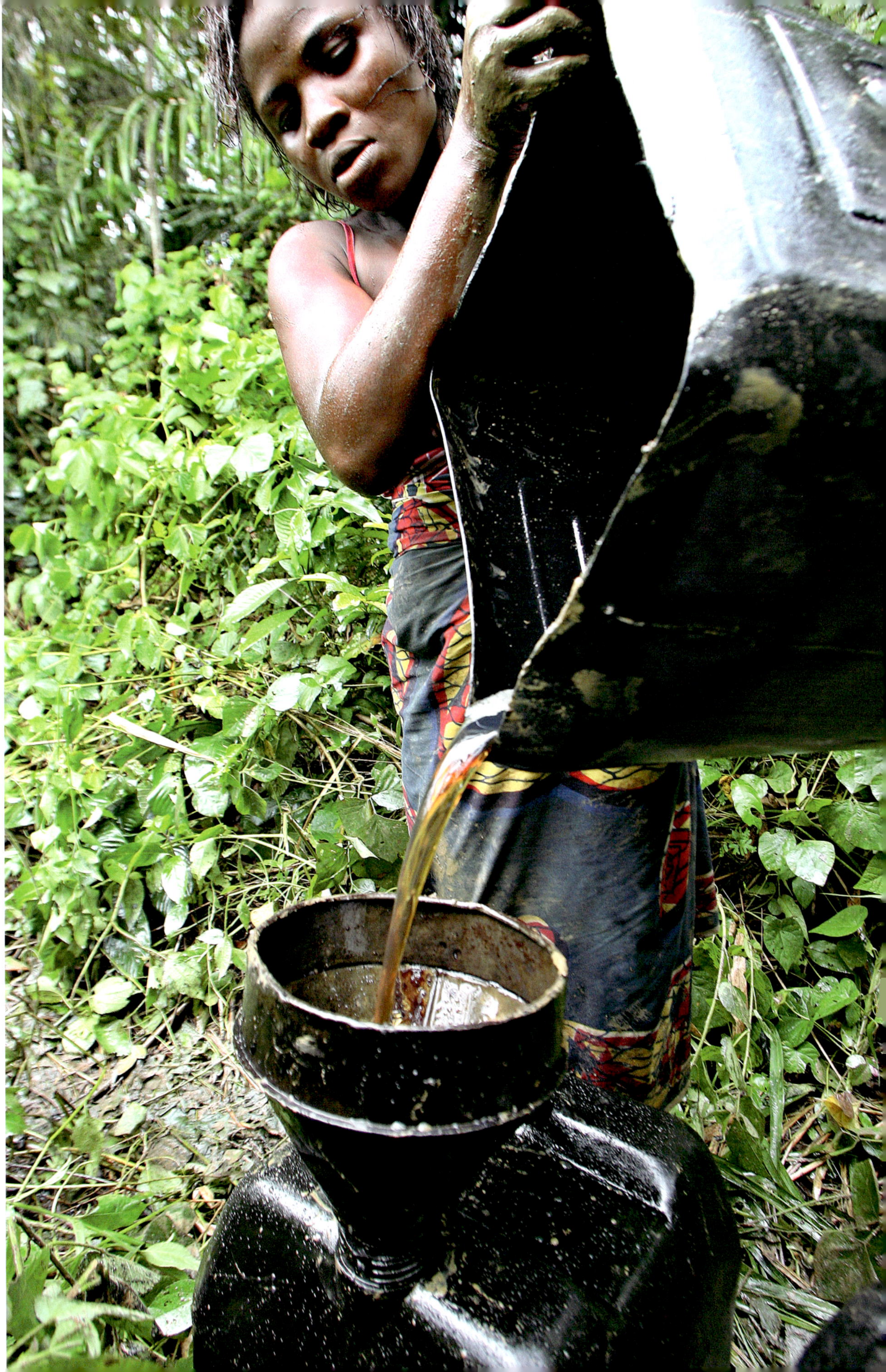

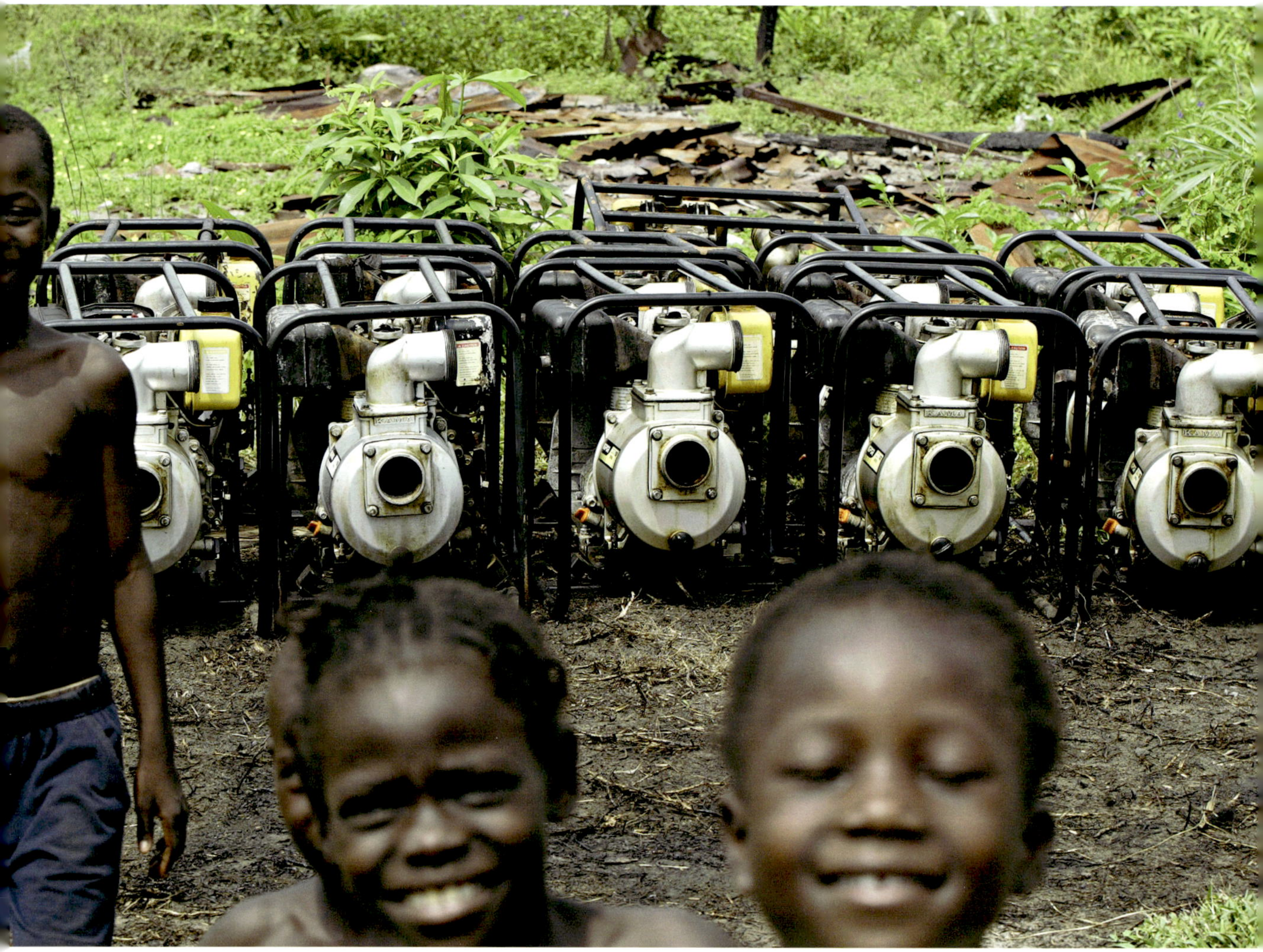

'Oil bunkering' is one of the main problems in the Niger Delta, fuelling greed, conflict and insecurity in the region. There have always been tensions between militant groups, the government security forces and some top powerful, politically influenced elites. Oil bunkering is a crime that enriches a few criminals, insurgents and corrupt officials, while still impoverishing many normal people. It undermines the rule of law, deepens corruption, pollutes the environment, violates human rights and depletes natural resources. It has

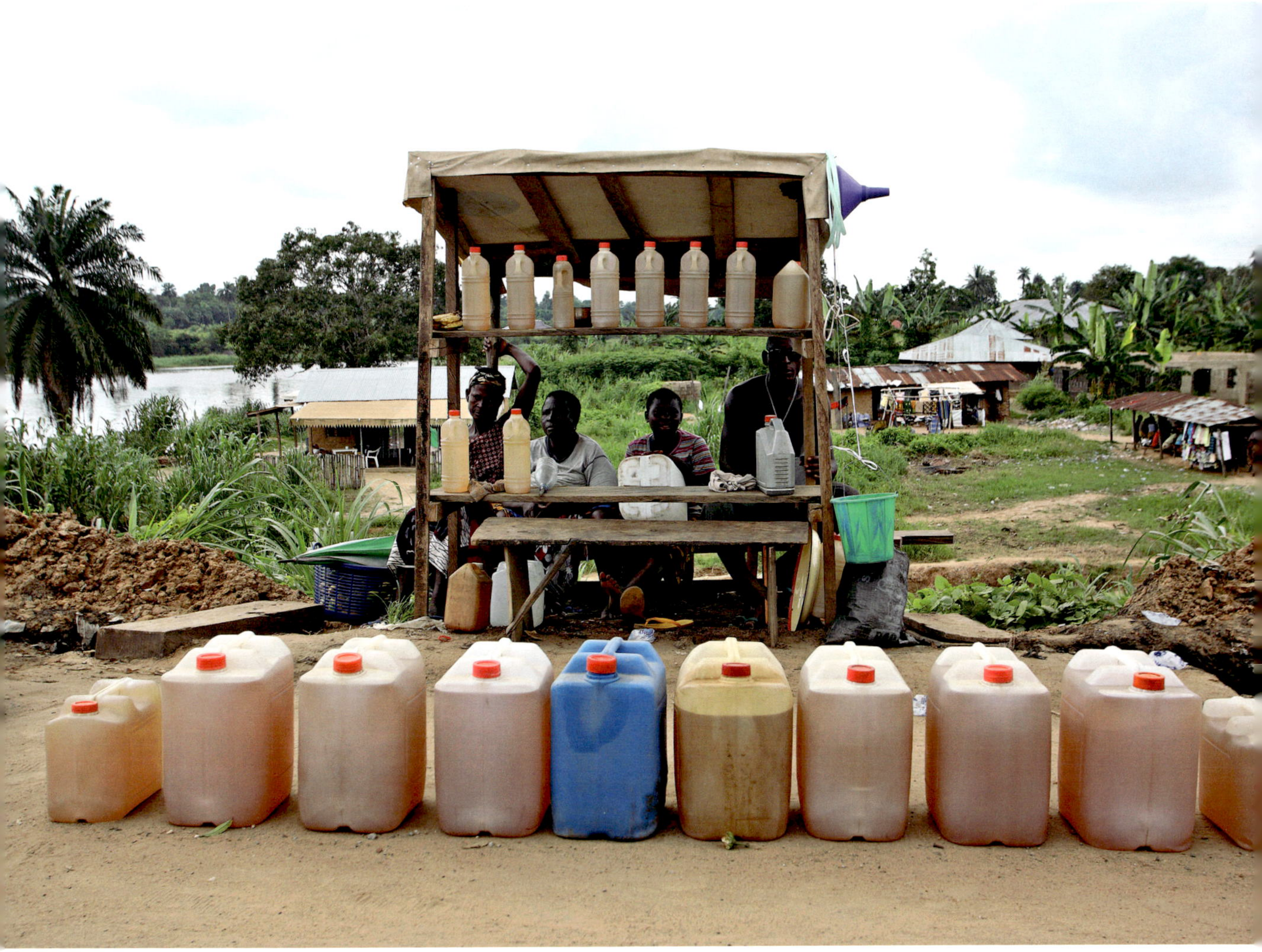

become a transnational criminal enterprise in its own right, and the violent political struggle provides a convenient smokescreen for those intent on personal enrichment. The wide range of players involved in bunkering and the complex web of their alliances makes it difficult to discern the victims from the perpetrators. The loss of oil revenues - perhaps one third of the income on which 80% of the Nigerian national budget is based - means that, aside from a few well-placed individuals, Nigerians are the net losers.

Above: Locals sell black market fuel in Mbiama, 2007.

Left: Oil pumping machines used for oil theft in the Delta region, 2005.

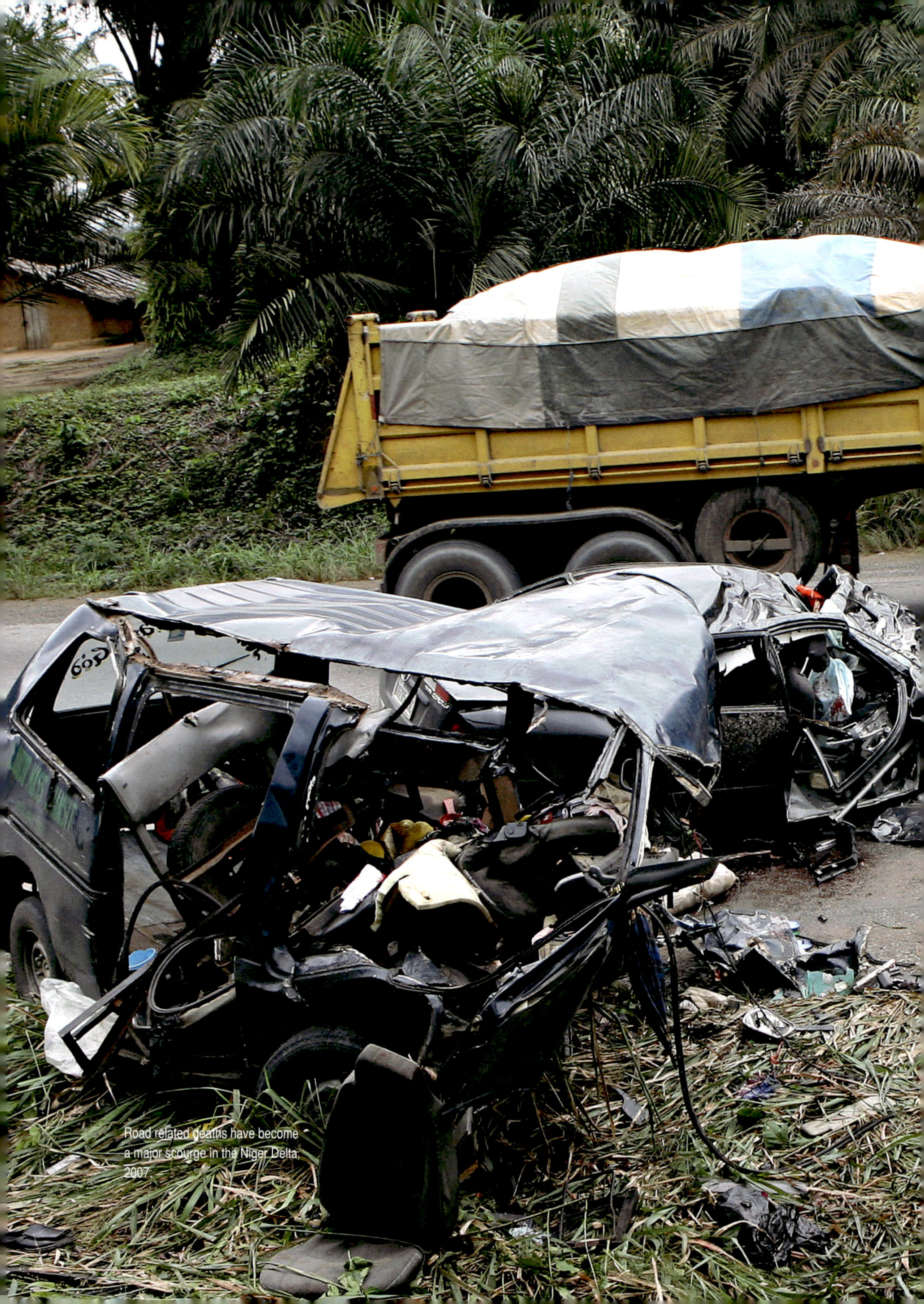

Road related deaths have become a major scourge in the Niger Delta, 2007

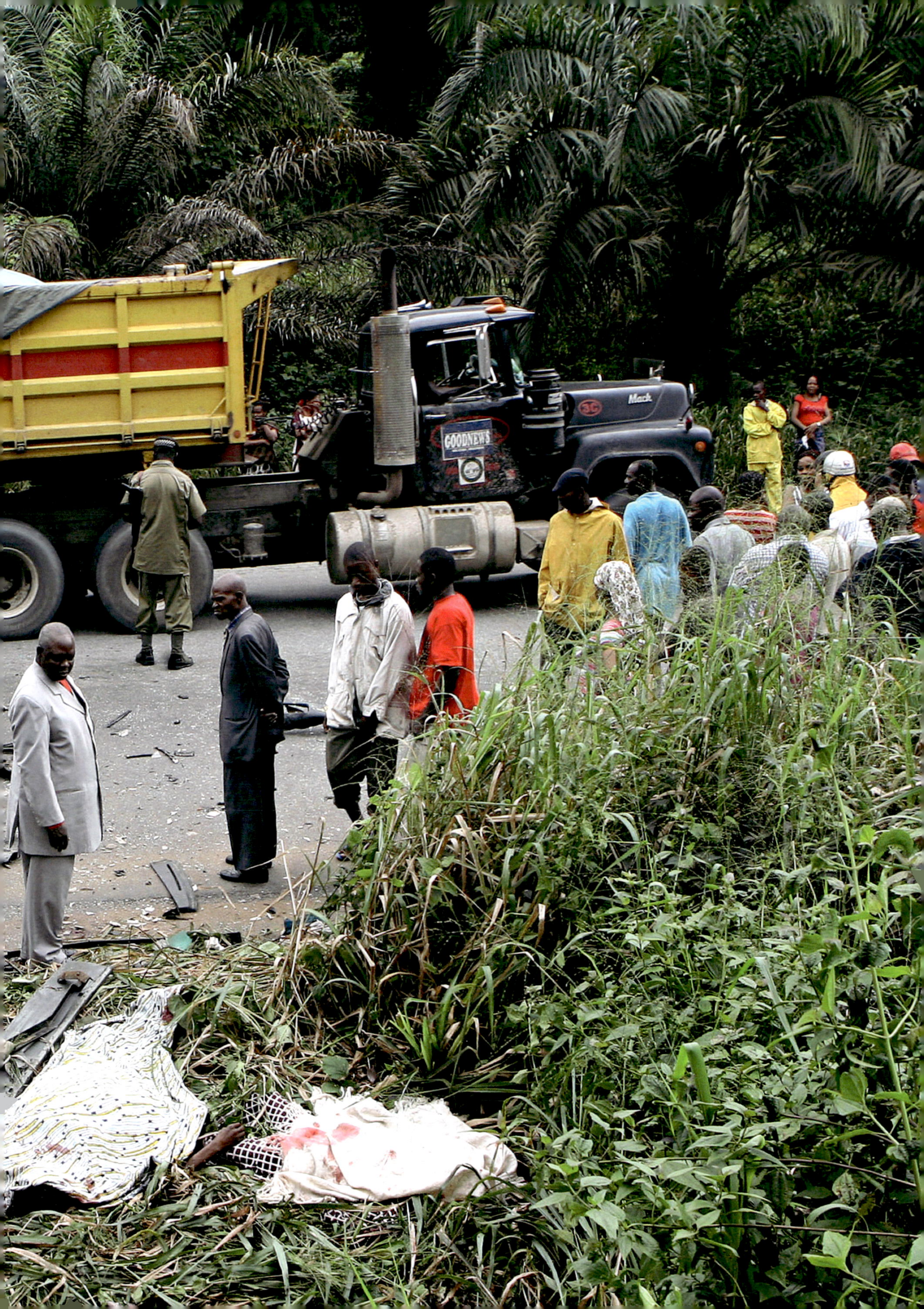
GOODNEWS
Mack

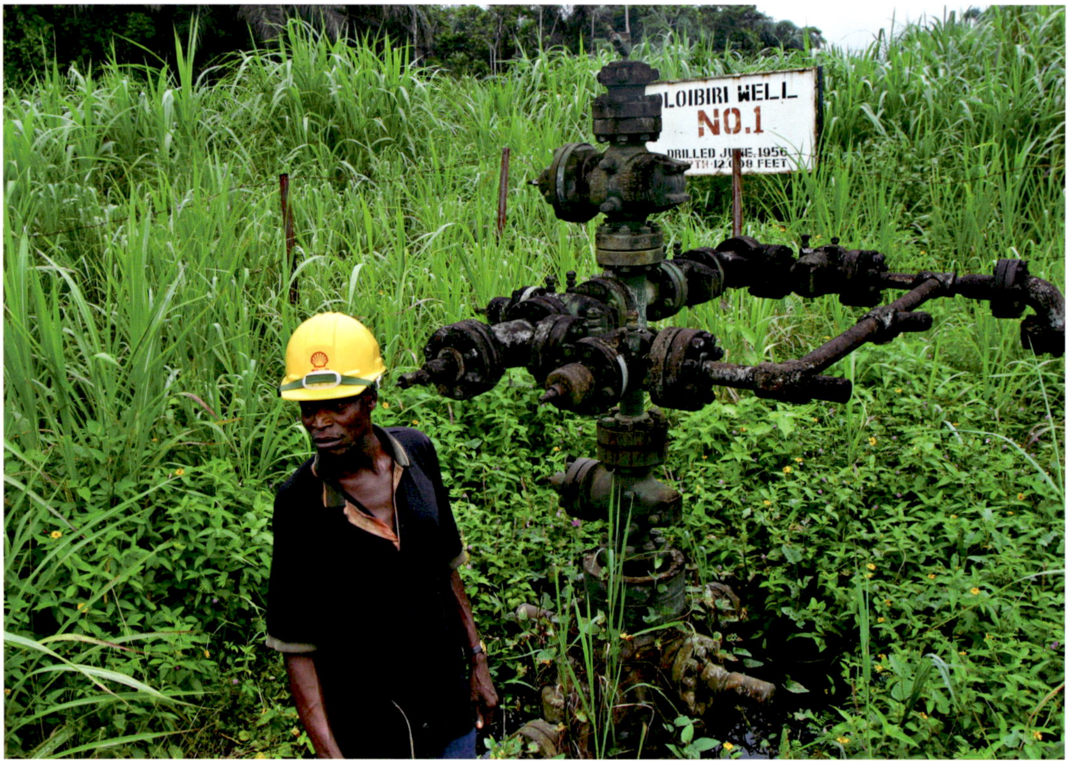

Above: The remains of the first successful oil well drilled at Otabagi, January 1956, in the Oloibiri district by Shell D'Arcy, the original name of Shell Petrol Development Company (SPDC).

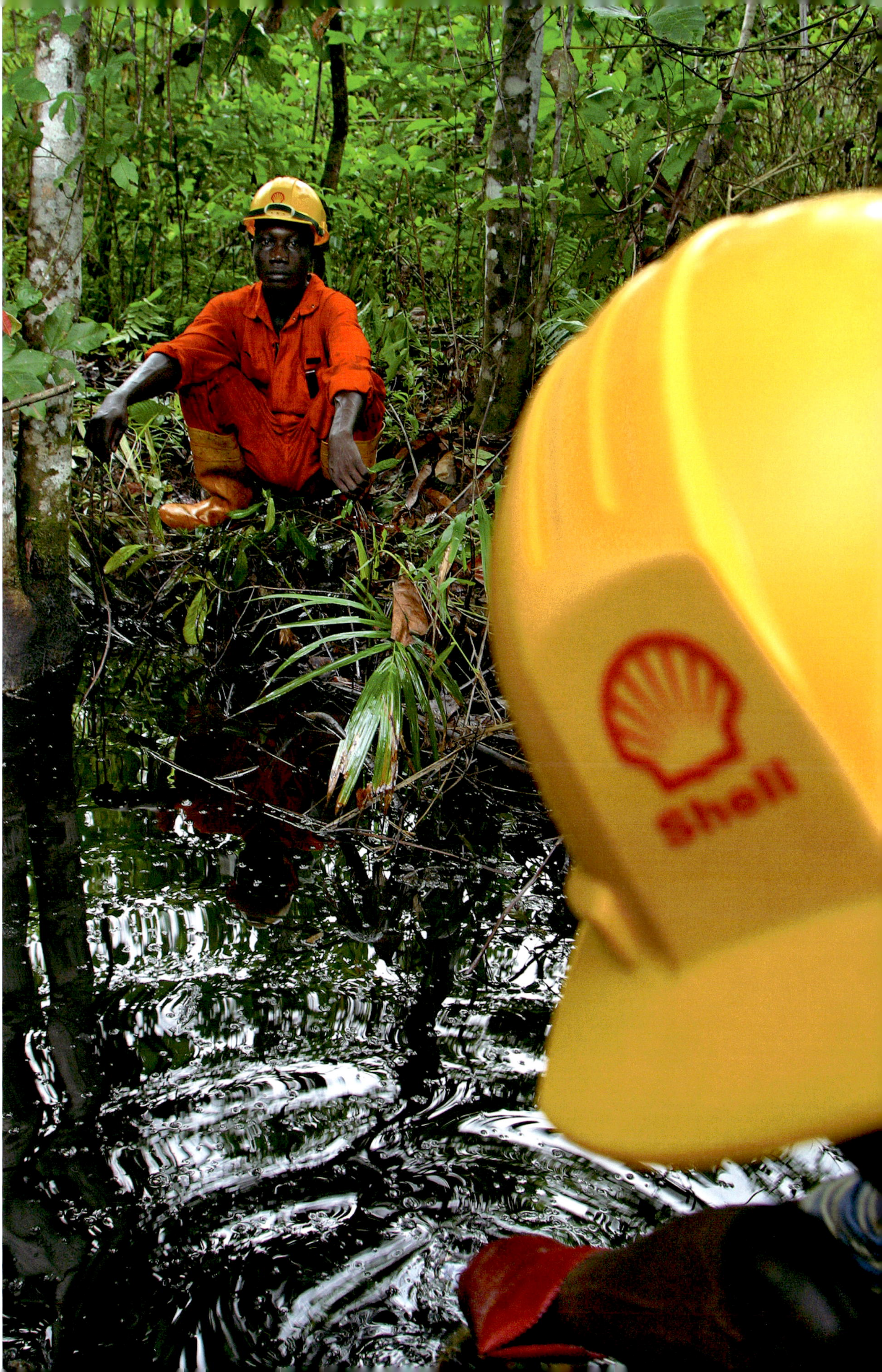
Shell

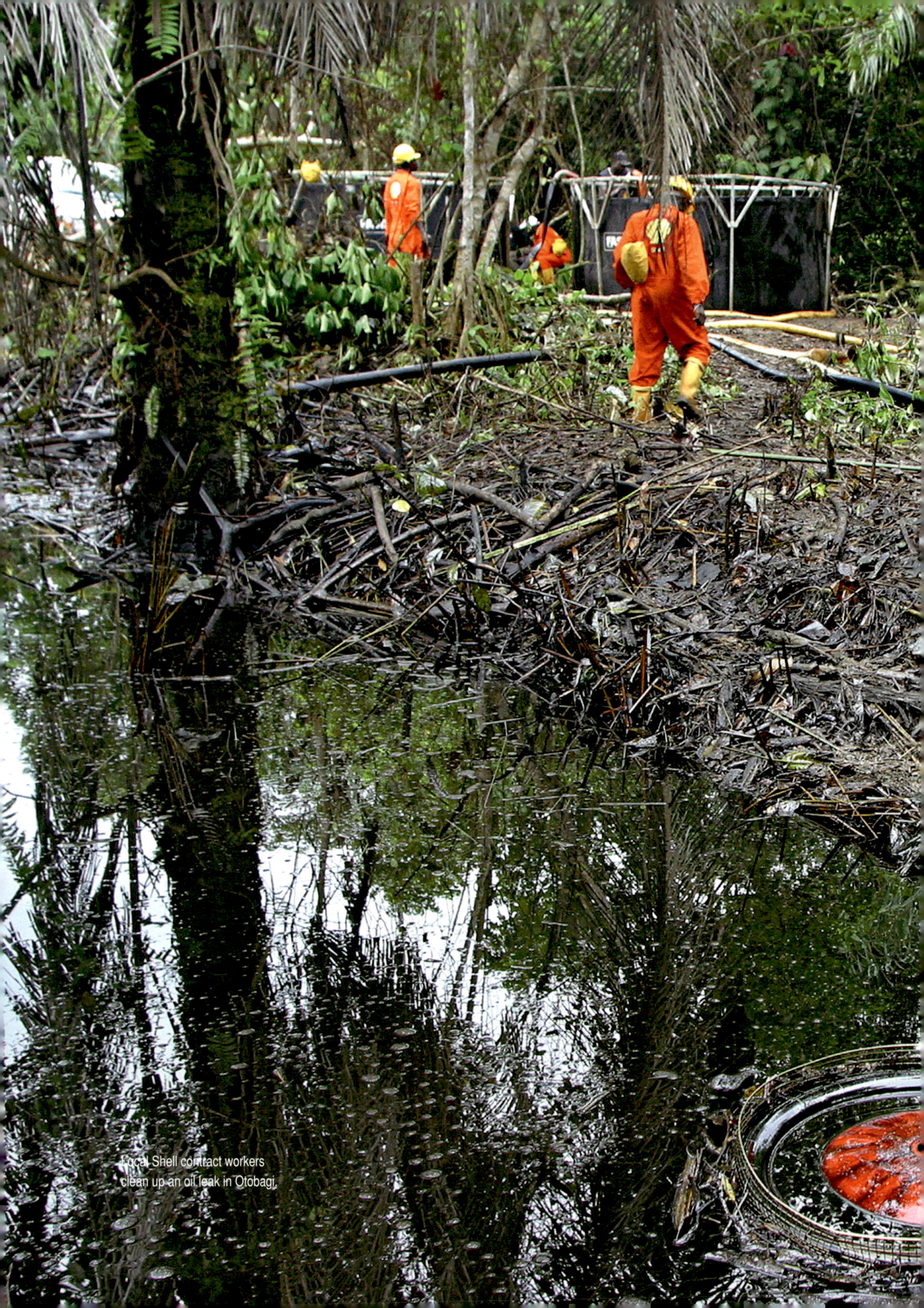

Local Shell contract workers clean up an oil leak in Otobagi.

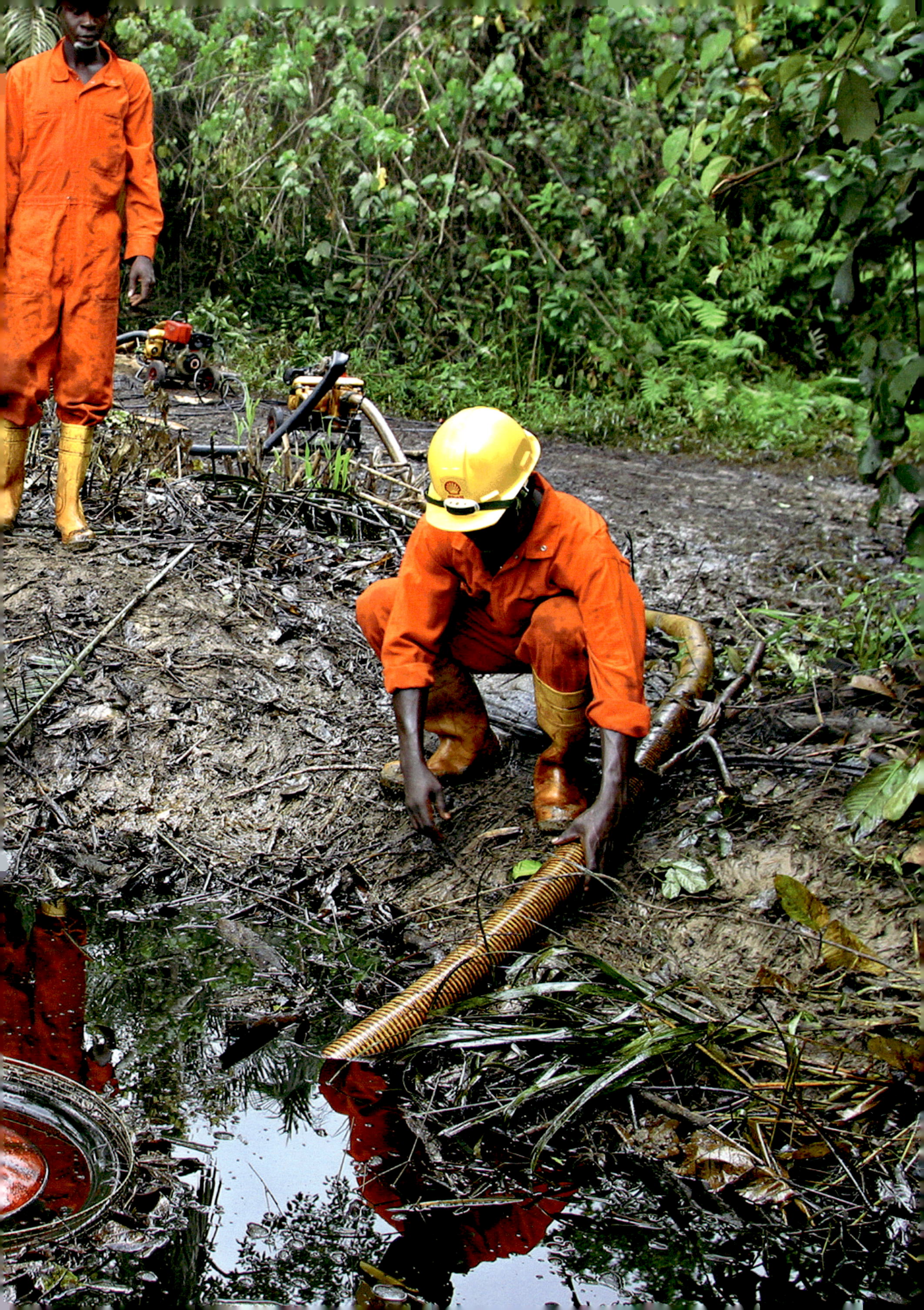

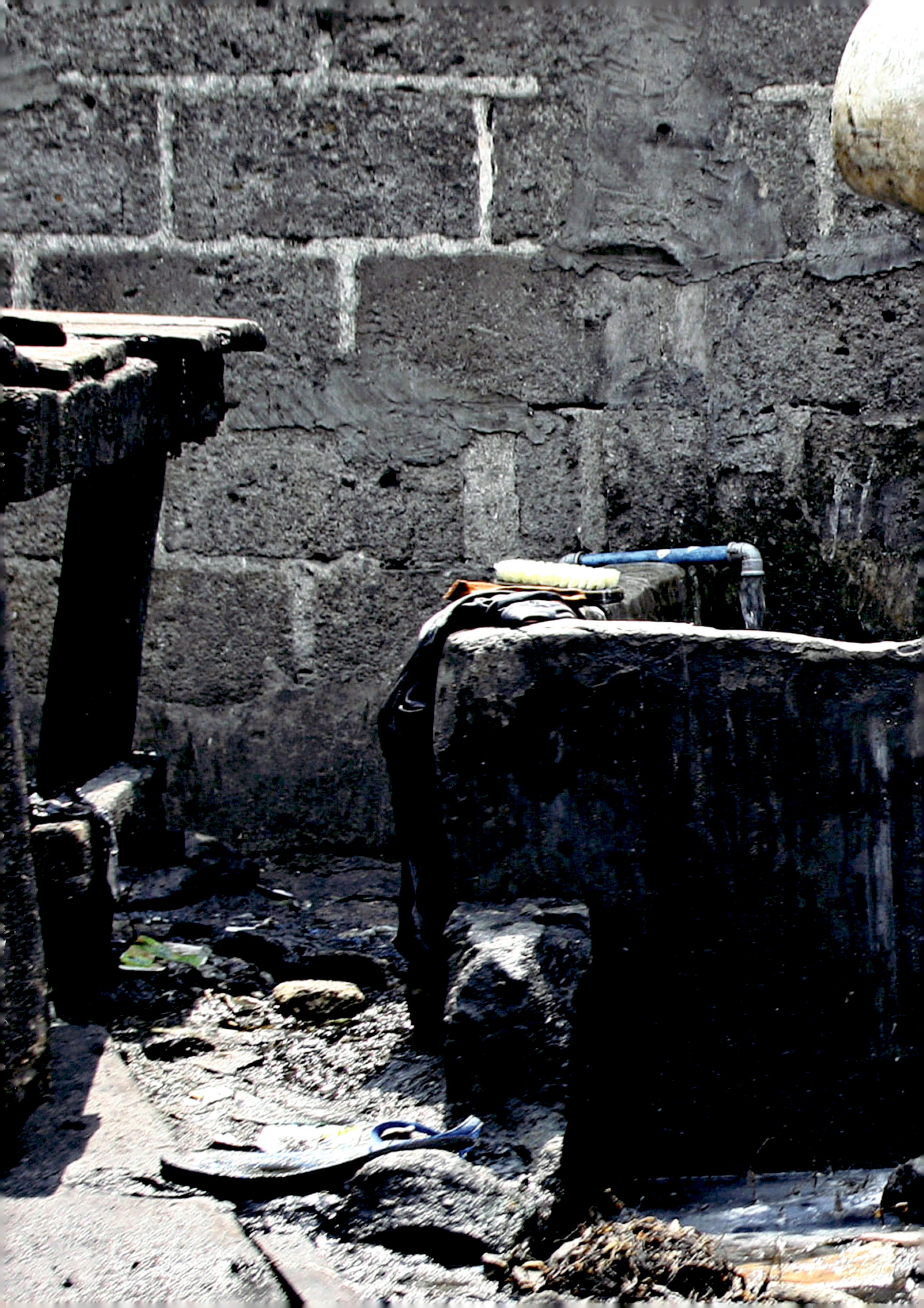

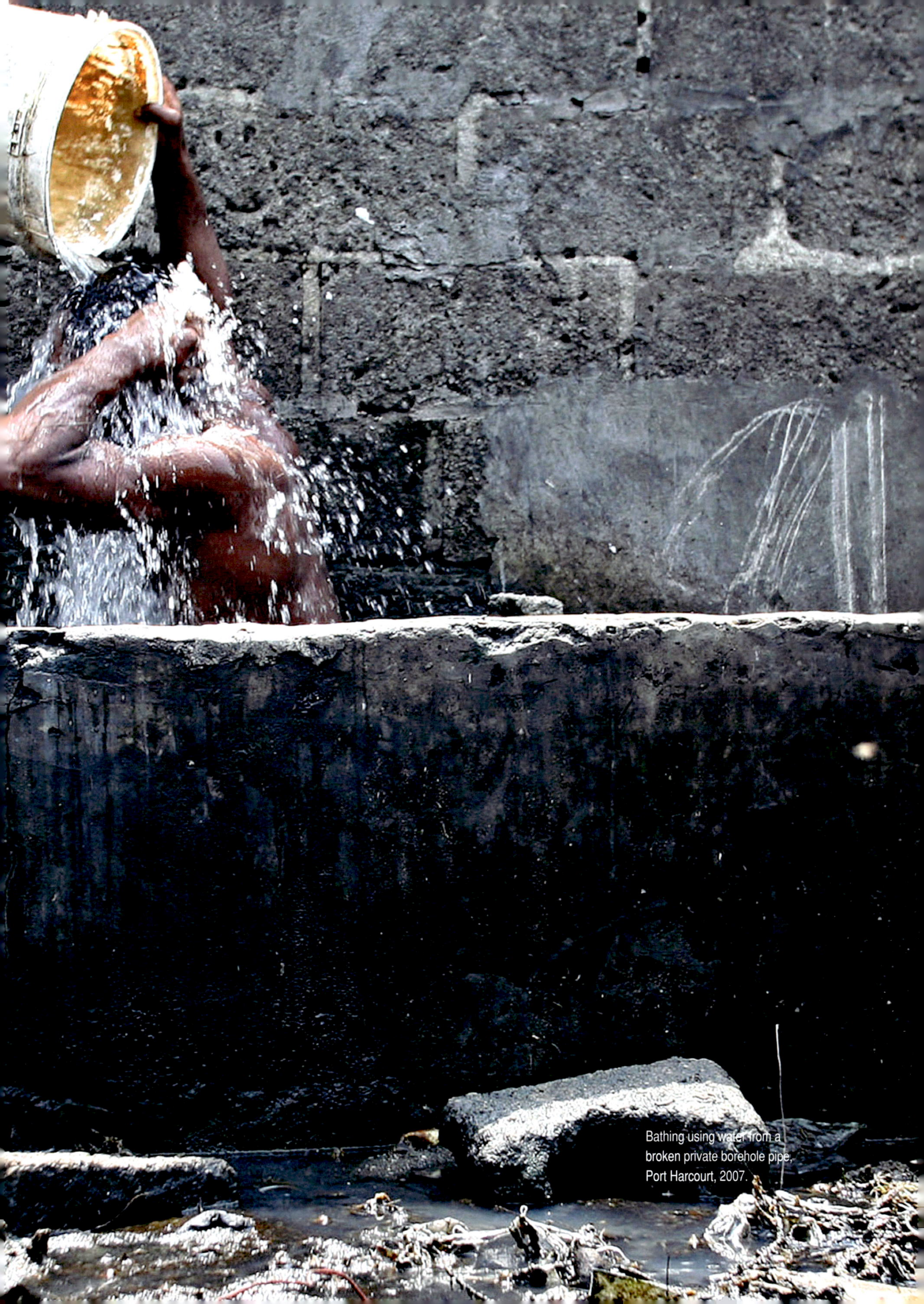

Bathing using water from a broken private borehole pipe, Port Harcourt, 2007.

EFFECT OF CLIMATE CHANGE

The rise in sea water levels has lead to erosion of land and has left thousands homeless in most of the coastal regions. The rich ecosystem there is also threatened by devastating floodwaters and rising air temperatures.

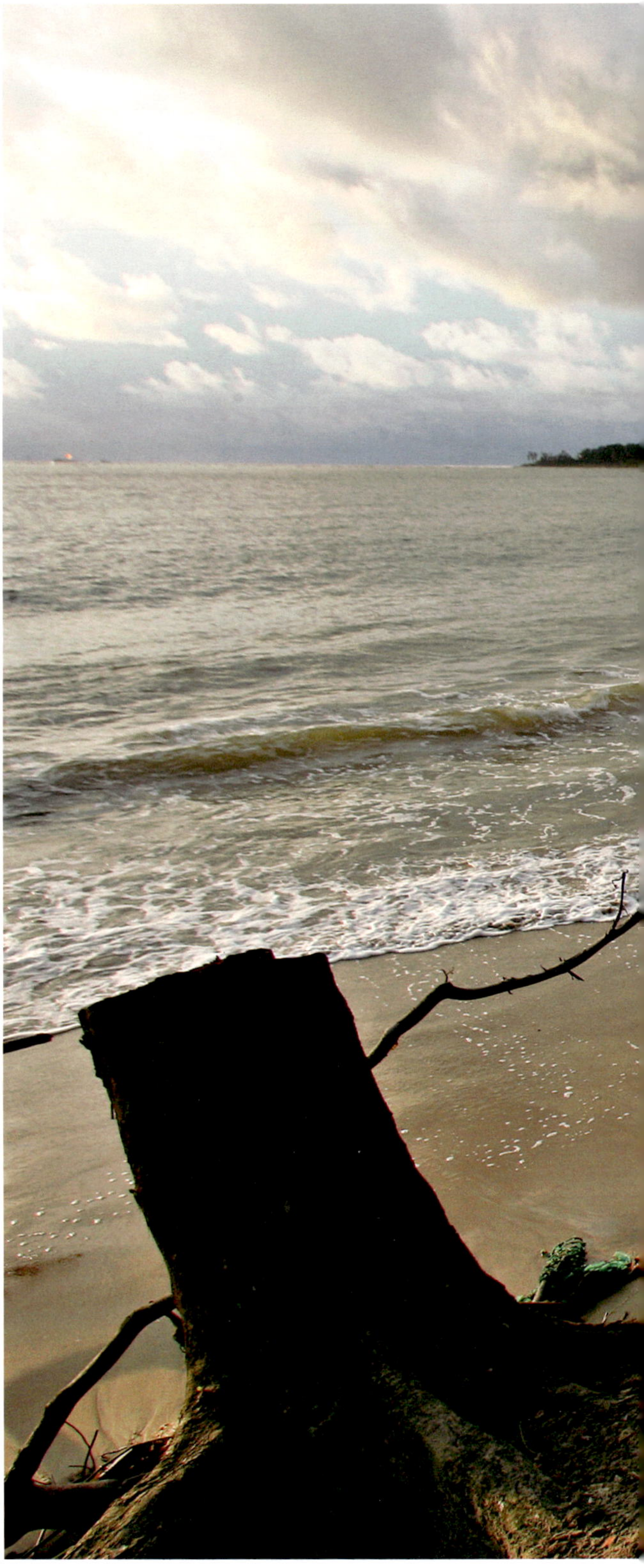

Coastline affected by sea erosion in Akassa, with oil rig in distance, 2007.

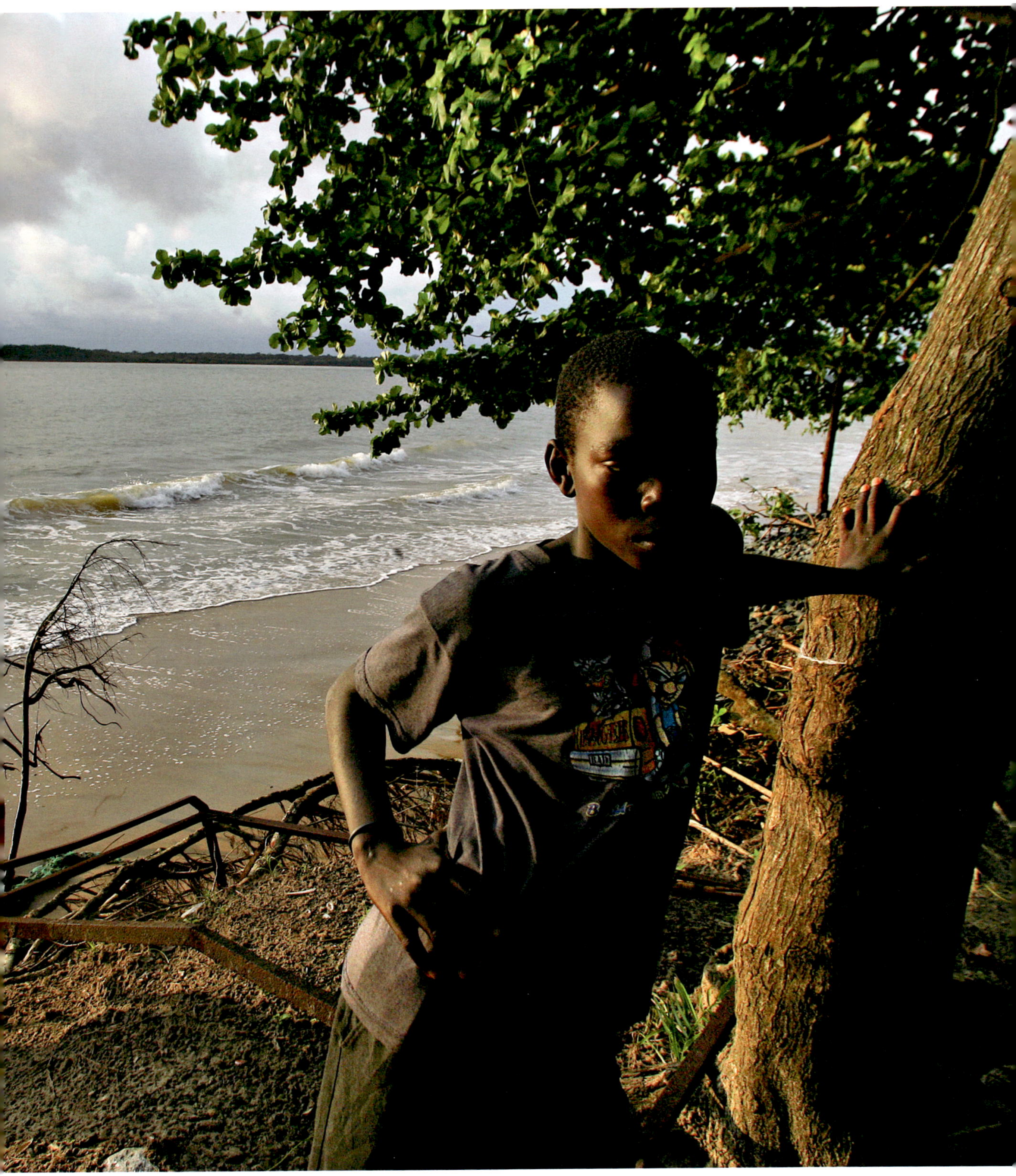

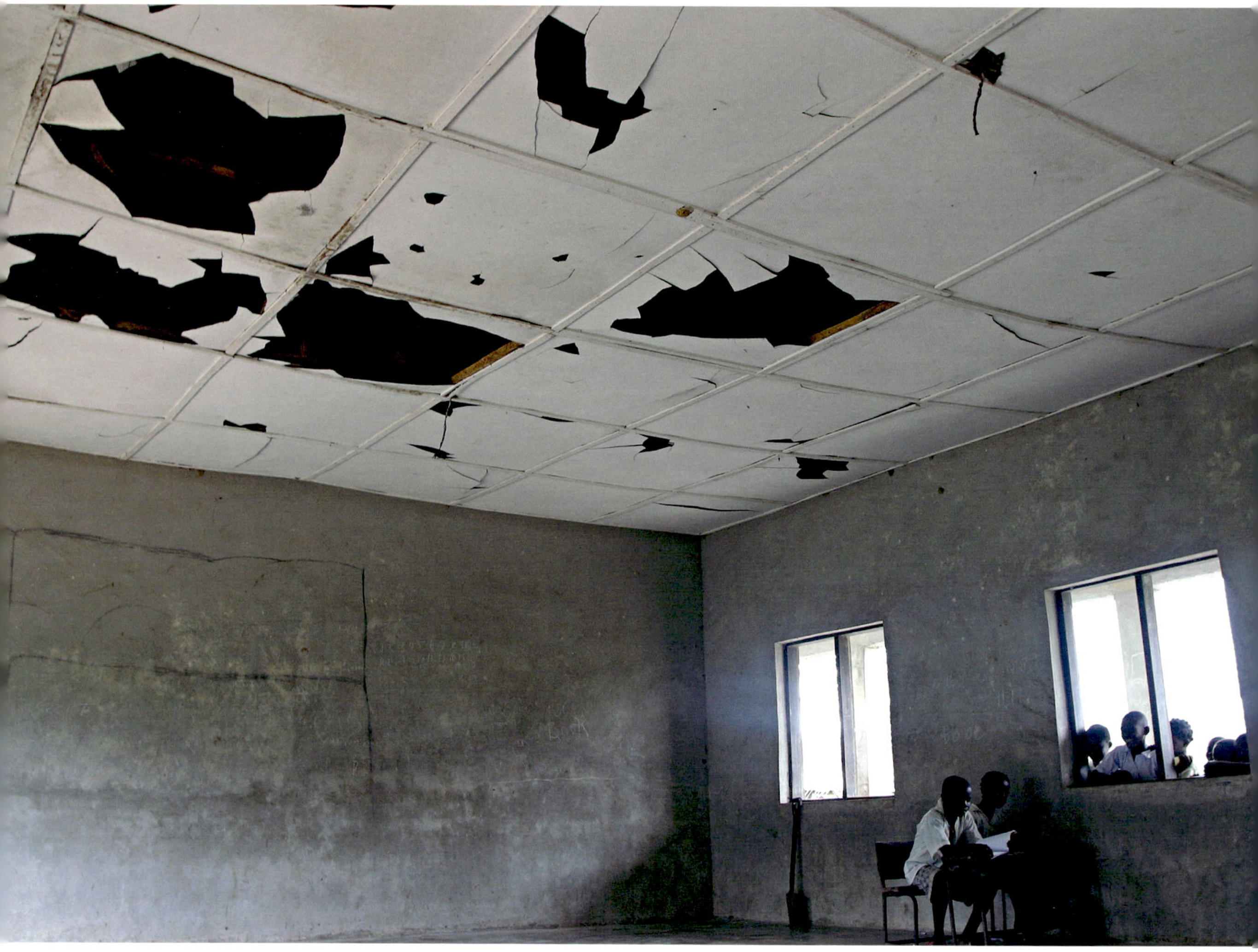

Above: Students sit inside their classroom at the government-run primary school in Oshie village. Many students learn on bare floors in some government schools in the Delta, 2004.

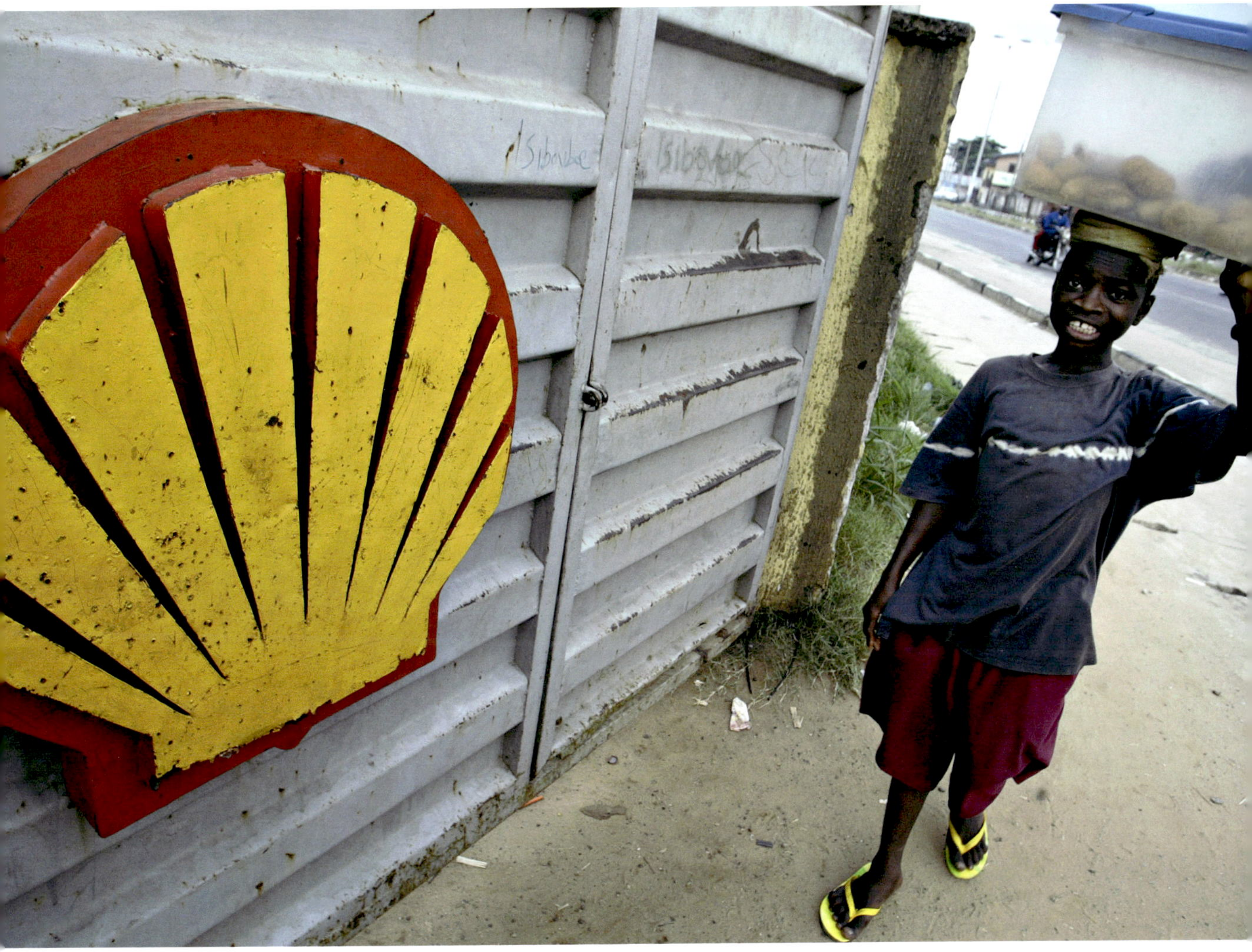

Above: One of the schools built by Shell, Warri, Delta State, 2005.

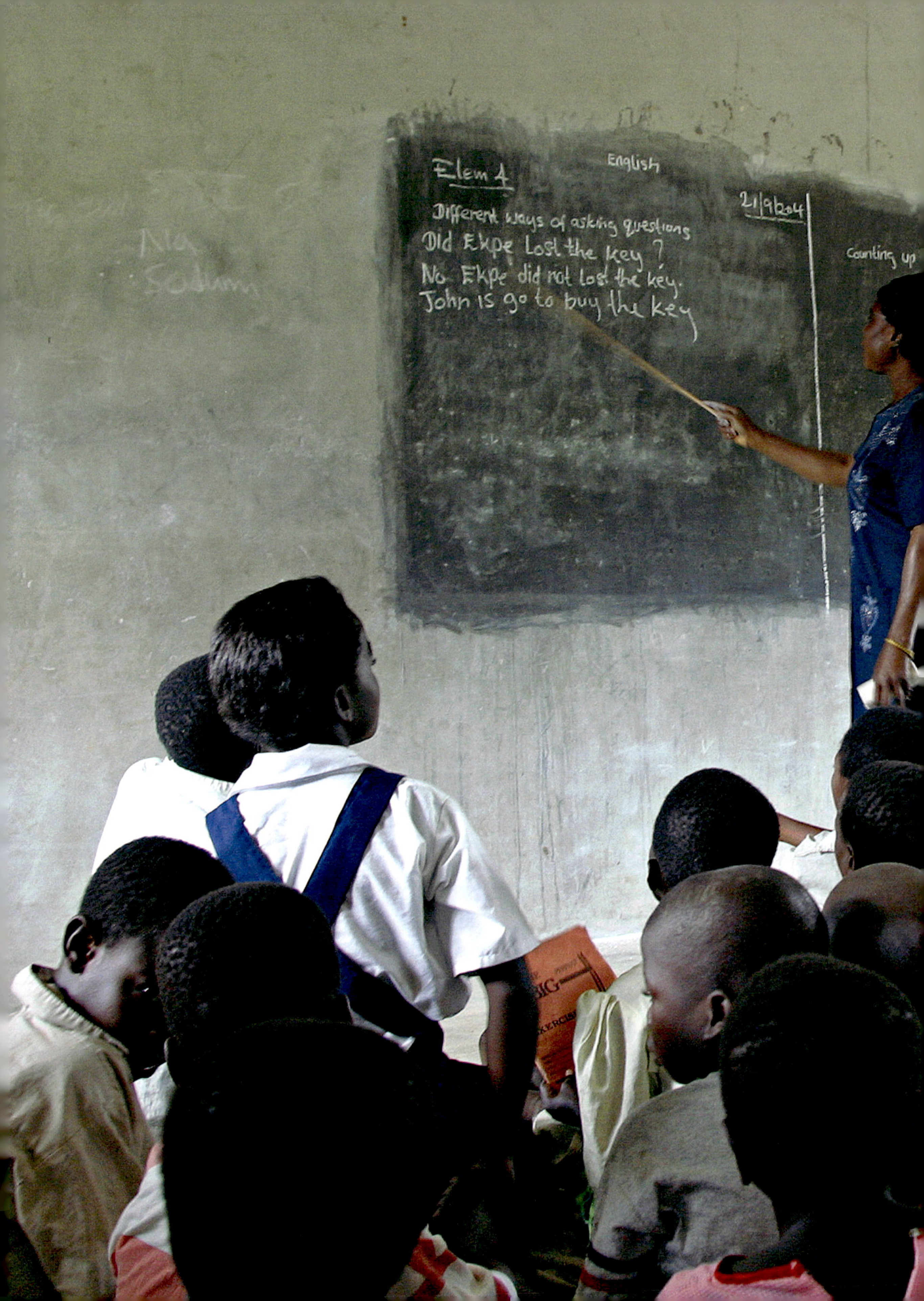
Elem 4
English
21/9/04
Different ways of asking questions
Did Ekpe lost the key ?
No. Ekpe did not lost the key.
John is go to buy the key
Counting up

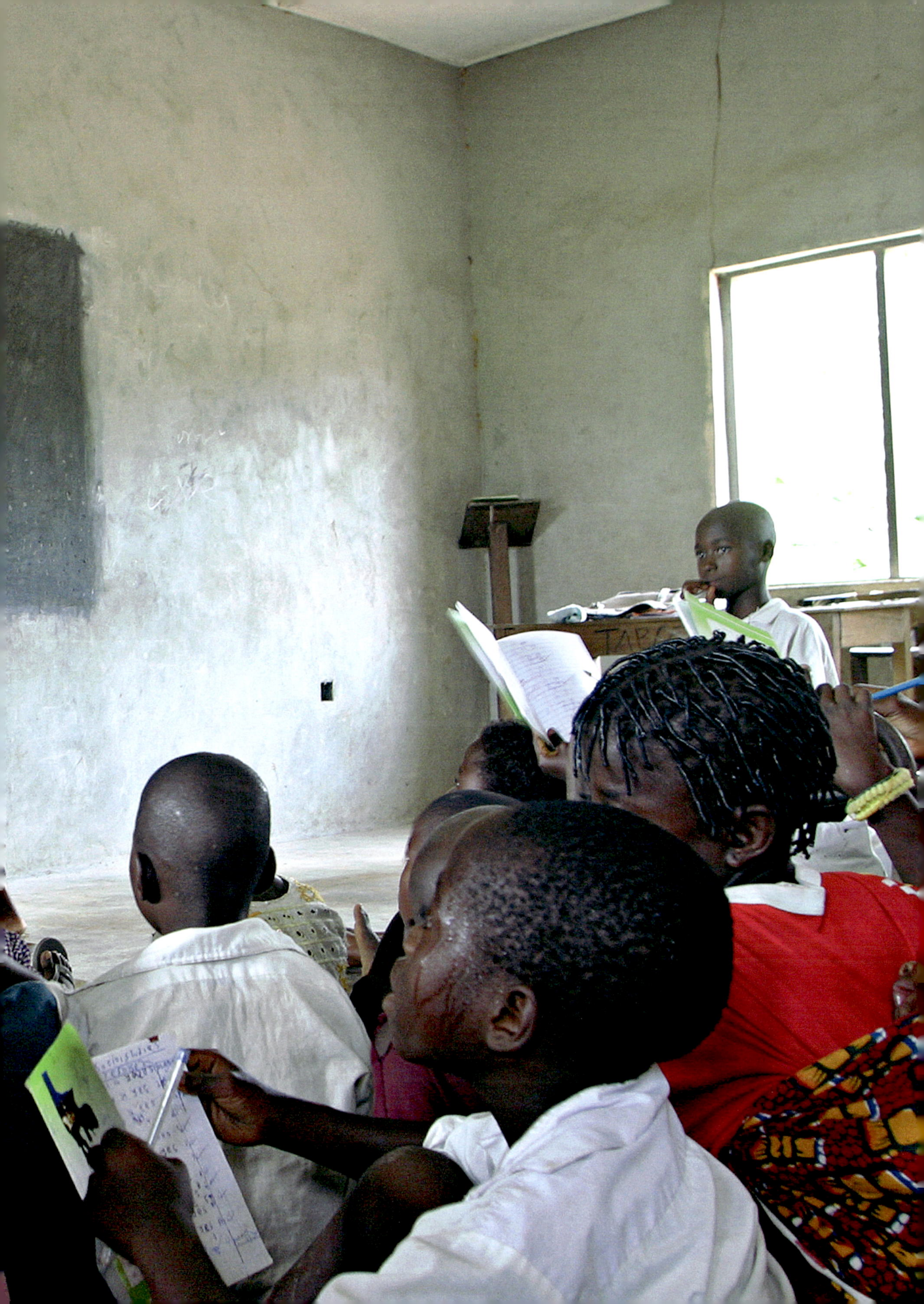

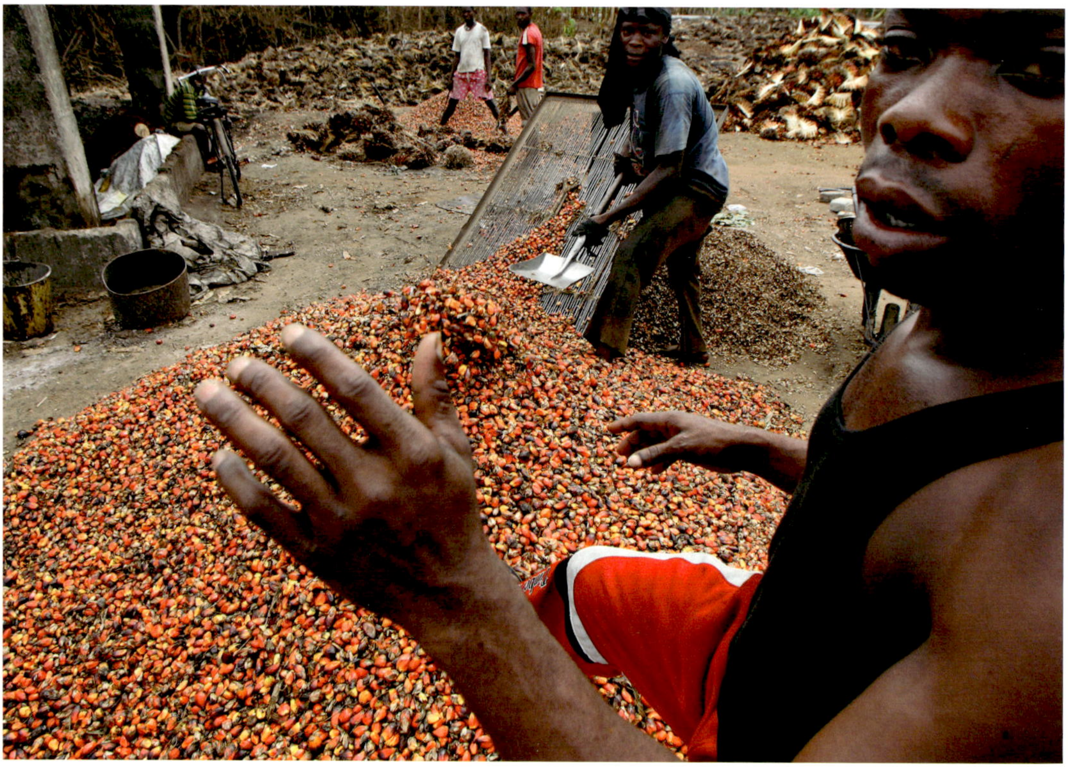

Above: Processing palm fruit to extract the palm oil, an industry abandoned for crude oil, 2007.

Right: A lady selling antelope , (bush meat) in Agbor, 2006.

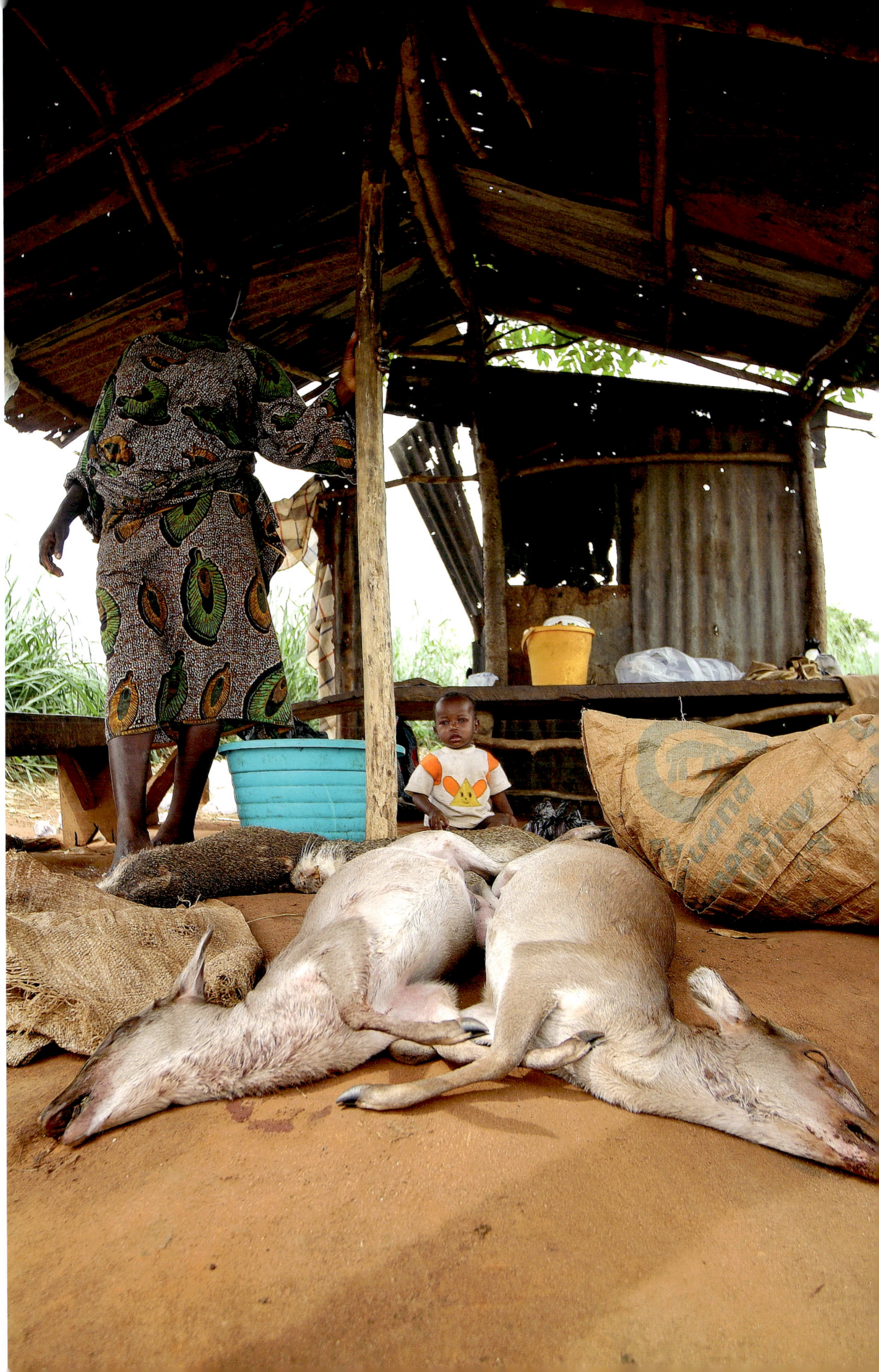

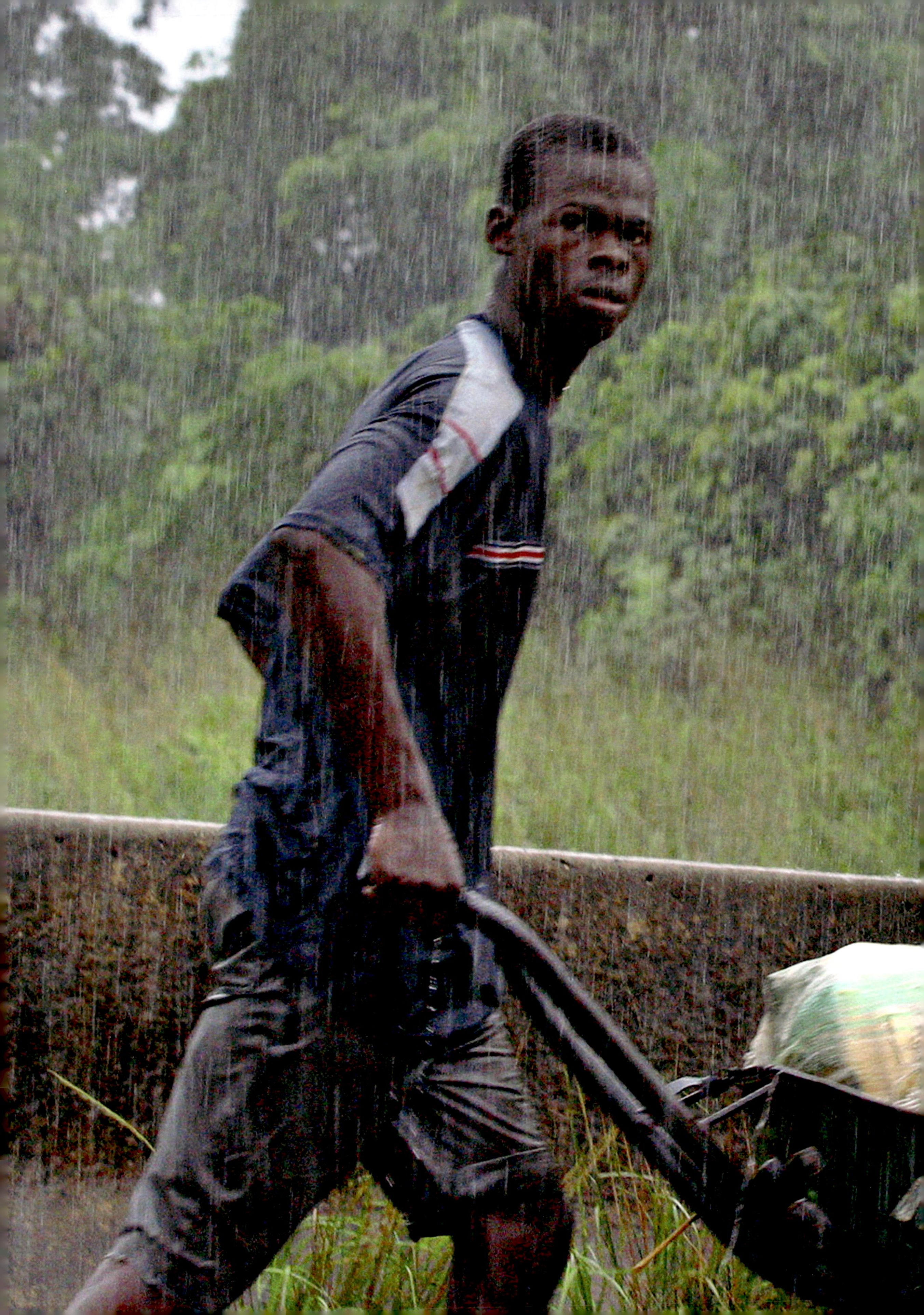

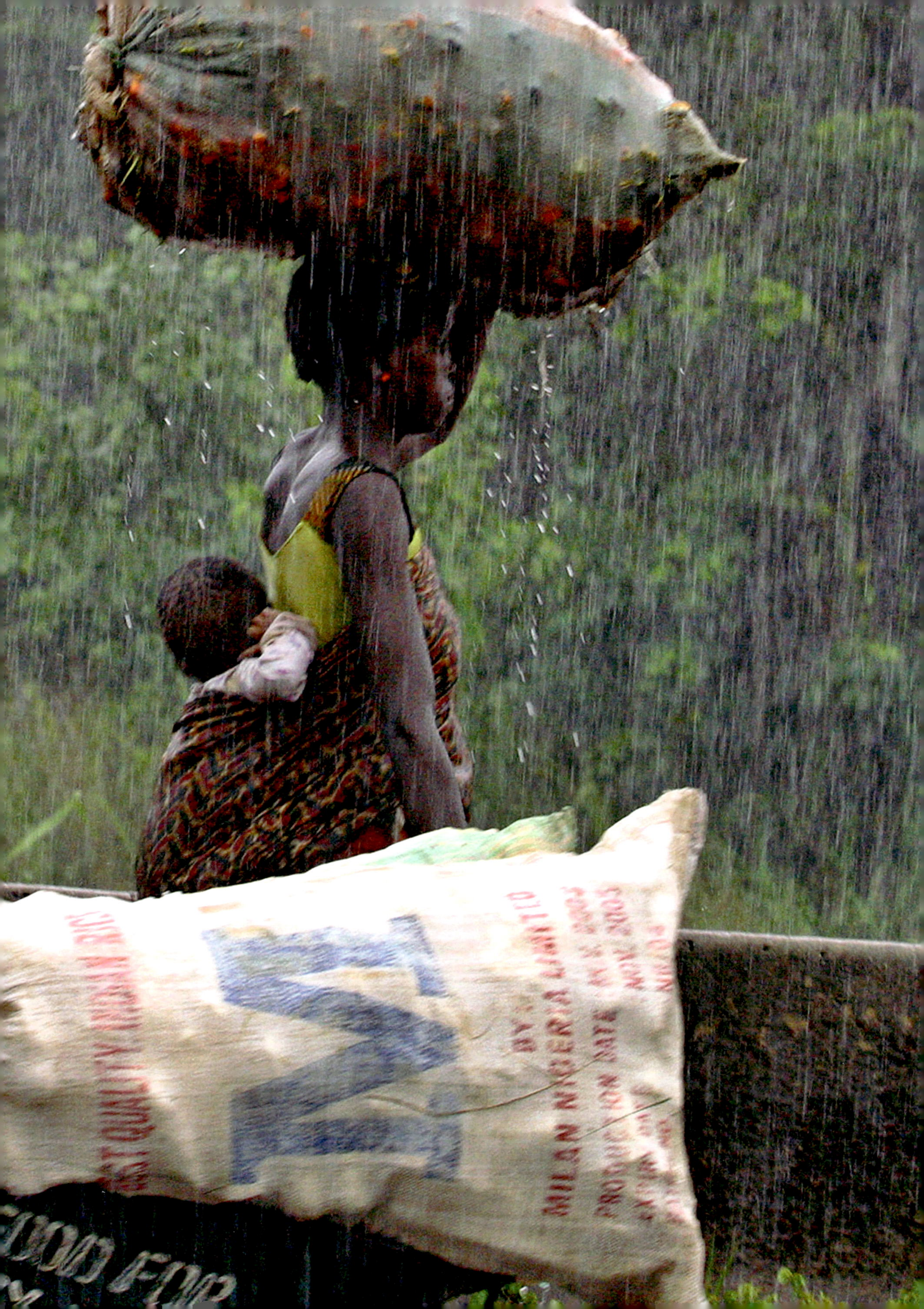
MILAN NIGERIA LIMITED

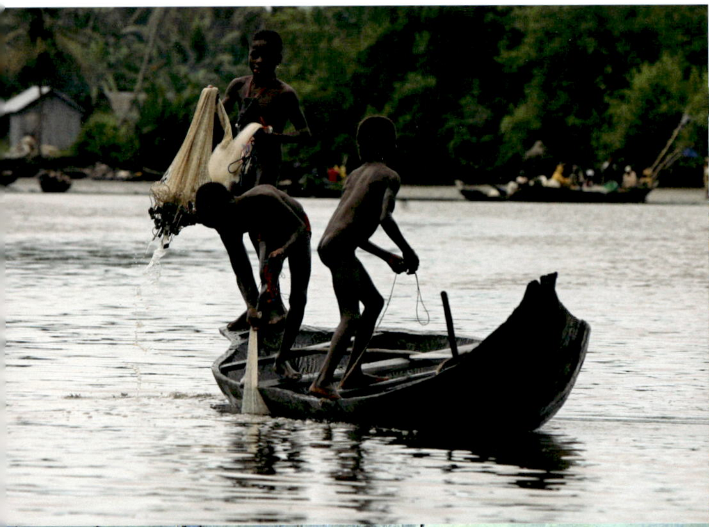
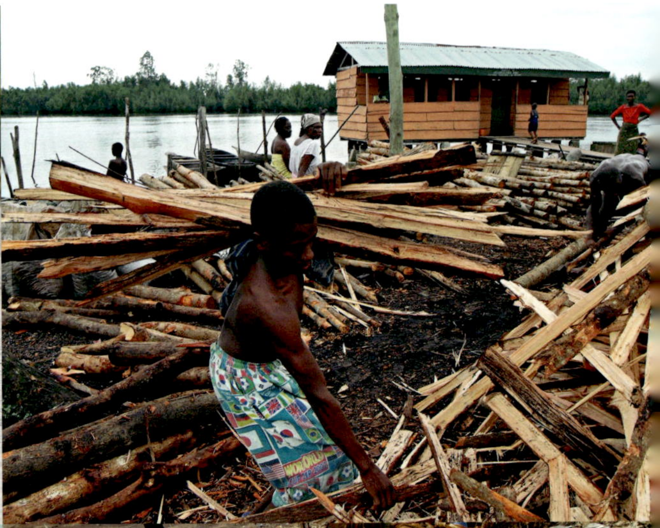
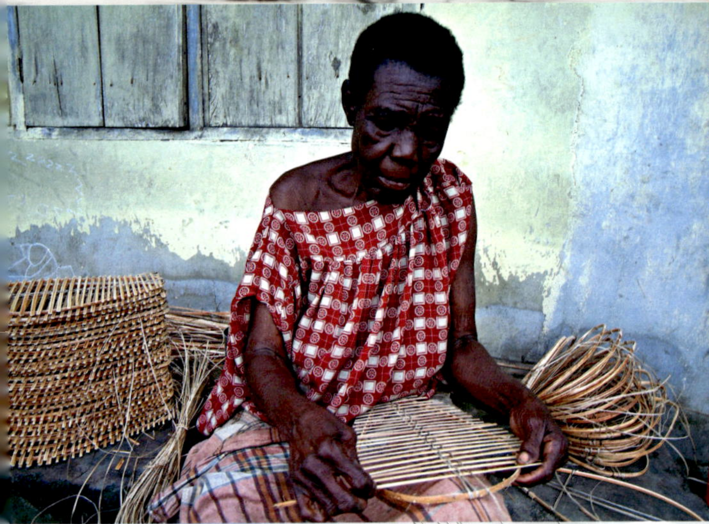
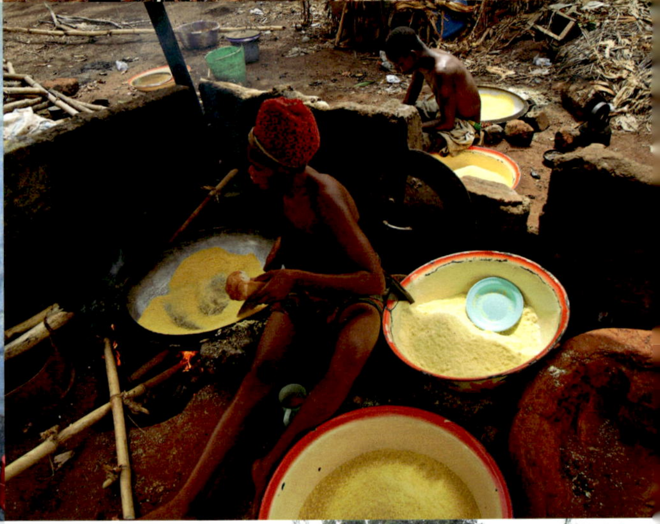
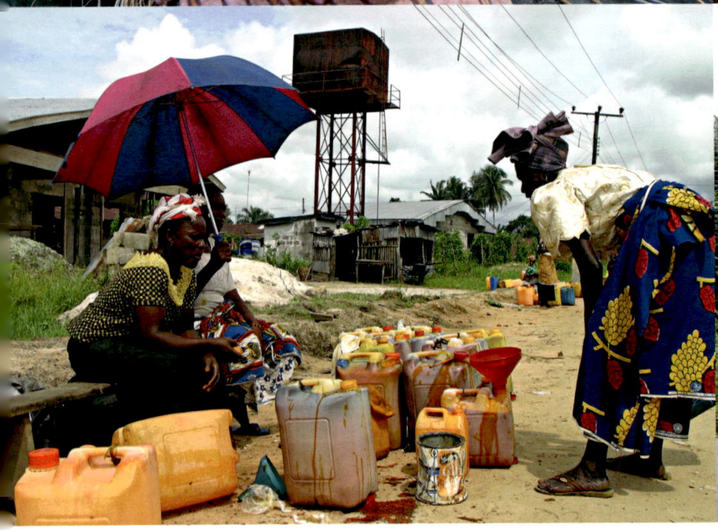
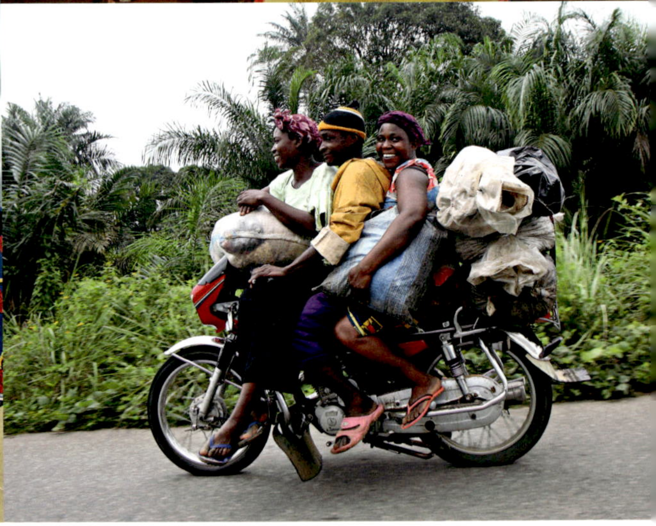

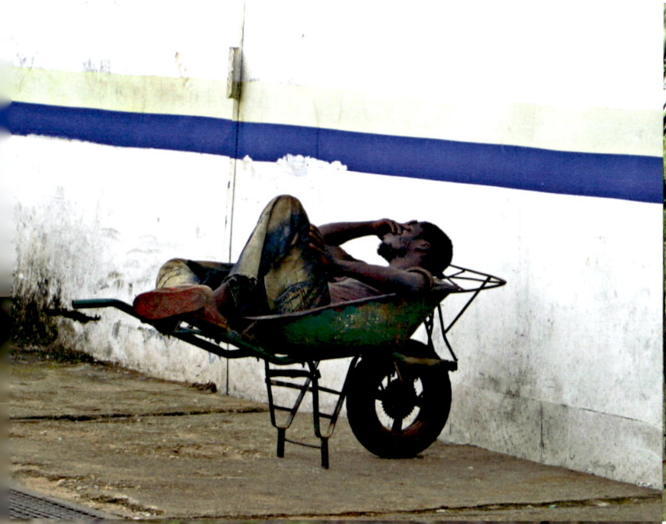
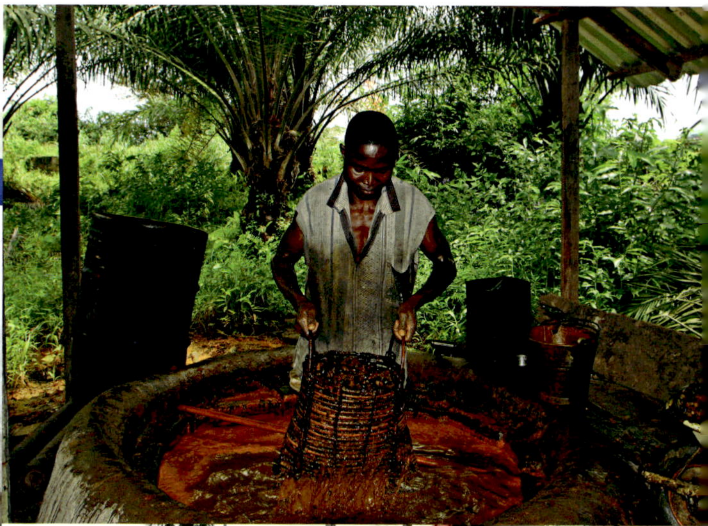
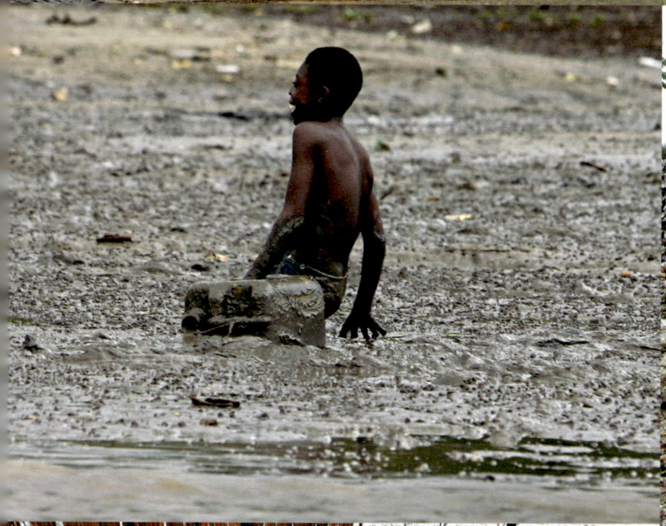
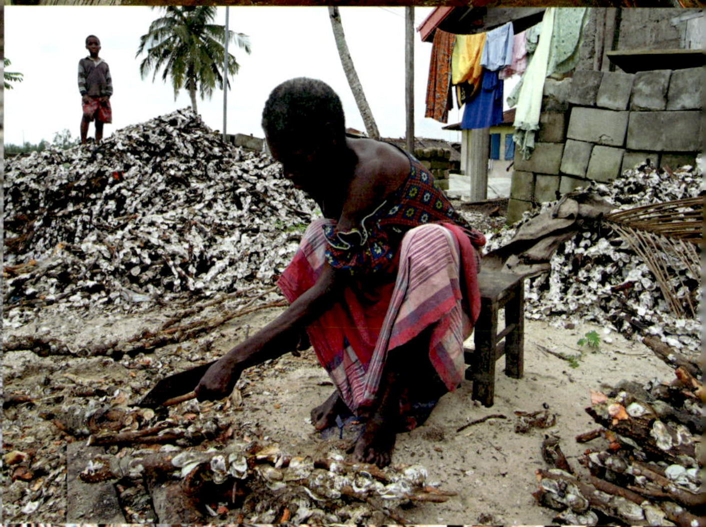
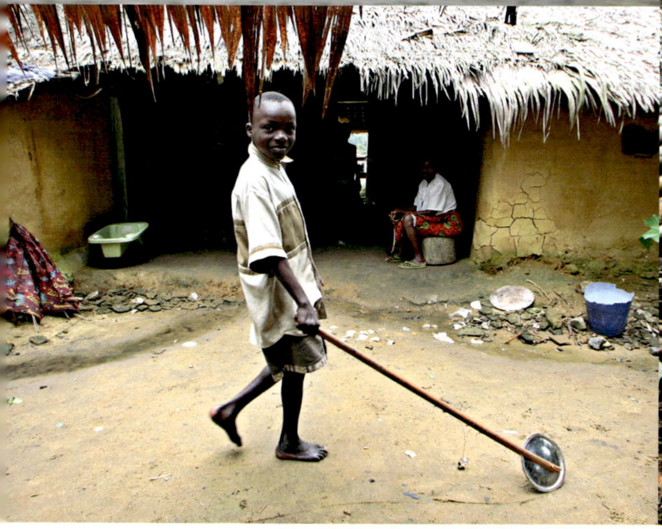
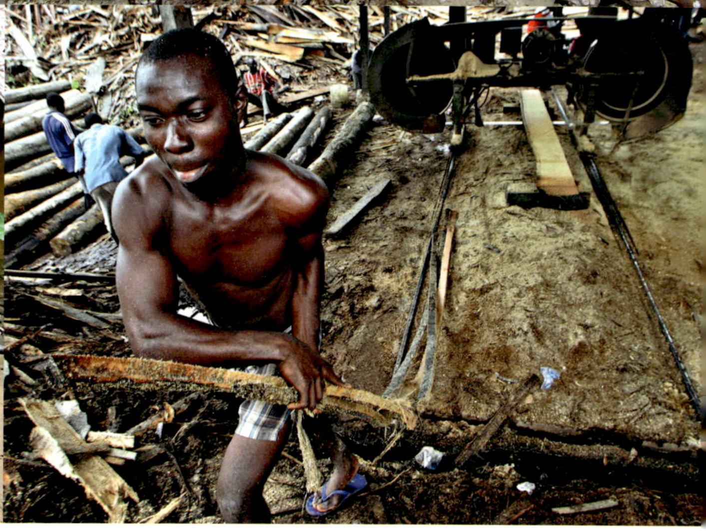

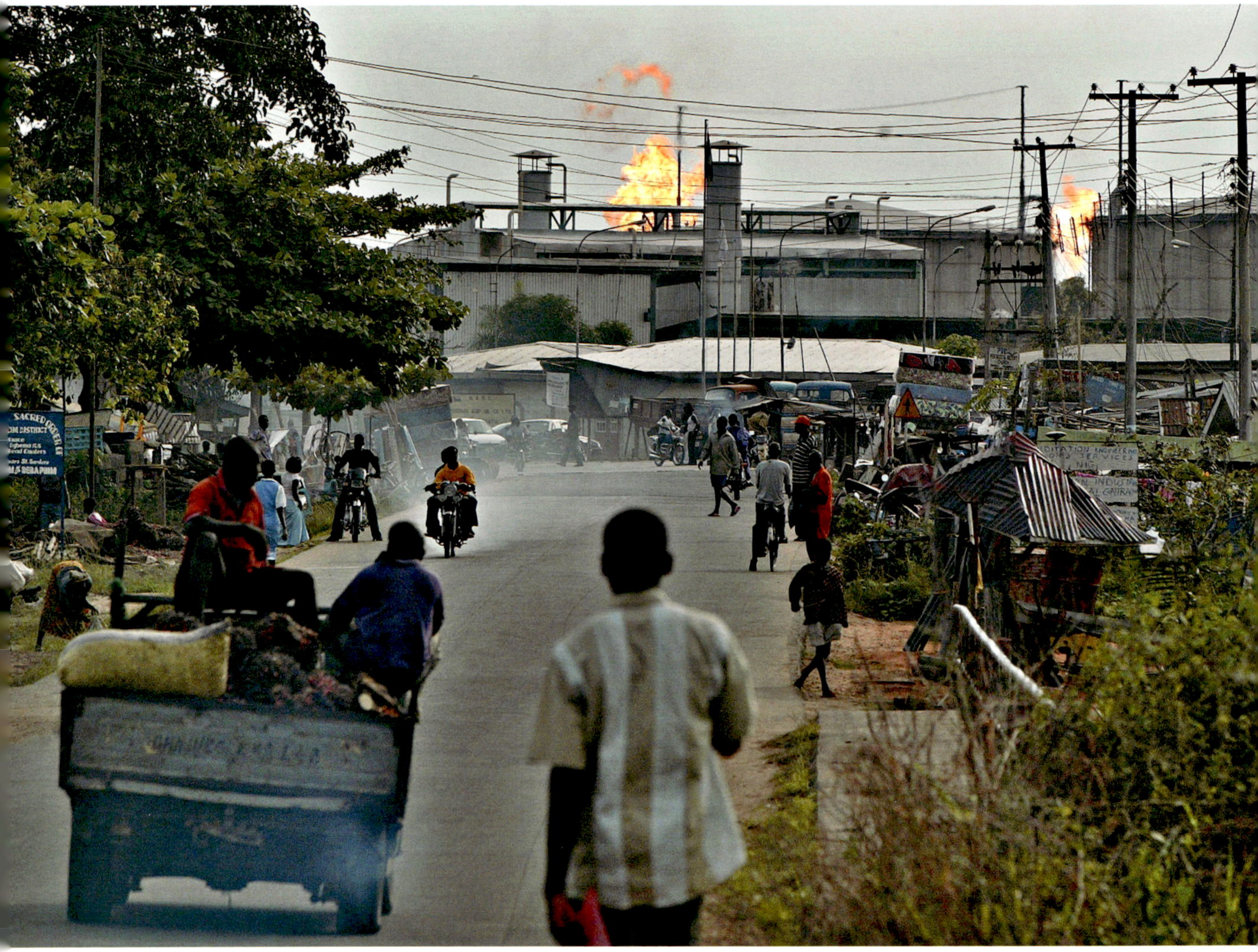

Above: Agip oil gas flare,
Ebocha, 2007.

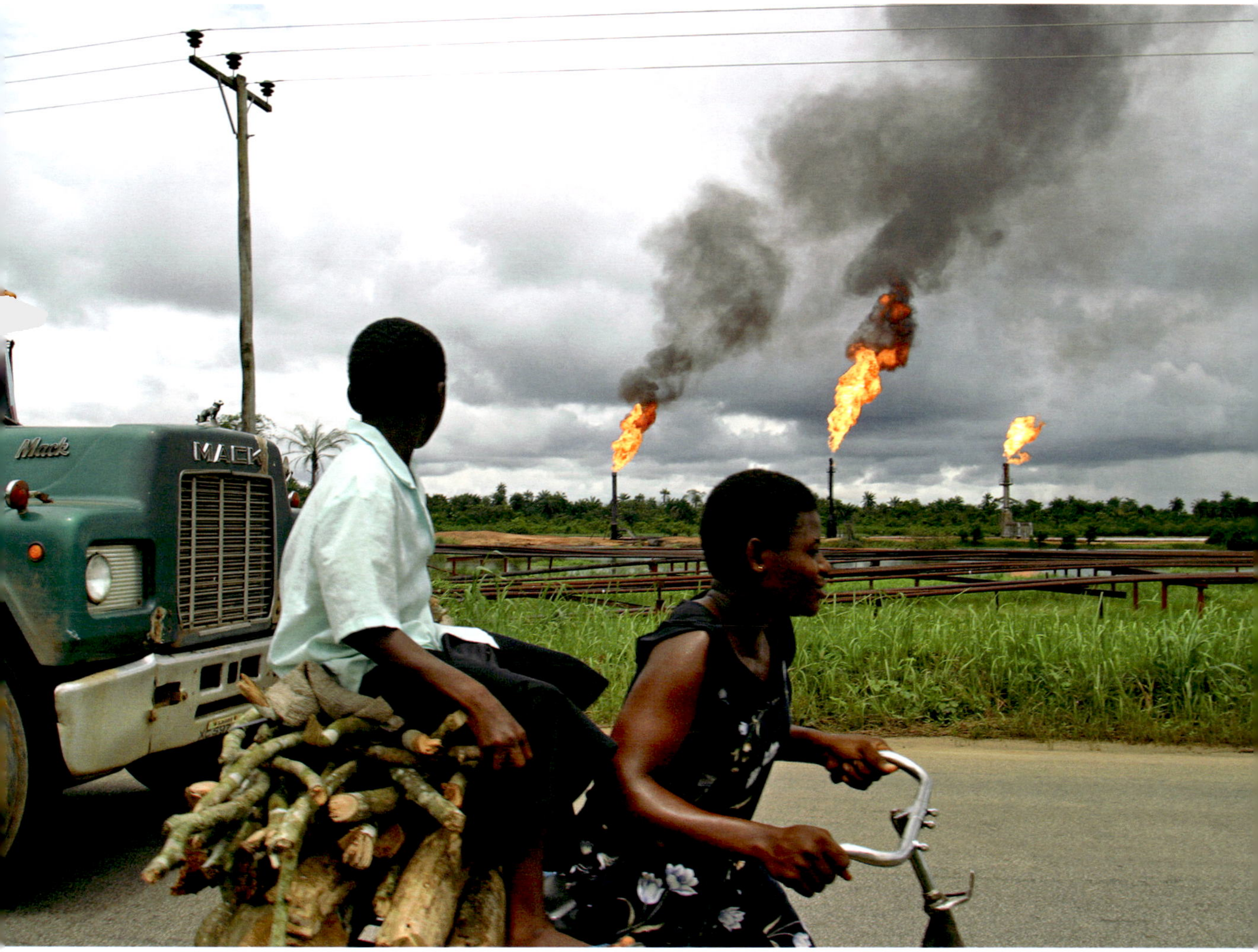

Gas flaring is common practice throughout the Niger Delta. After Russia, Nigeria has more gas flares than anywhere else in the world. It is considered a safe way of burning off excess gas typically released during oil production. In reality it leaves a terrible effect on health and the environment and is extremely wasteful of natural gas resources.

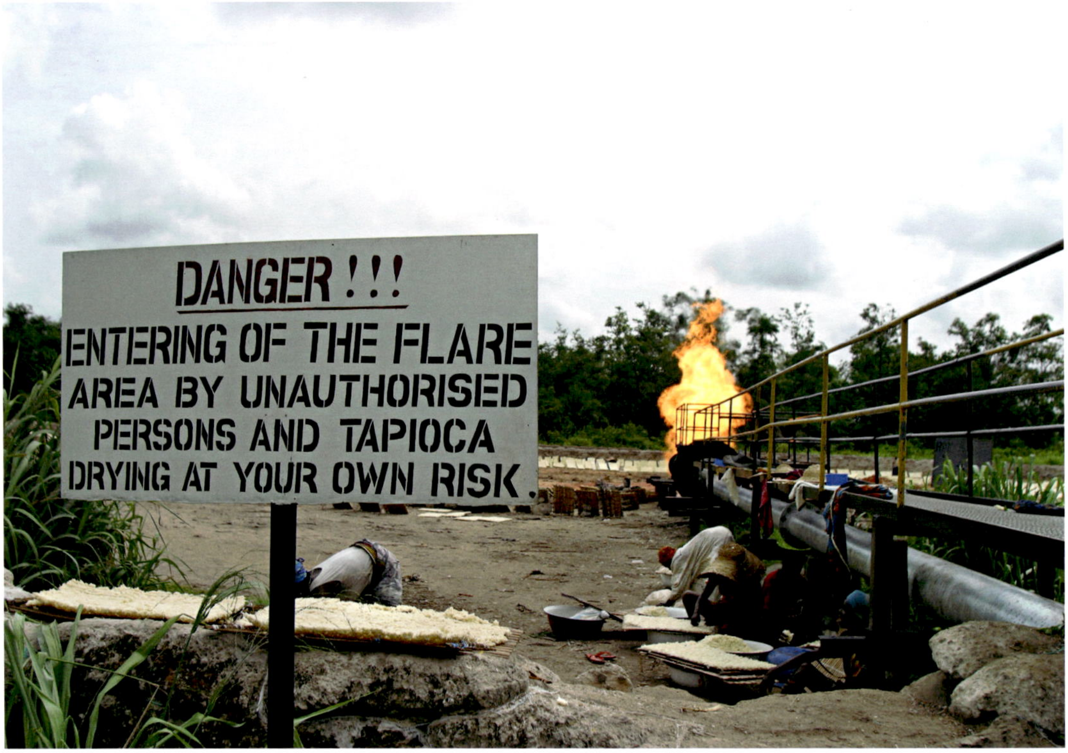

Tapioca drying is very common in the Delta, especially among women and children. Originally cassava cake would dry naturally in the sun over several days. With the advent of gas flares the tapioca could dry beneath it in a matter of hours. However the flares emit poisonous fumes which the women and children inhale, unaware and impotent against the risks. The cassava is then put into bags and sold at market. Tapioca is a major way for them to earn a living.

Right: Playing football near Oshie Agip gas flare, 2004.

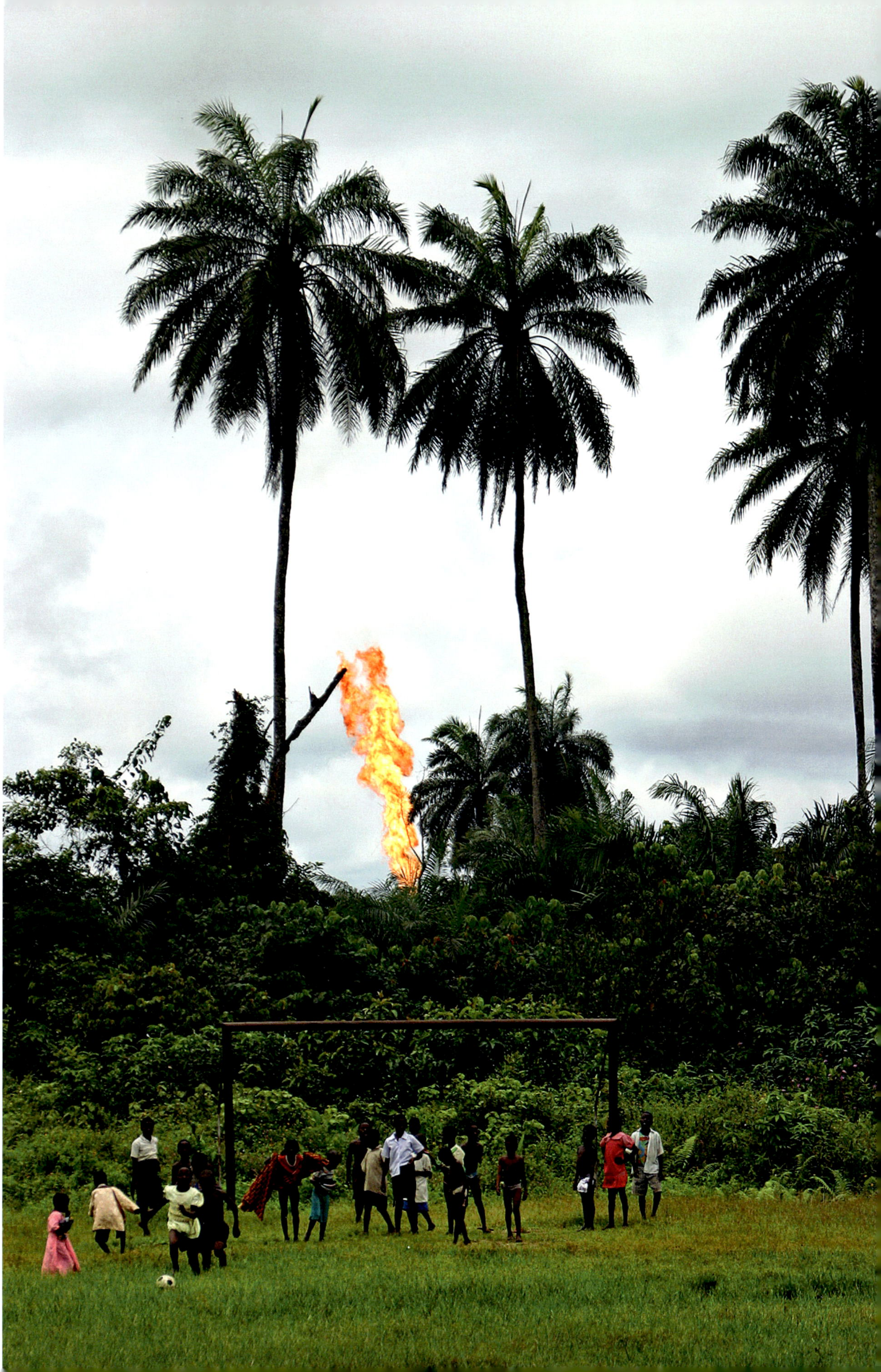

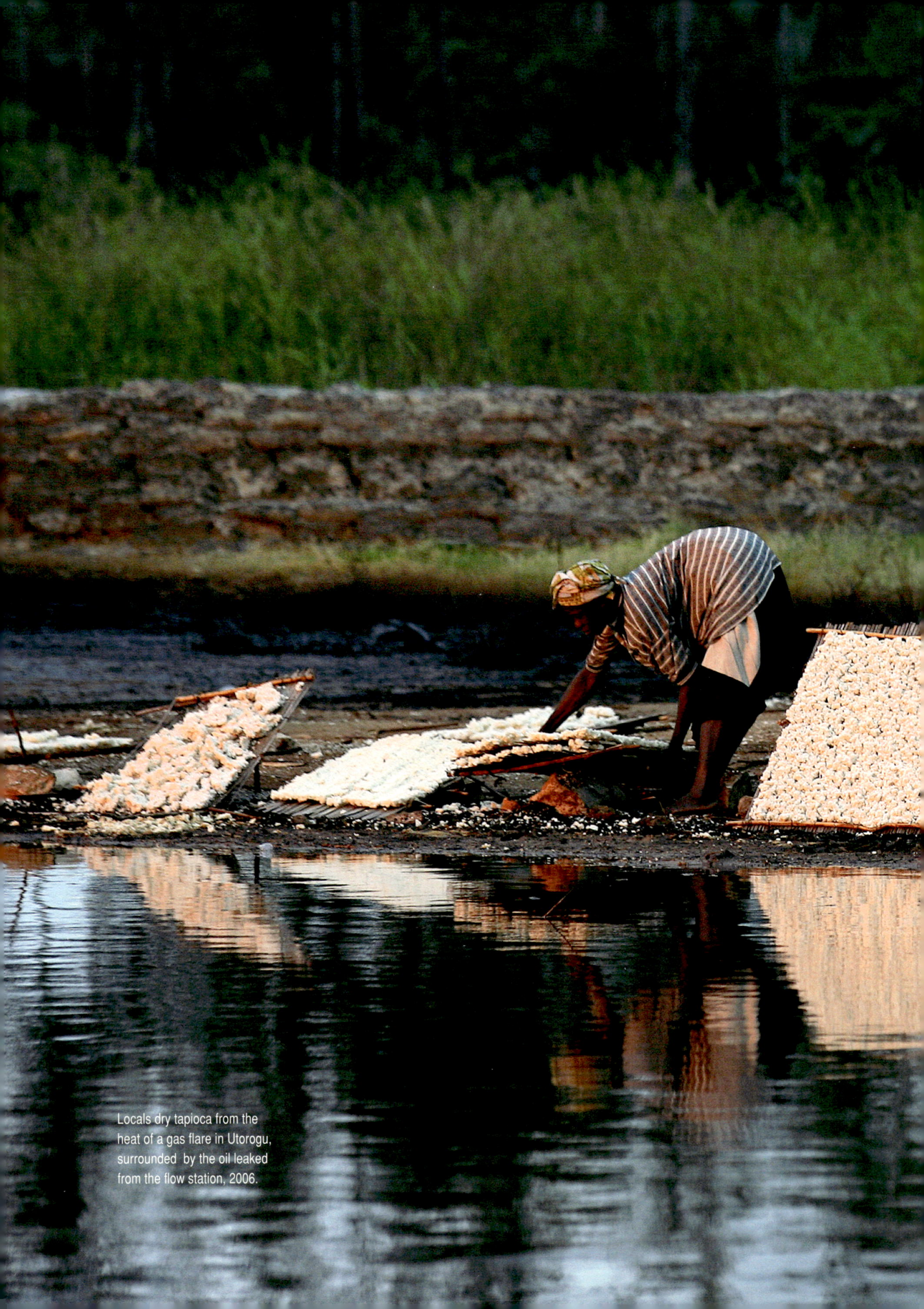

Locals dry tapioca from the heat of a gas flare in Utorogu, surrounded by the oil leaked from the flow station, 2006.

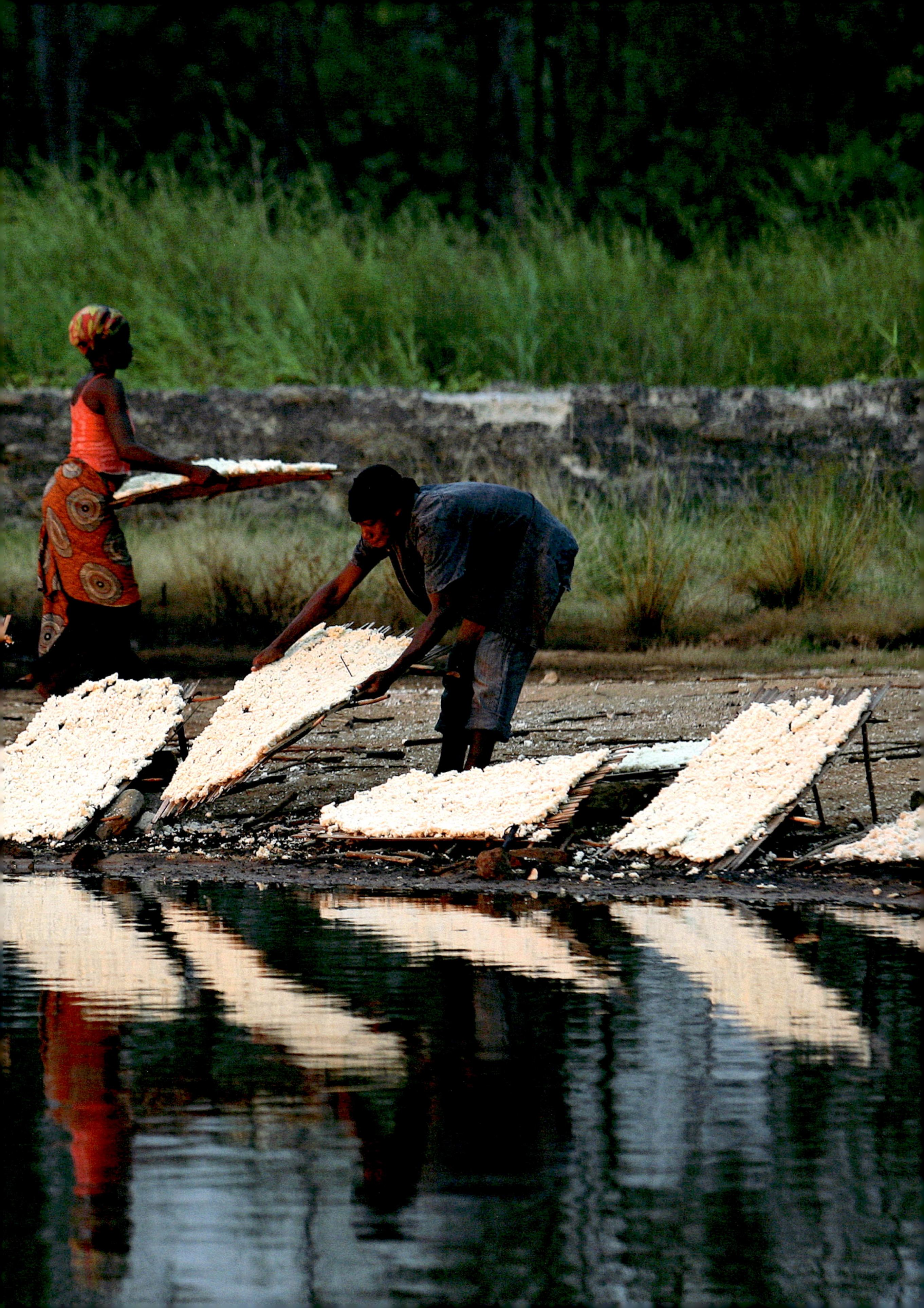

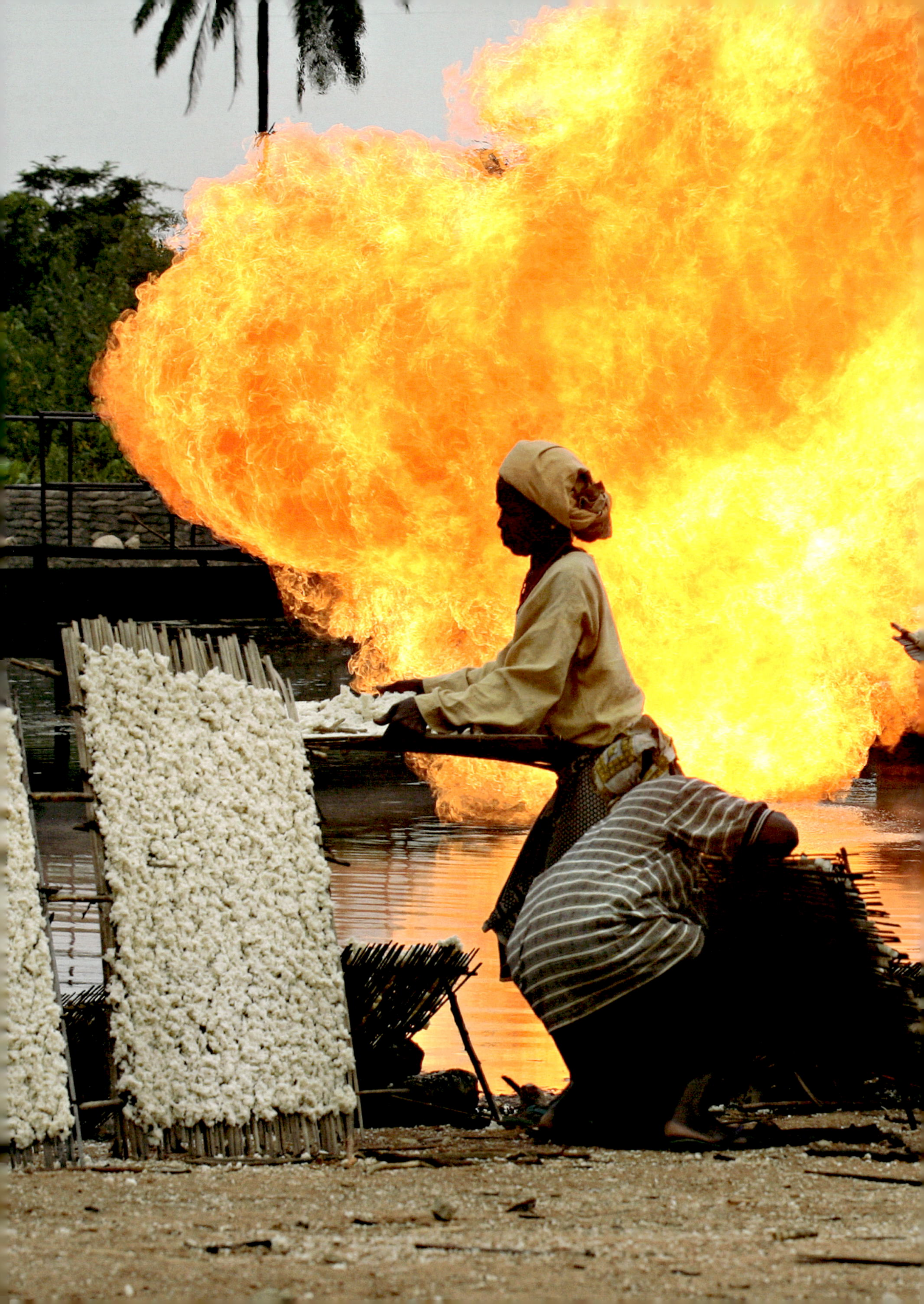

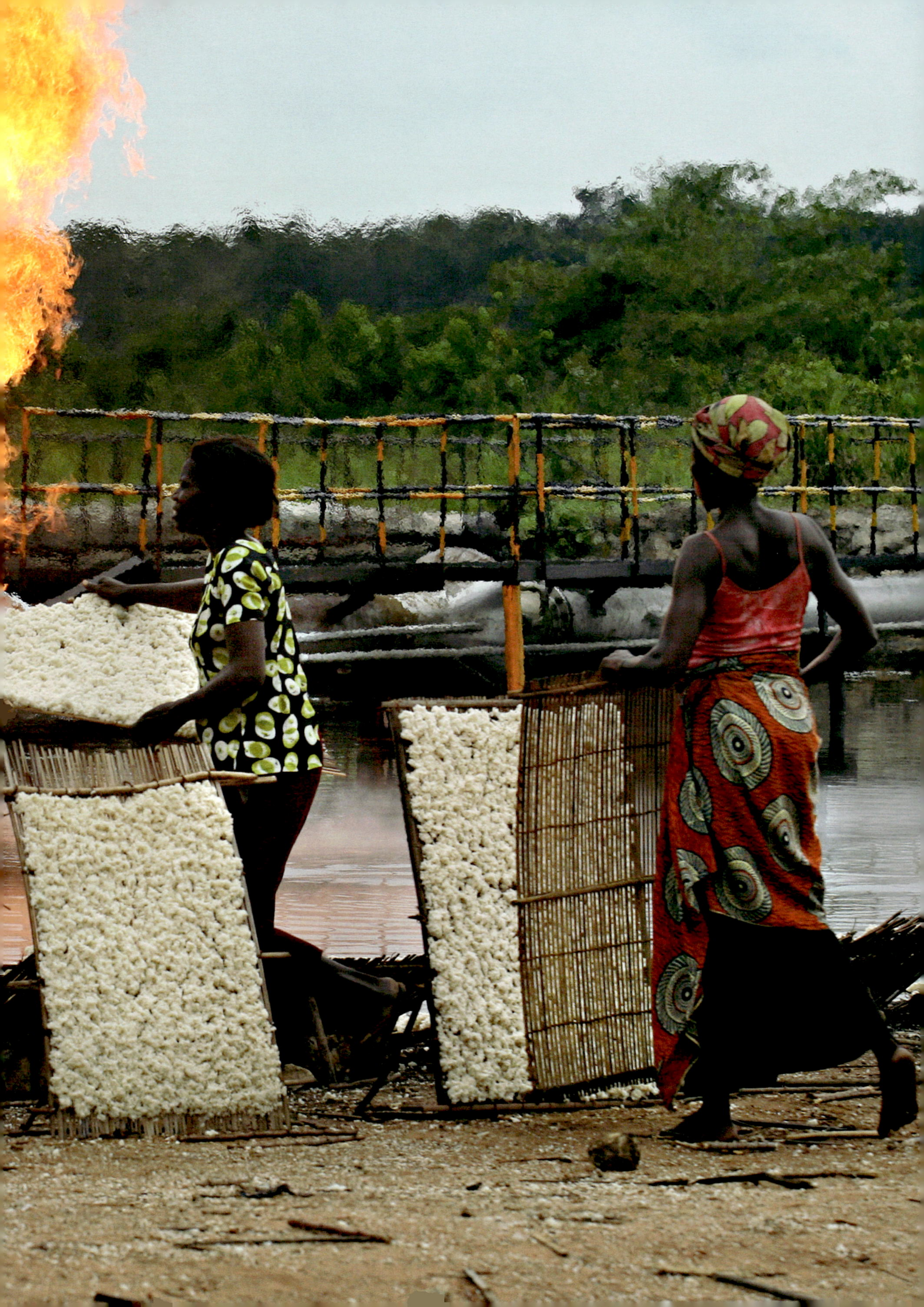

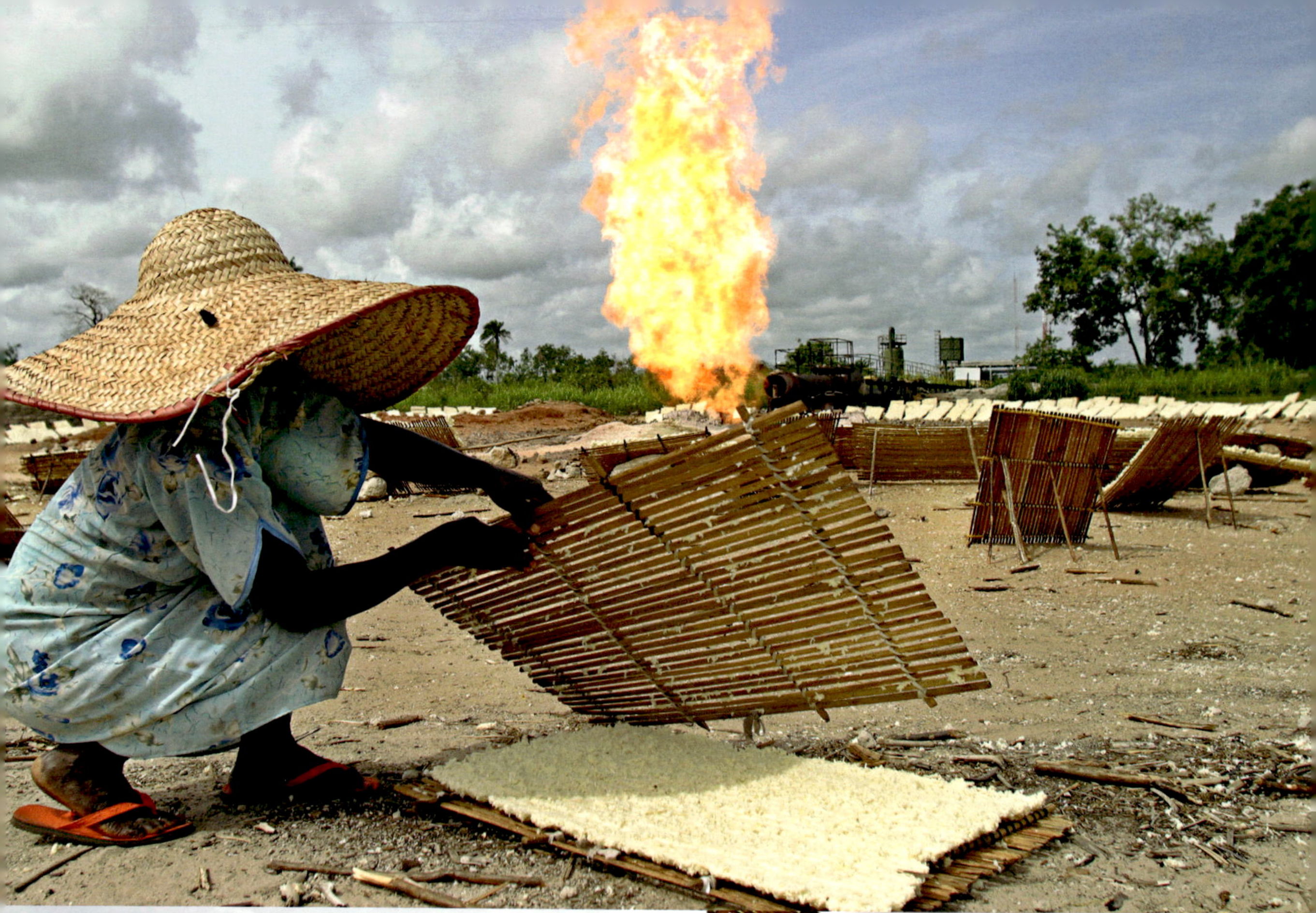

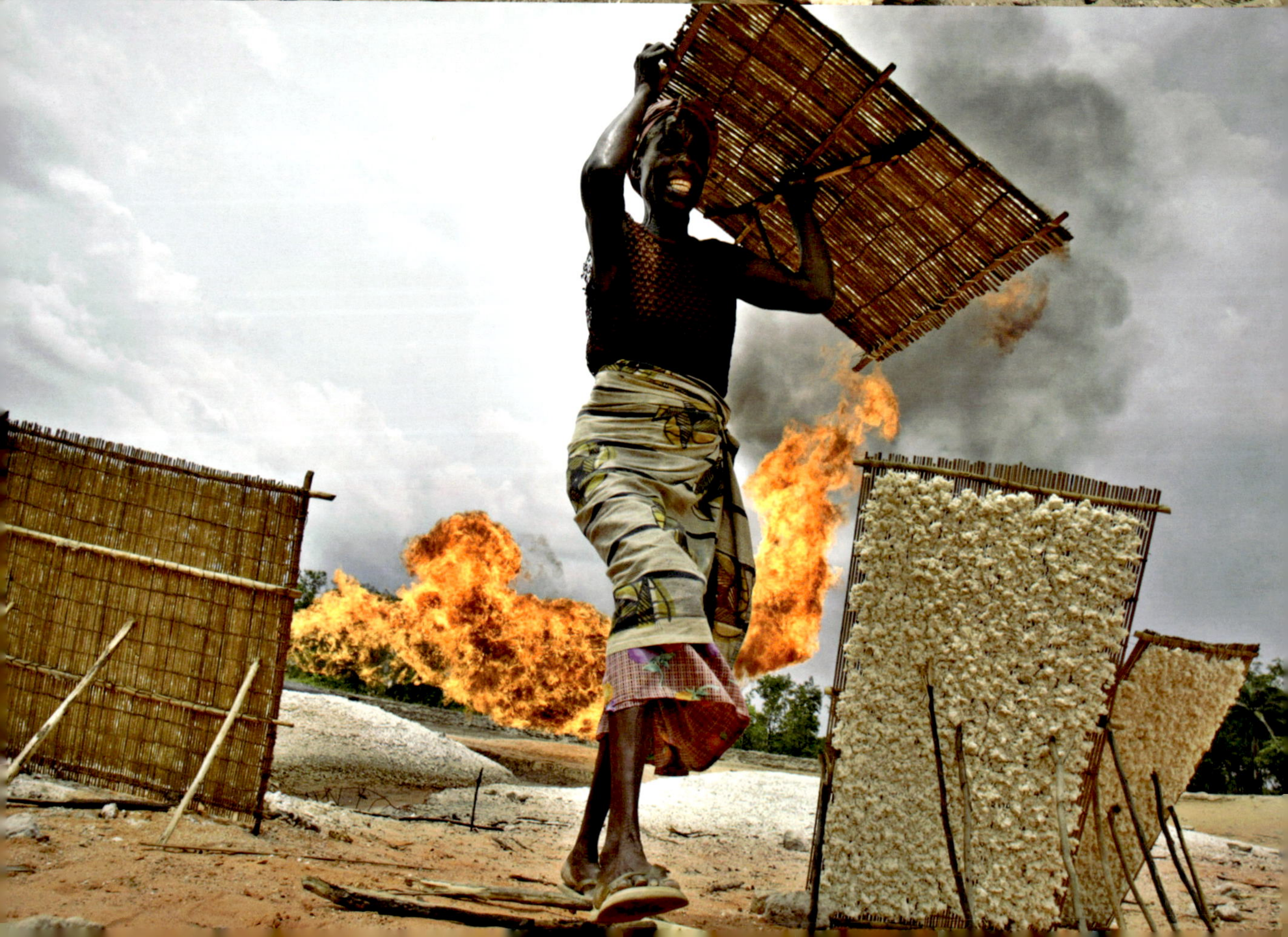

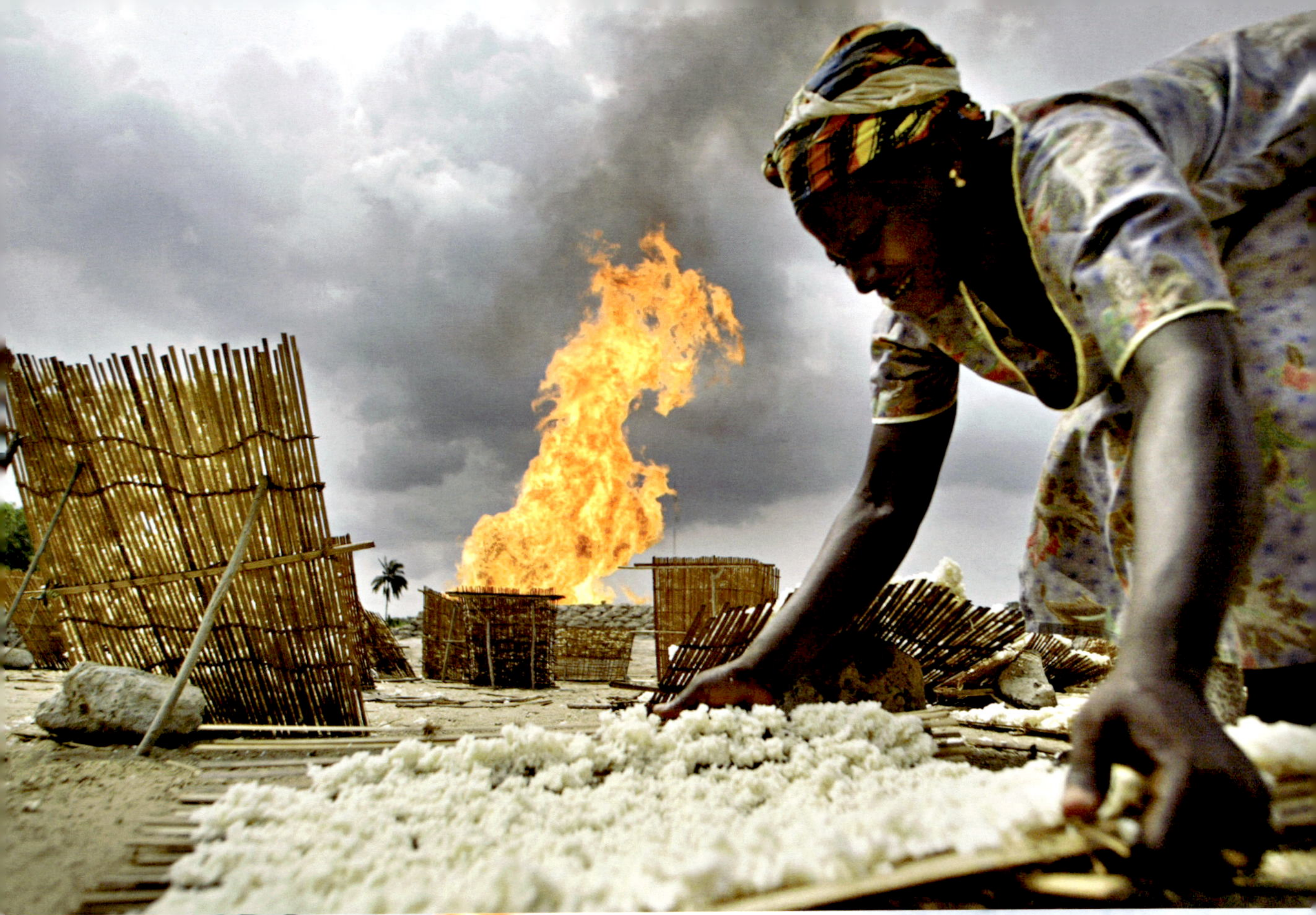

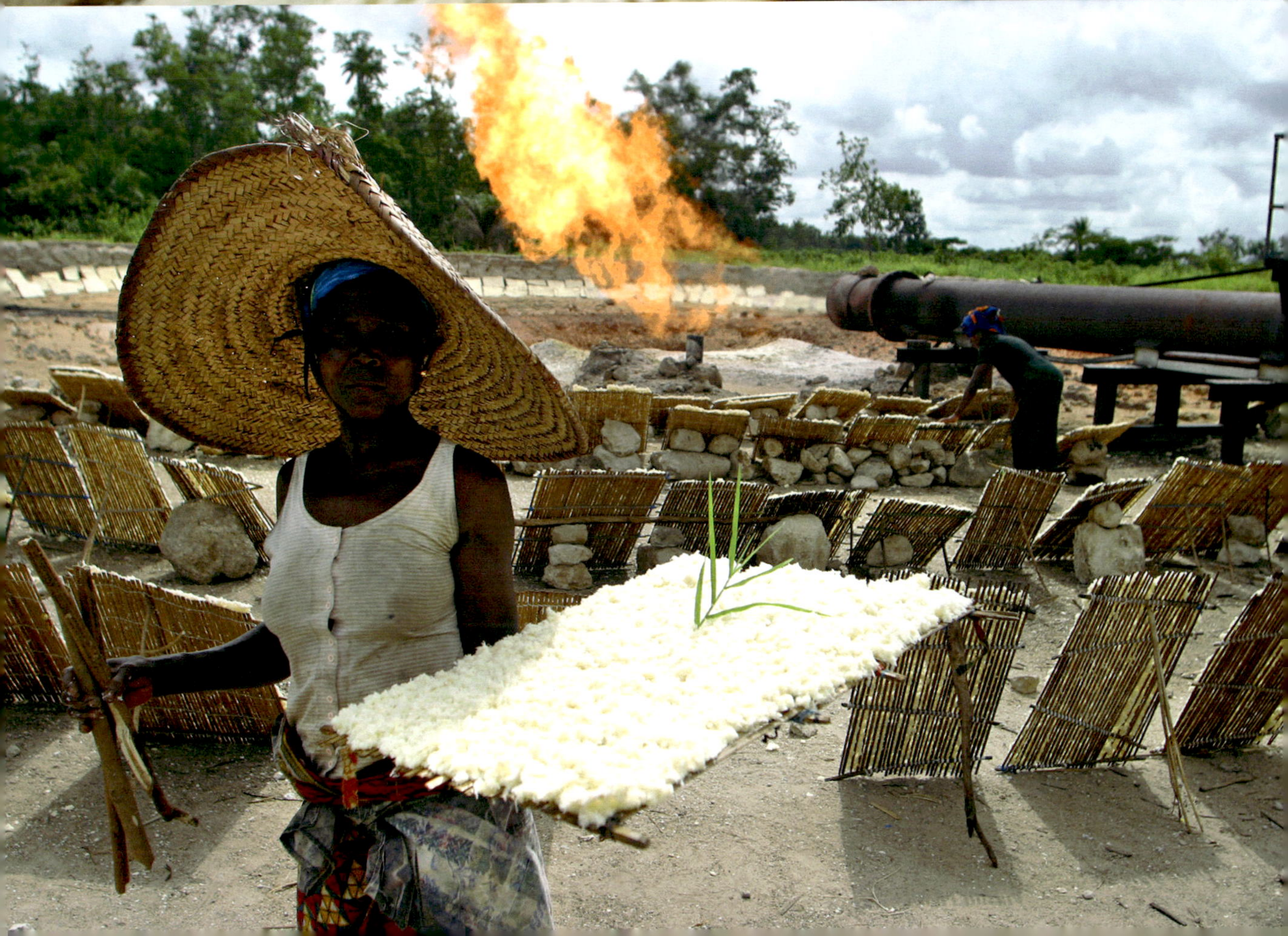

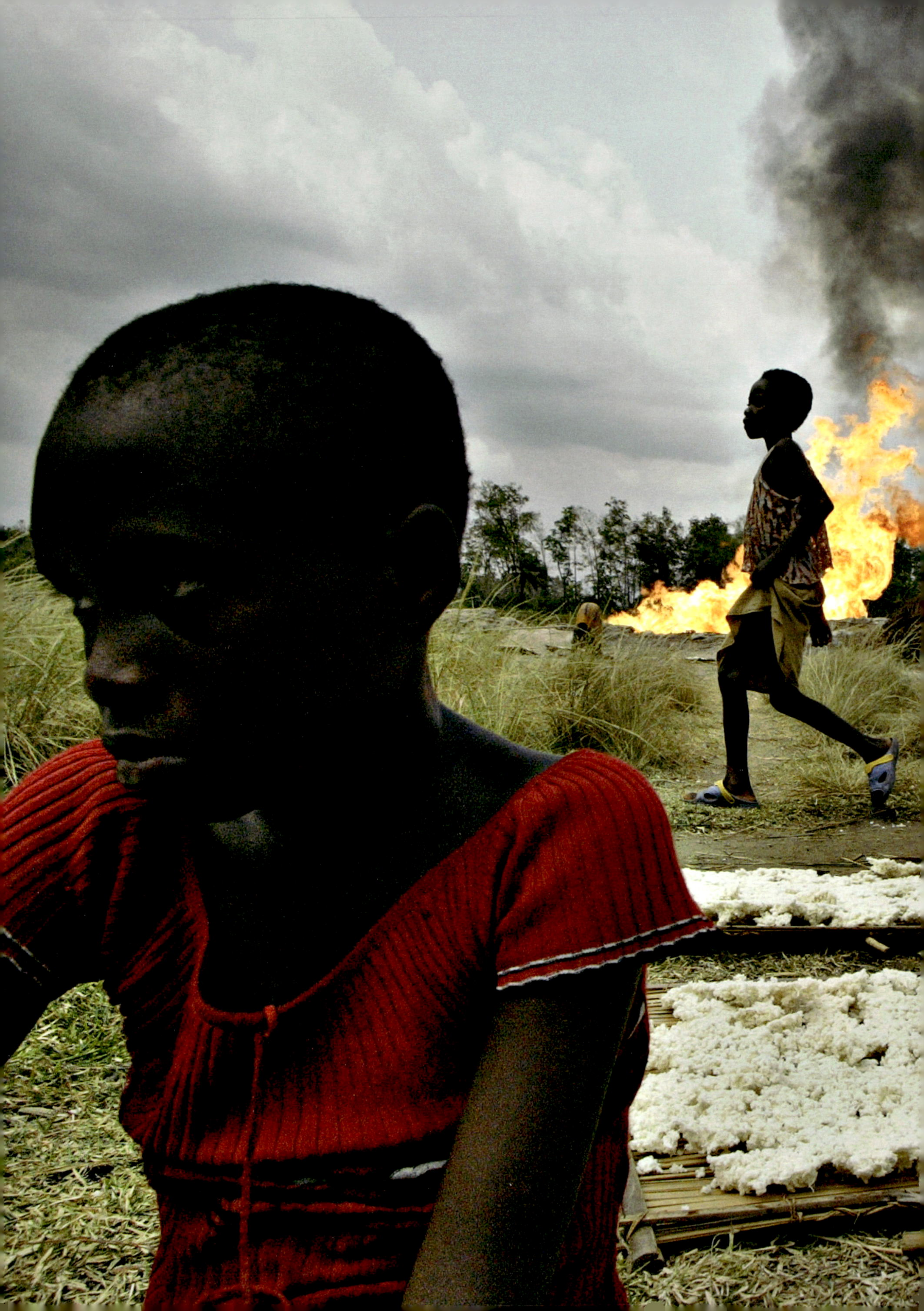

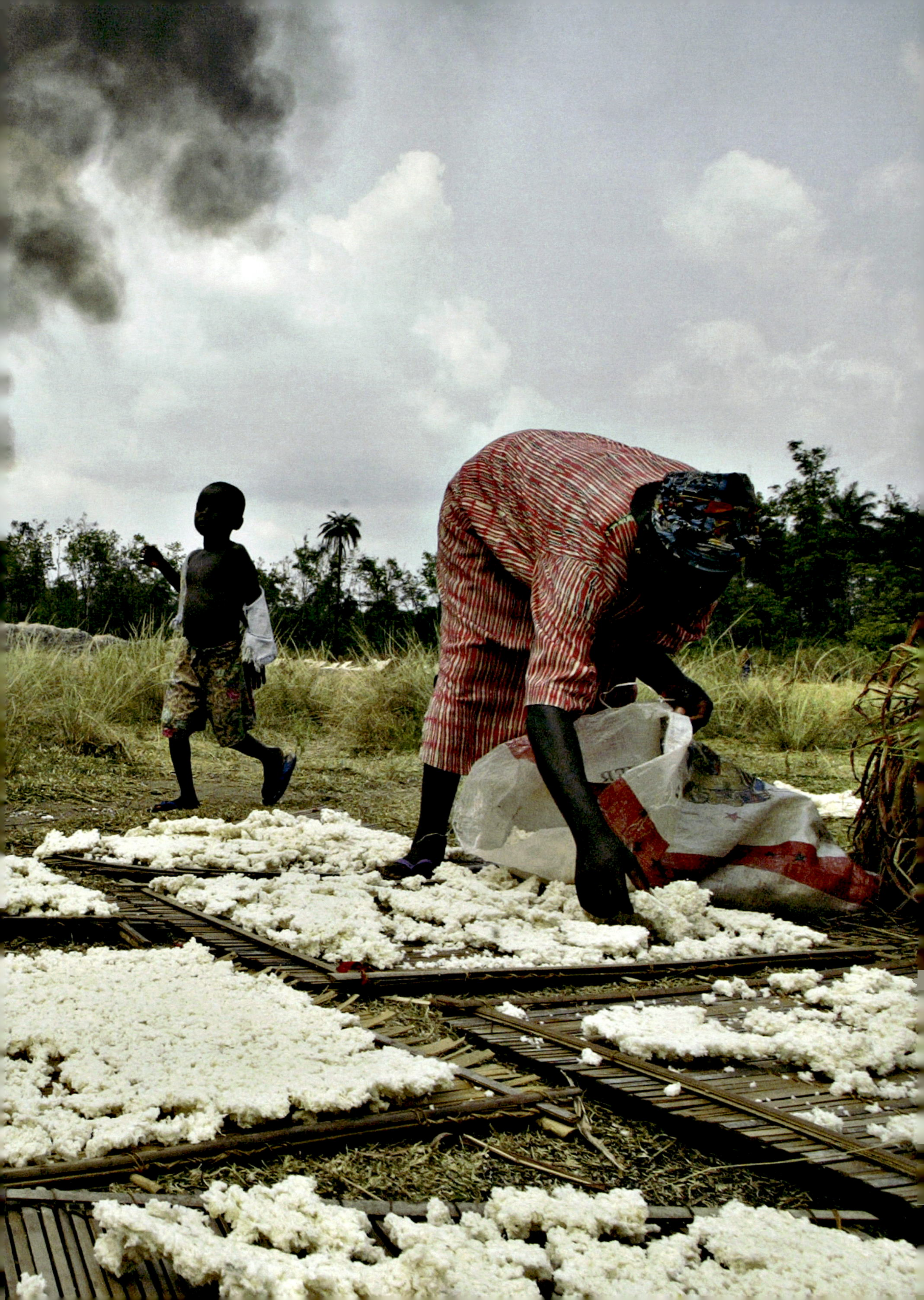

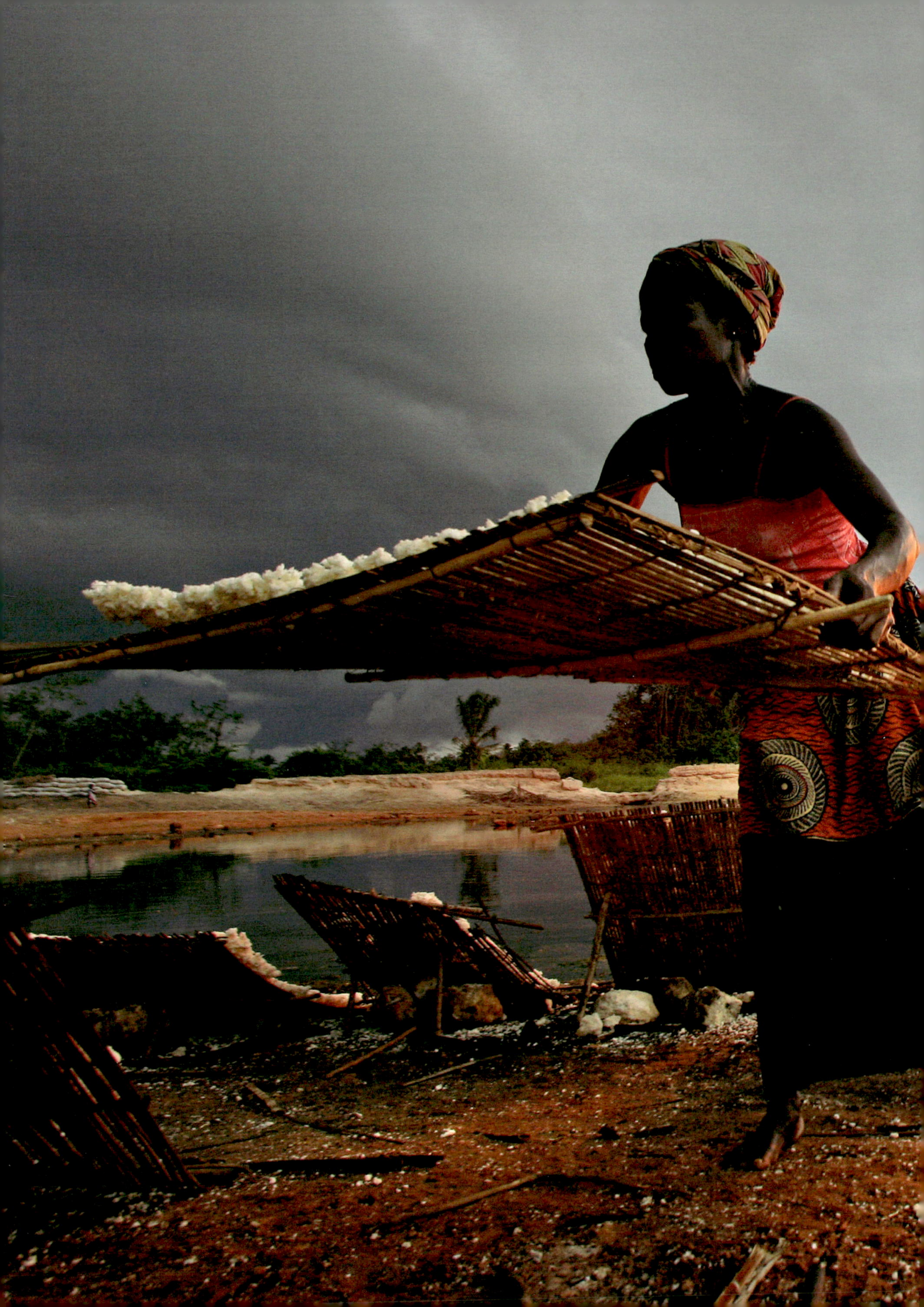

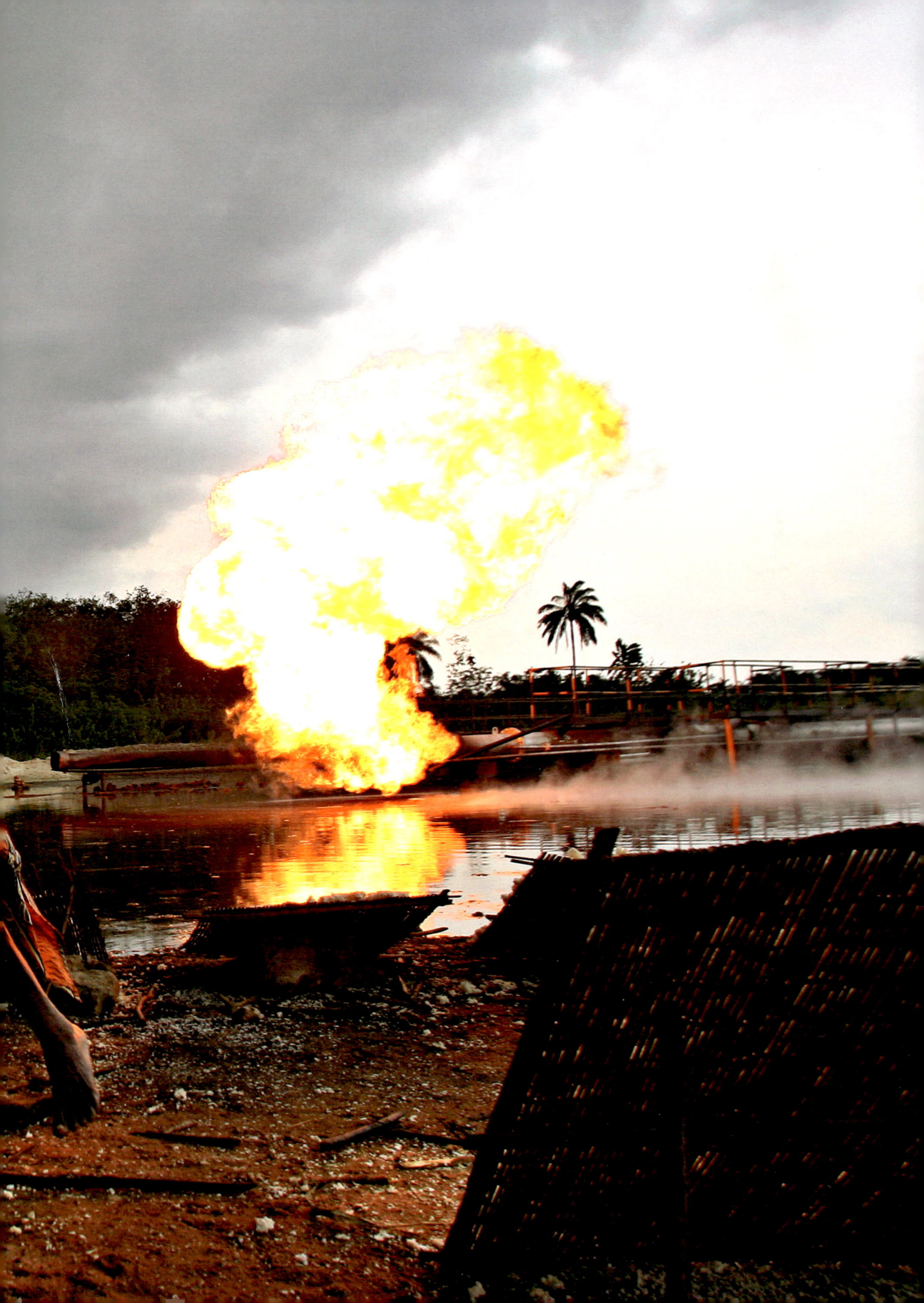

FIRE IN THE DELTA

BY TAM FIOFORI

"There's fire on the mountain / Run, Run, Run," exhort the lyrics of a popular Nigerian nursery school rhyme.

There is definitely fire in the crude oil-producing host communities in the Niger Delta, but unfortunately, there is nowhere the inhabitants of these communities can run to or hide.

The discovery of crude oil in vast commercial quantities in Nigeria, over fifty years ago, and the continued exploitation and export of millions of barrels of crude oil everyday, has been an unqualified economic blessing for Nigeria but a malignant curse for the Niger Delta, the source of Nigeria's crude oil reserves.

Nigeria's crude oil boom has been accompanied by the continued doom of the environment and inhabitants of the Niger Delta. Since crude oil was first produced in the late 1950s, the associated gas has simply been separated and flared off by the oil companies. This associated gas, well over 58 million standard cubic feet (mscf), is daily flared with impunity in thousands of oil fields (each containing multiple oil wells) dotted all across the Niger Delta.

These gas flares remain a big danger to both the environment and inhabitants of the Niger Delta. Seen from the air, these gas flares give the Niger Delta the eerie semblance of an open-air chemistry laboratory with thousands of Bunsen burners.

It would seem that the 'mad scientists' responsible for this deadly combustion are permanently away without leave. On the ground, some flares assume awesome heights of well over twenty metres; dwarfing houses and humans, as they hungrily lick the sky and continuously emit audible hisses of sizzling heat.

At night, they create an unusual phenomenon of daylight around rural riverine communities that have no electricity supply. For these communities the gas flares create a dangerous 24 hour artificial light with fumes of deadly poisons.

The flares cause acid rain that in turn poisons both the soil and the creeks; stunting the growth of agricultural produce and killing most marine life. Buildings are affected and roofs soon rot away. The gases burn the skin of humans and cause skin cancer. They also affect the eyes and the digestive system. Simply put,

It would seem that the 'mad scientists' responsible for this deadly combustion are permanently away without leave.

gas flaring continues to be a big daily health hazard to the inhabitants of the Niger Delta as well as the entire ecosystem.

This 50-year old daily and deadly health and environmental hazard, placed side by side with the reality of the multi-trillion dollars that Nigeria has earned from the crude oil of the Niger Delta, more than confirm that the Nigerian military and political class since independence are uncaring, callous and plain wicked.

Can racism and apartheid be any worse than this slow sure death? Can our former colonial masters be blamed for this mean streak in this privileged class who continue to exploit the golden-goose-laying minority in their same country? Harsh words you might say; give them time, you might equally plead.

But time is definitely not on the side of the inhabitants and environment of the Niger Delta. Also, evidence points to the worst-case scenario of cynical, deliberate and calculated neglect.

Twenty-five years ago (26 years after the first gas flares), the federal military government decreed that all gas flaring must end by January 1, 1984.

Corruption and collusion between the NNPC and the oil companies continue to make nonsense of this order. From a purely hard-nosed economic argument, the type favoured by our military/political class, Nigeria is now earning good money exporting gas from its NLNG complex in Bonny.

So, progress has indeed been made, whatever the continued cost to the Niger Delta. It is also some progress that Nigeria now has three crude oil refineries. Alas, when they operate (in spurts), they too are deadly to the environment and inhabitants of the immediate communities, as the Eleme/Okrika refinery (the oldest) has proven for over four decades.

On February 22, 1966, Jasper Isaac Adaka (Lion) Boro told his men, before they embarked on their 12-day revolution to carve out a Niger Delta Republic from Nigeria, "Today is a great day in our lives and the history of the Niger Delta. Before today, we were branded robbers, bandits, terrorists and gangsters but after today, we shall be heroes of our land." How true!

Boro's Niger Delta Volunteer Service was crushed by the federal army. He and his colleagues were sentenced to death, later pardoned, and they fought to liberate the old Rivers State from Biafra. Boro's revolution has, in the last decade, spawned all manner of militants in the Niger Delta. Some have even keyed in on his name, and fight for justice and equity for the oil-producing Niger Delta.

Others have decided to cash in by becoming the robbers and terrorists that Boro warned against. Sadly, it has become extremely difficult to separate the true militant 'freedom fighters' from the criminal enforcers.

All the same, their activities have introduced another raging and seemingly permanent fire in the Niger Delta. Gun fire! Ironically, the gun fire from both the militants and the federal Joint Task Force are provided from barrels of crude oil from the same Niger Delta, which has truly become Nigeria's killing fields for half a century!

All the same, their activities have introduced another raging and seemingly permanent fire in the Niger Delta. Gun fire!

As crude oil prices dramatically drop on the world market and our economic experts rightly panic, will they pause and suggest to the federal government that it is time to stem the tide of fire, destruction and death in the Niger Delta in double quick time?

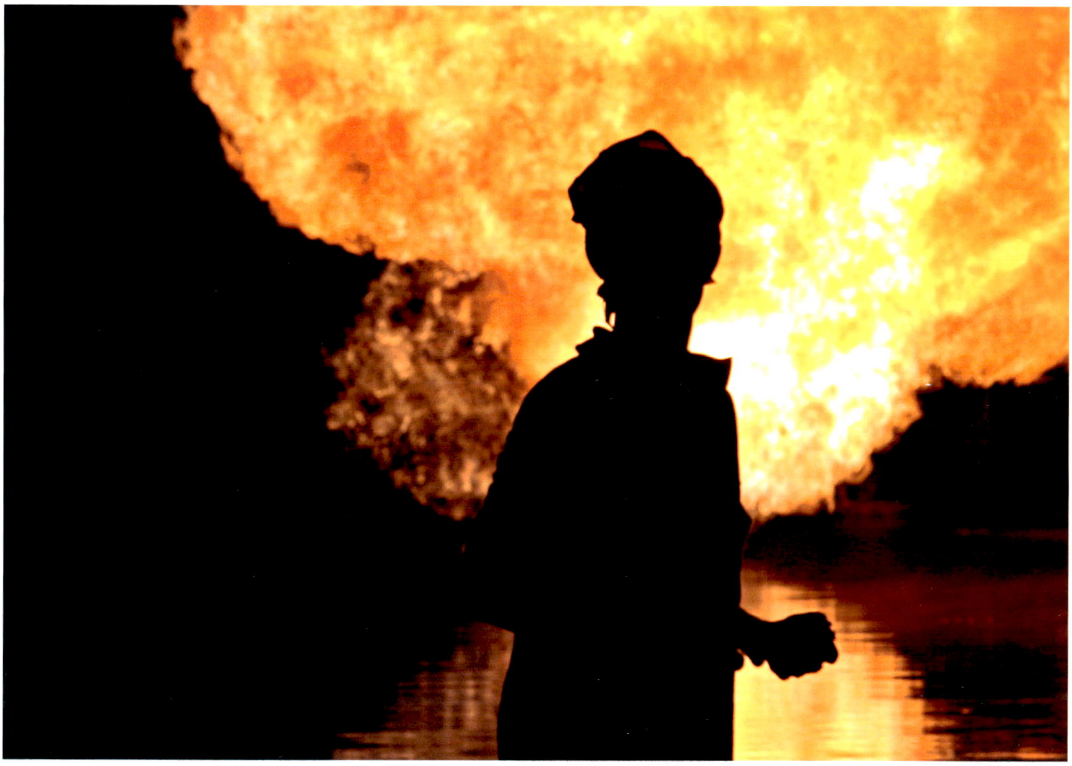

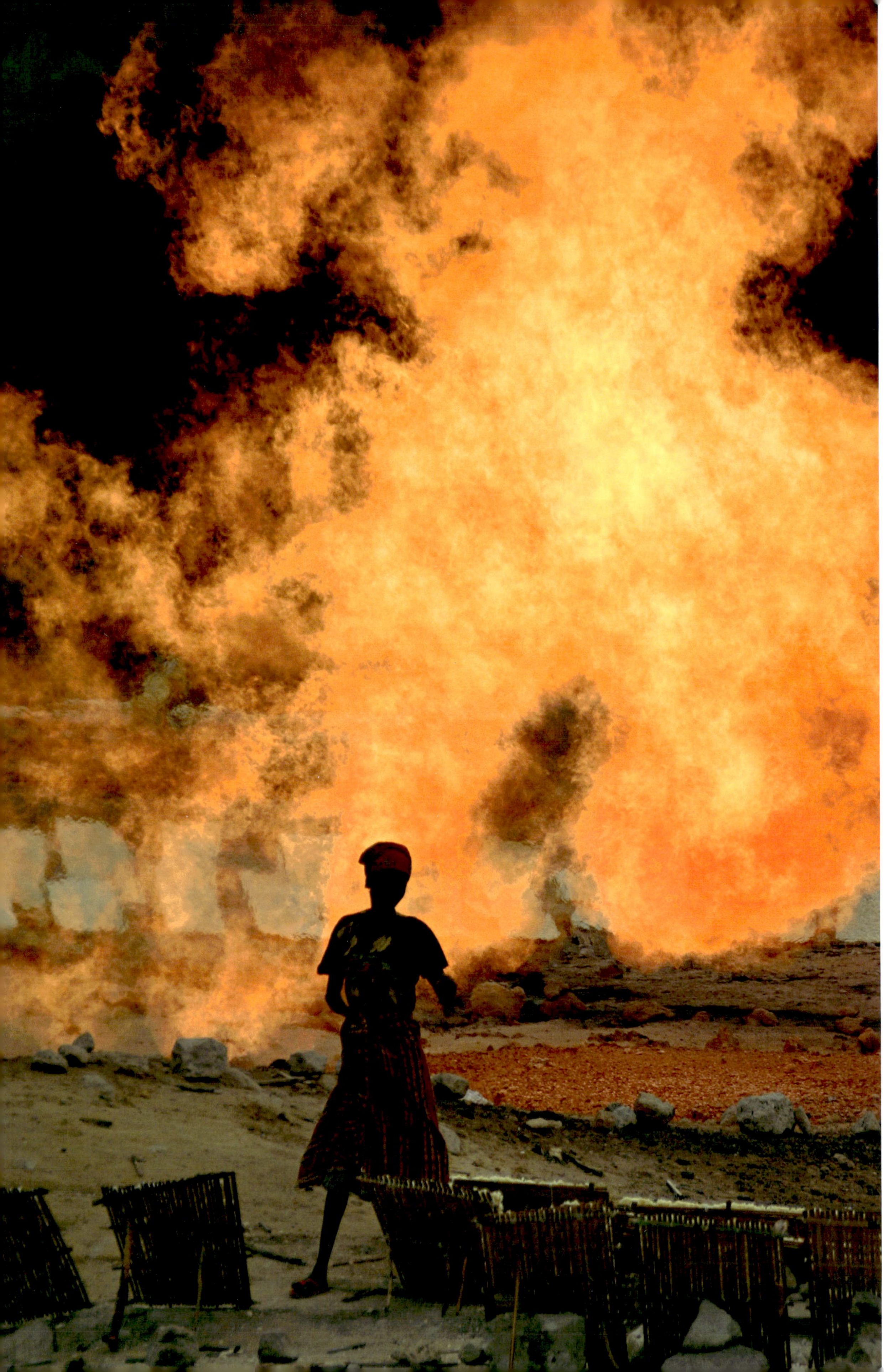

ECONOMIC AND HEALTH COSTS OF GAS FLARING

The economic loss to Nigeria from gas flaring is immense, as natural gas released through oil extraction is wasted instead of being harnessed as energy for the country. Figures released by the Nigerian government and industry officials estimate that Nigeria is losing between US$2.5- $3 billion a year through flaring, or over US$70 billion from 1970 - 2006. When the wider environmental, economic and social consequences are factored in, a ballpark figure could be in the order of US$150 billion. Part of this cost has been the impact on human health. A groundbreaking study by Environmental Rights Action and the Climate Justice Programme in 2005 attempted to quantify the damage done by the toxic cocktail of pollutants, including benzene and dioxins, emitted by gas flaring. The study estimates that in Nigeria's Bayelsa State alone, flaring is statistically likely to cause 49 premature deaths, 5,000 child respiratory illnesses, some 120,000 asthma attacks, as well as 8 additional causes of cancer each year. (Friends of the Earth)

Shell gas flare, Eriemu.

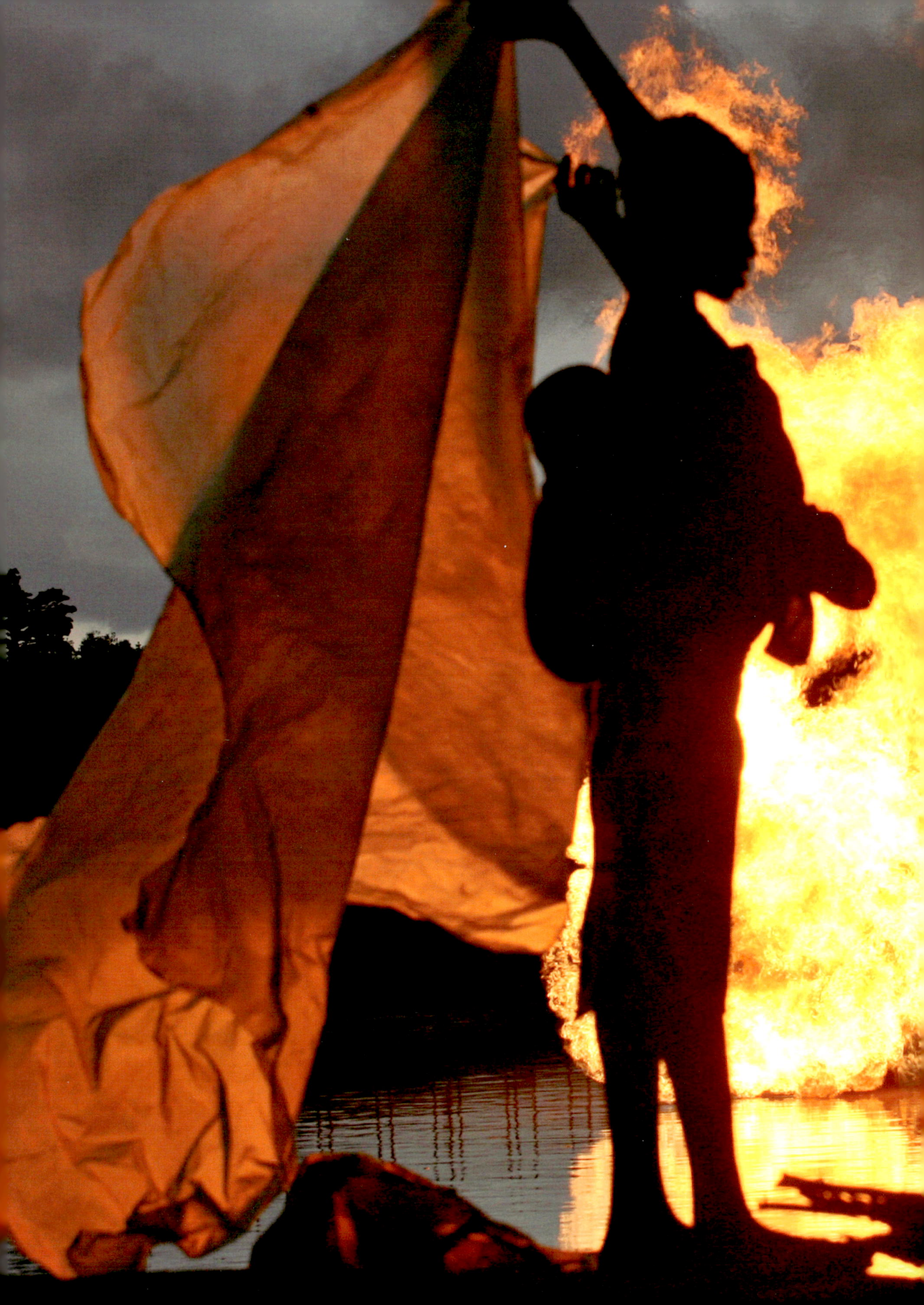

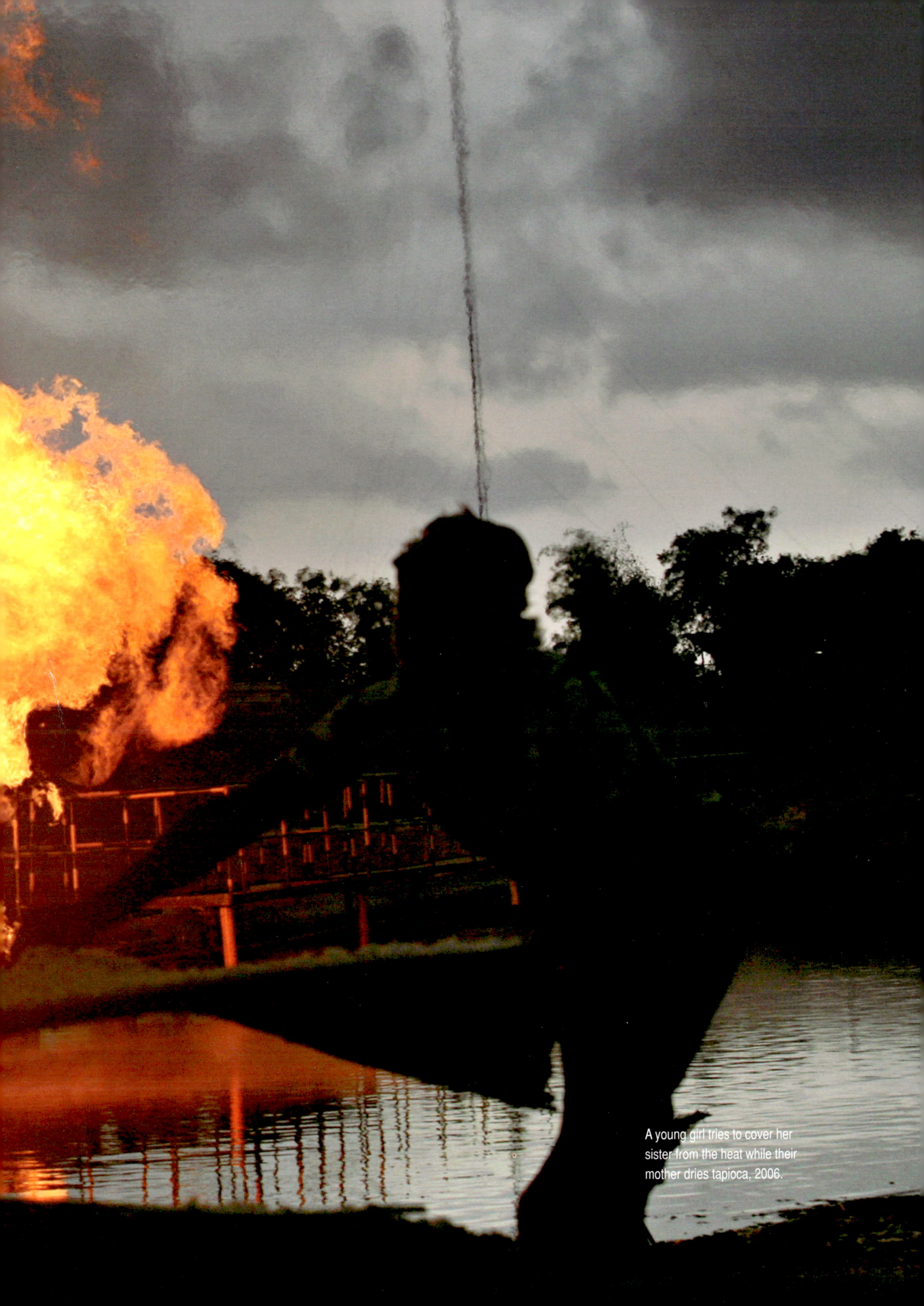

A young girl tries to cover her sister from the heat while their mother dries tapioca, 2006.

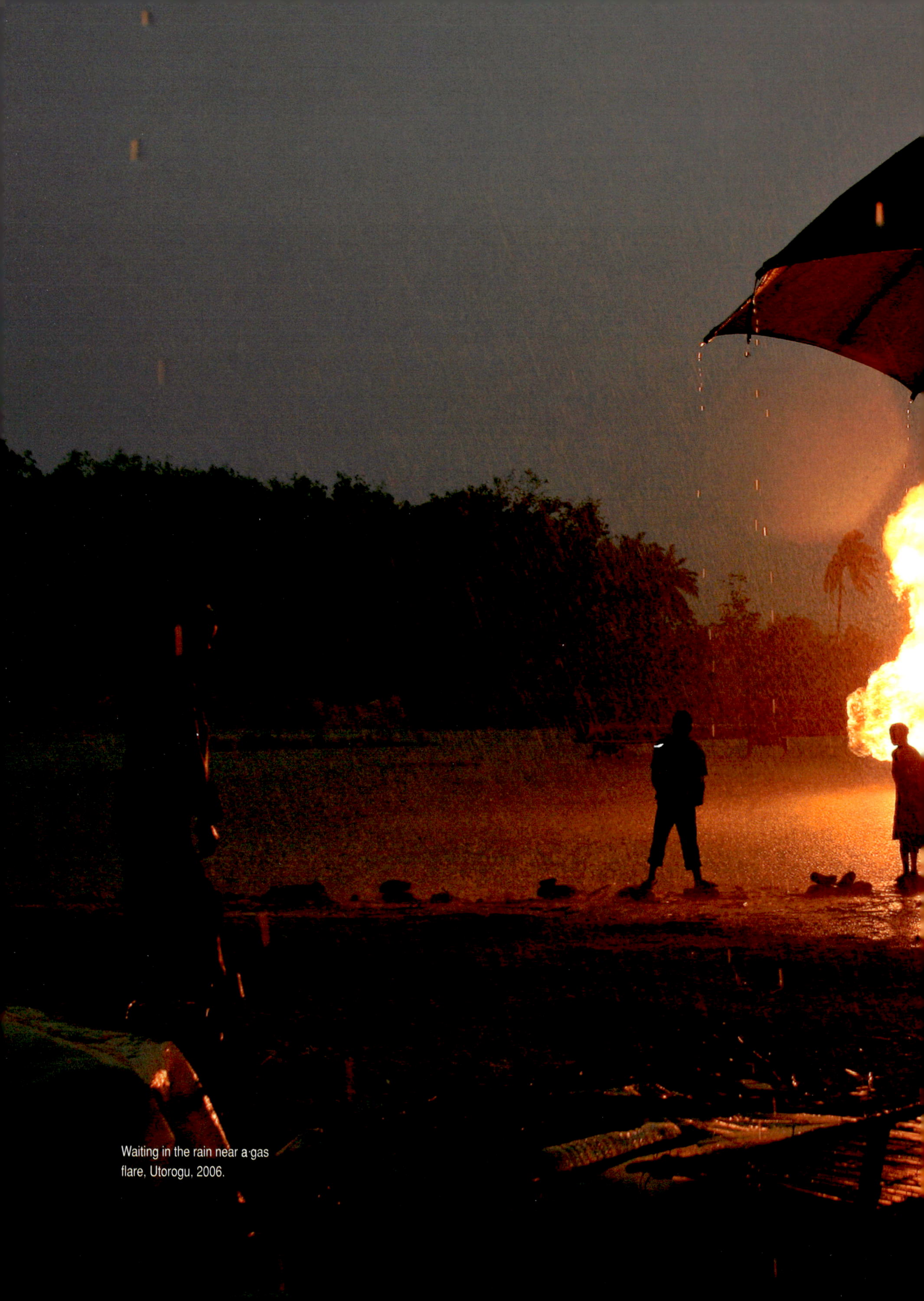

Waiting in the rain near a gas flare, Utorogu, 2006.

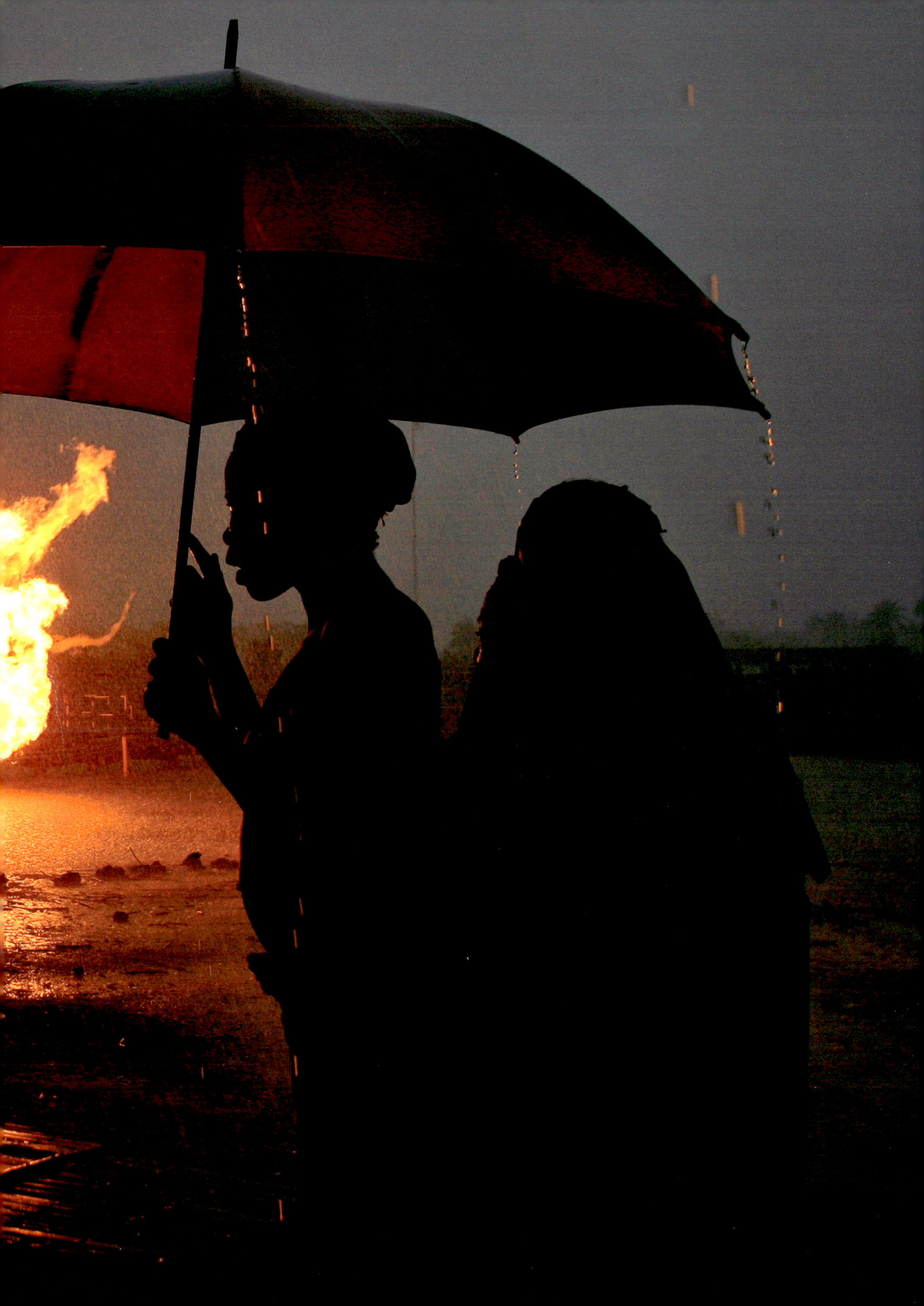

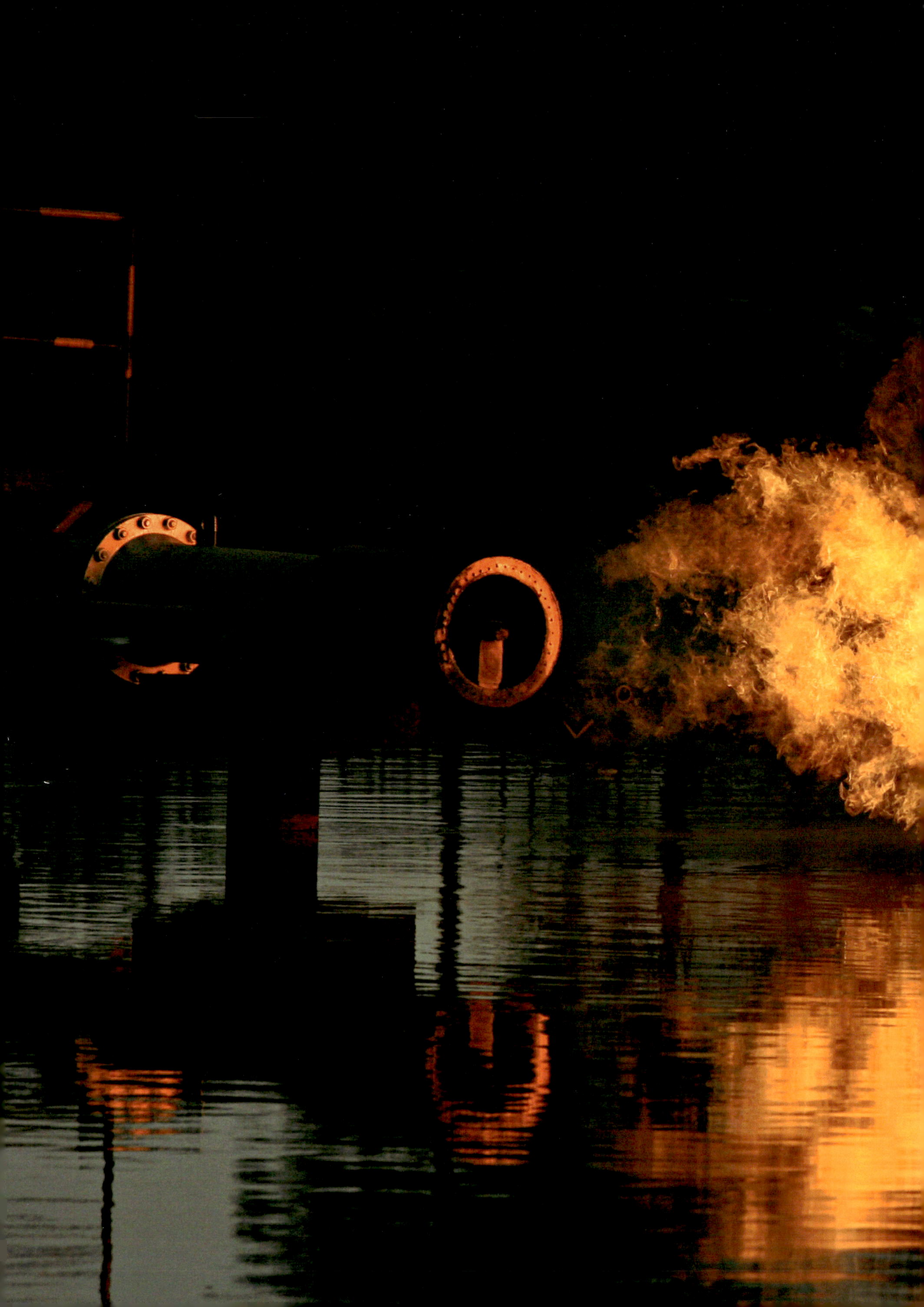

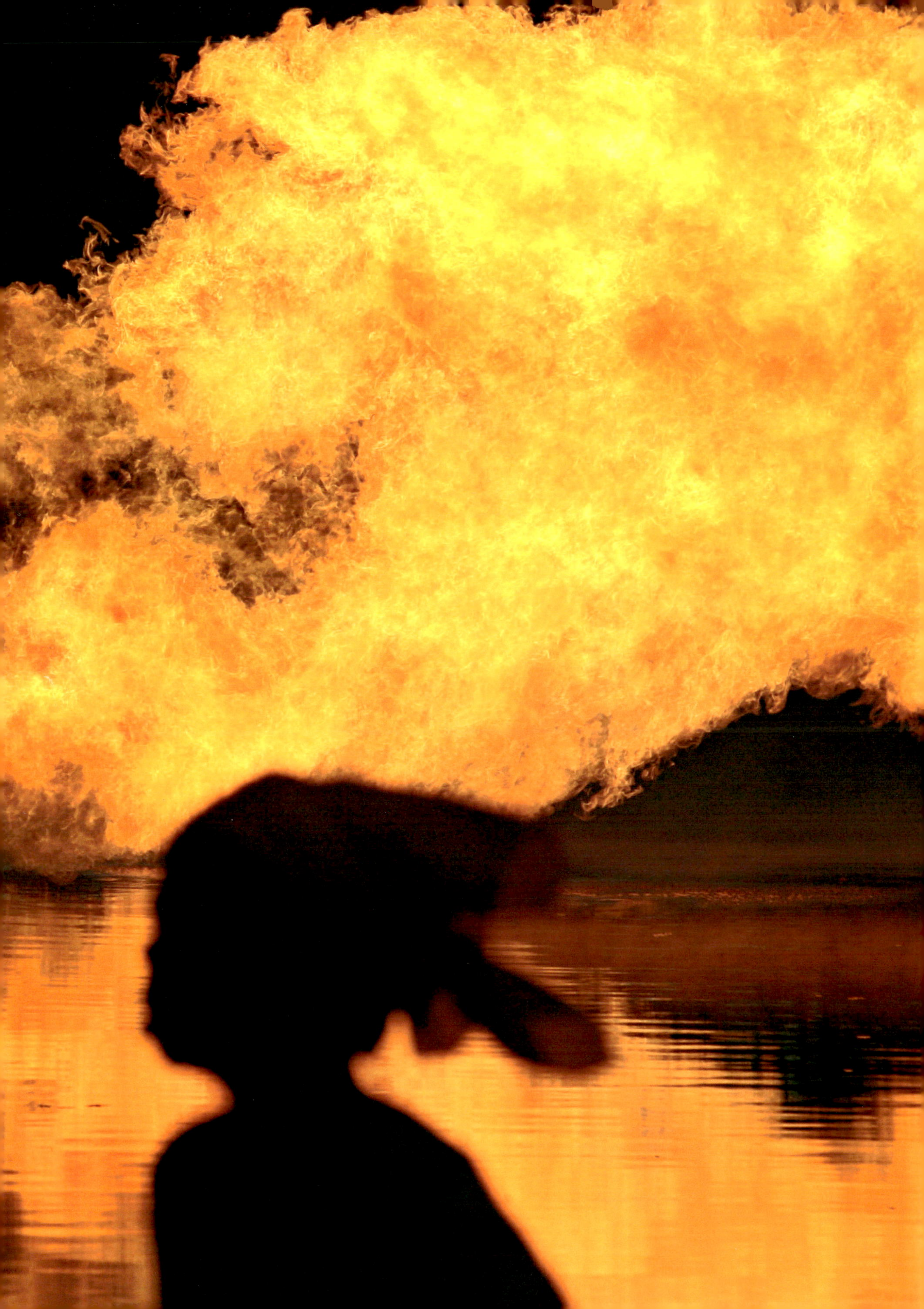

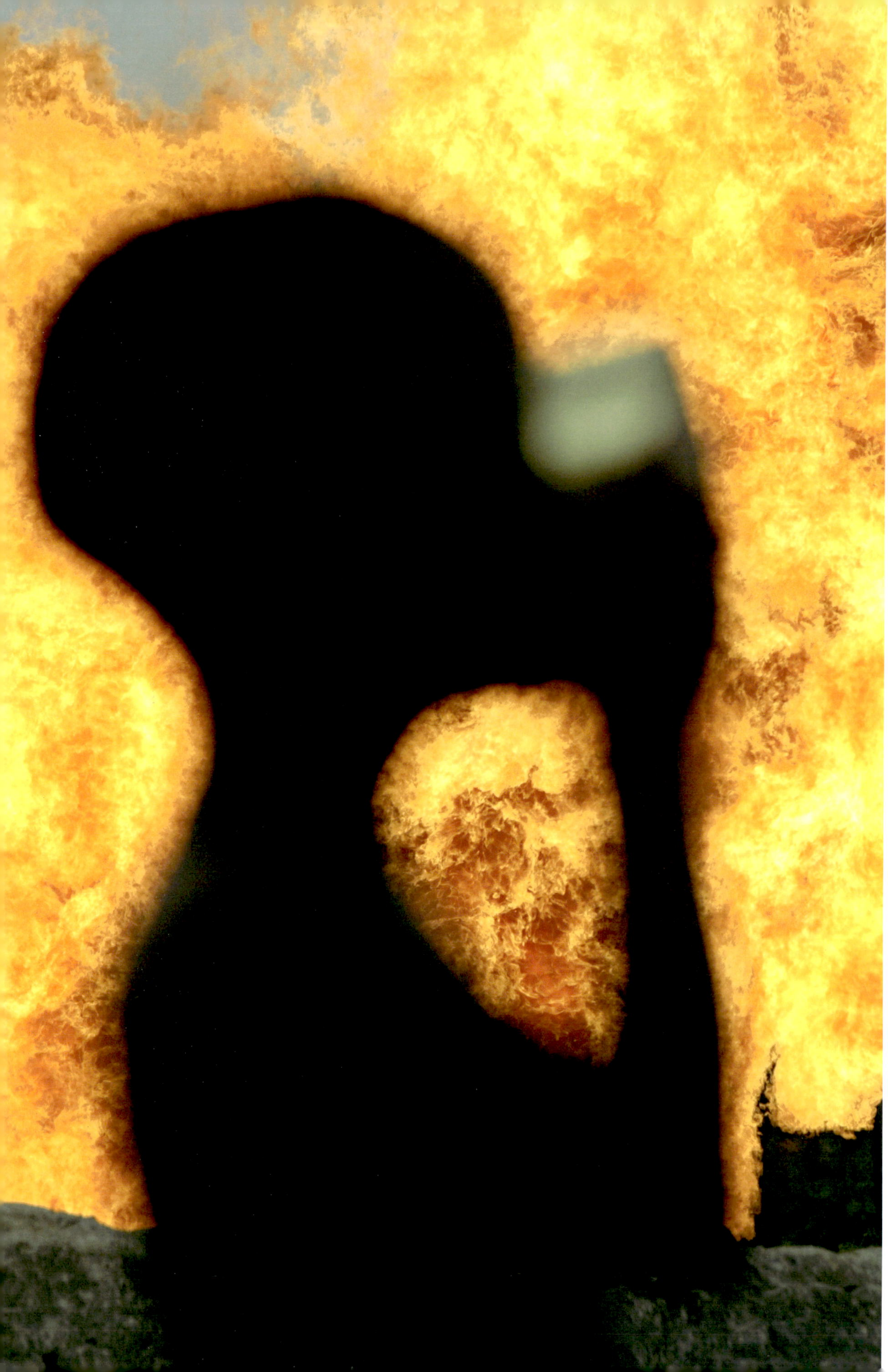

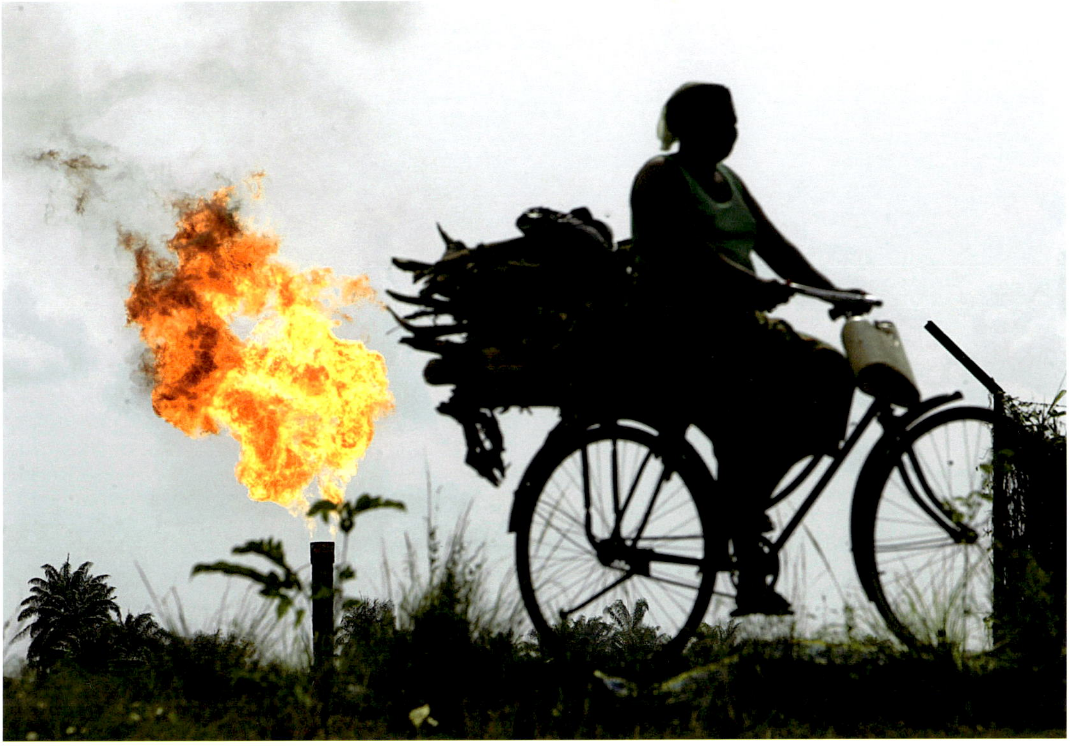

Above: Carrying firewood on a bicycle past a gas flare, Ebocha, 2006.

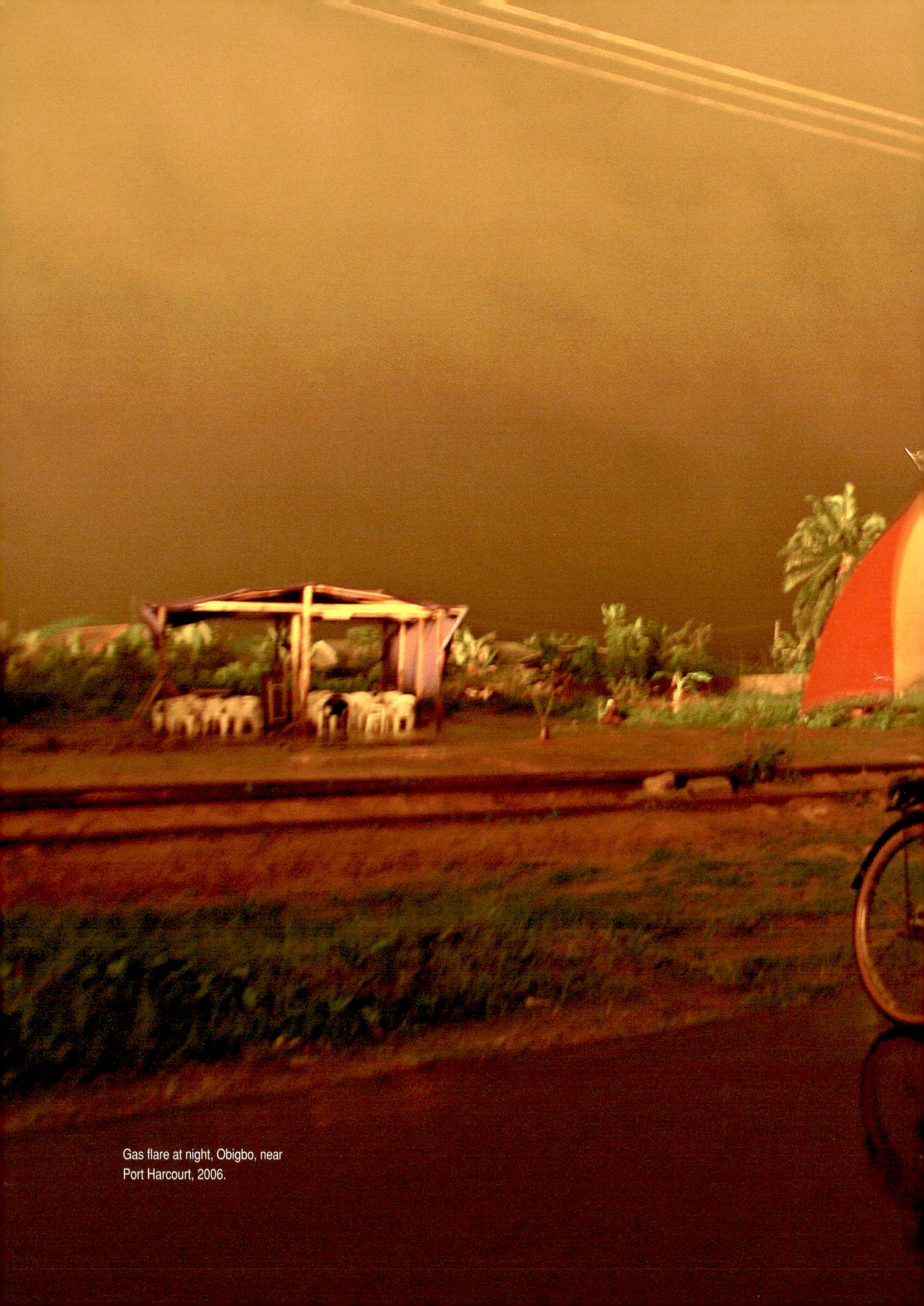

Gas flare at night, Obigbo, near Port Harcourt, 2006.

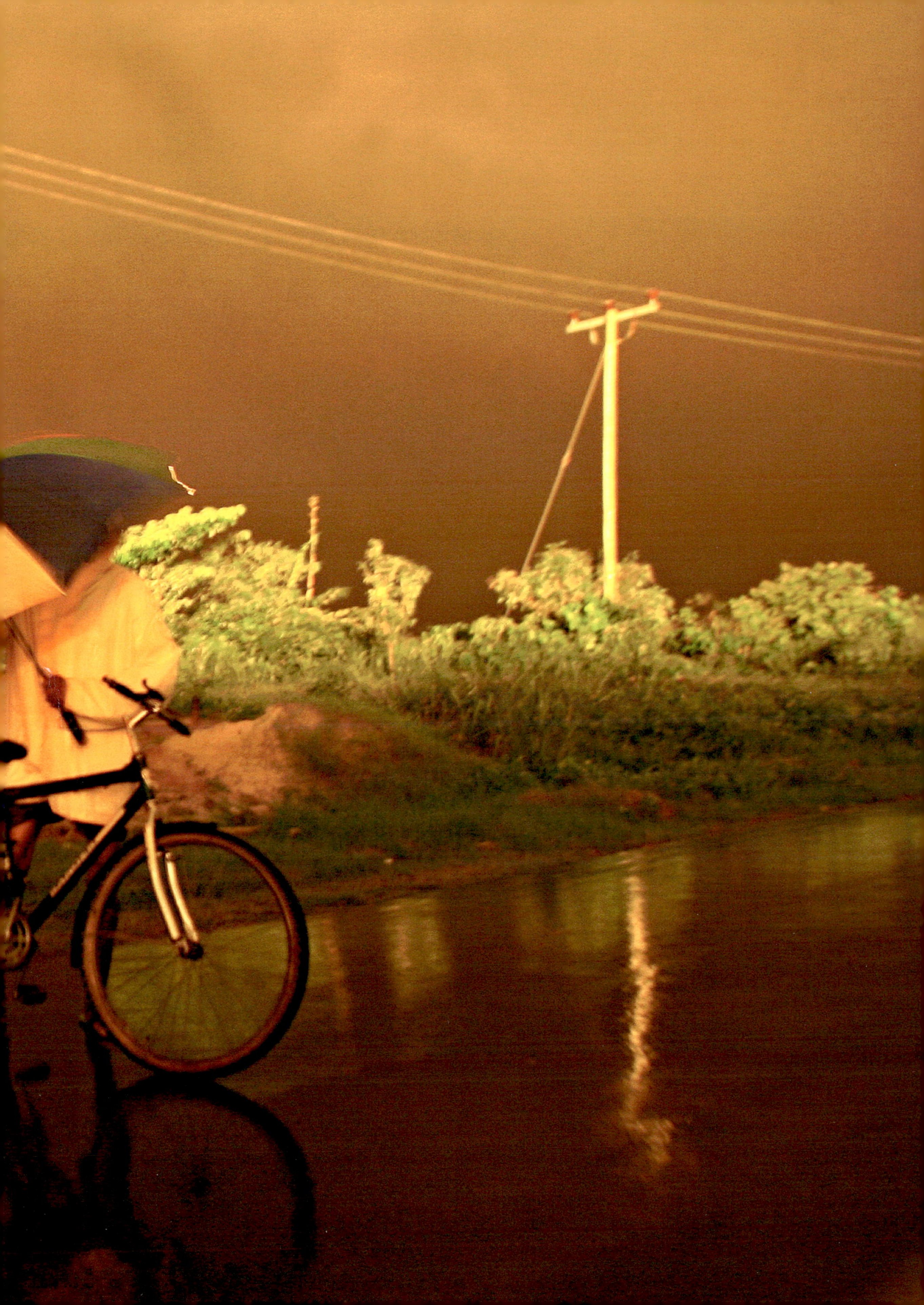

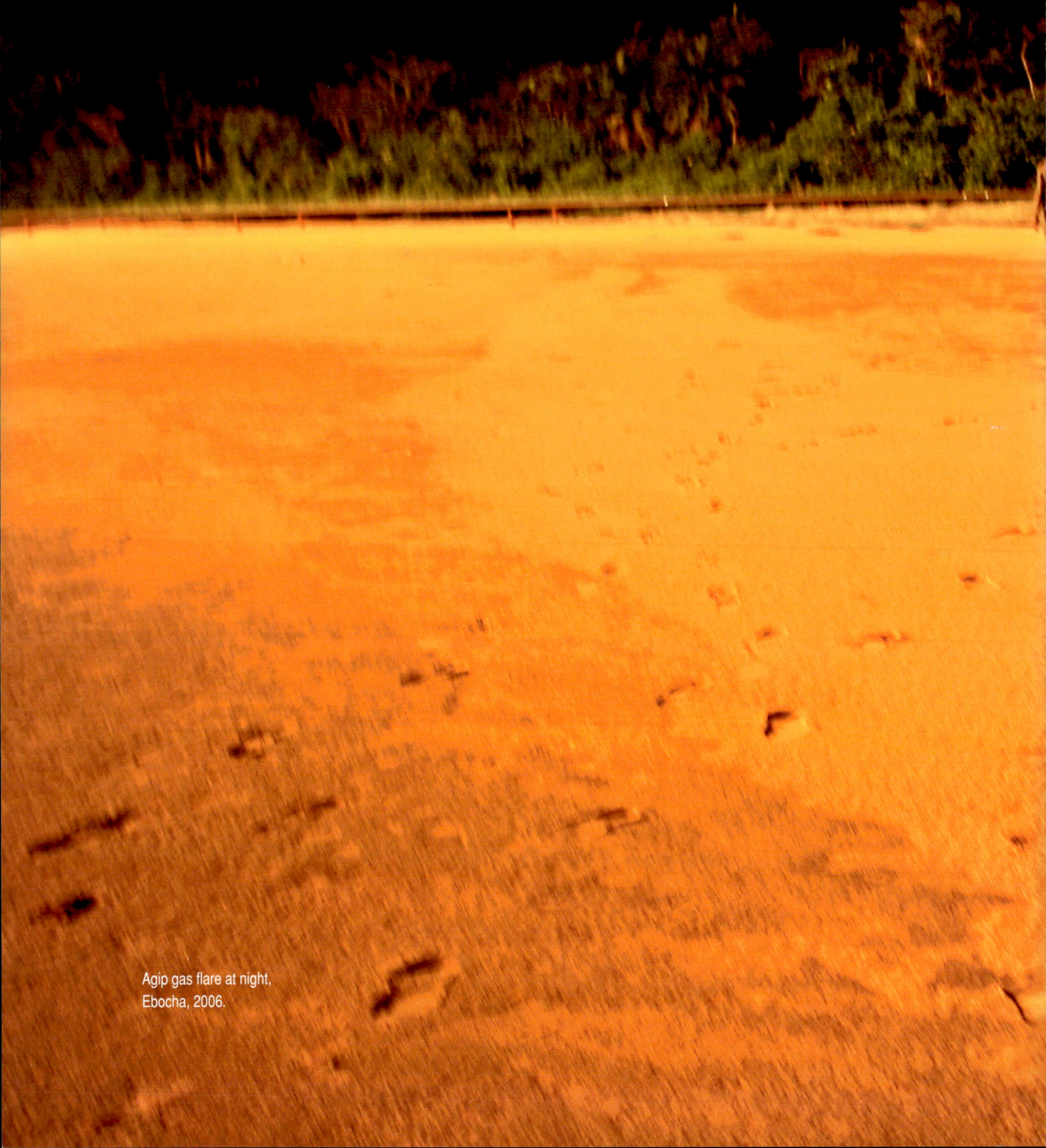

Agip gas flare at night,
Ebocha, 2006.

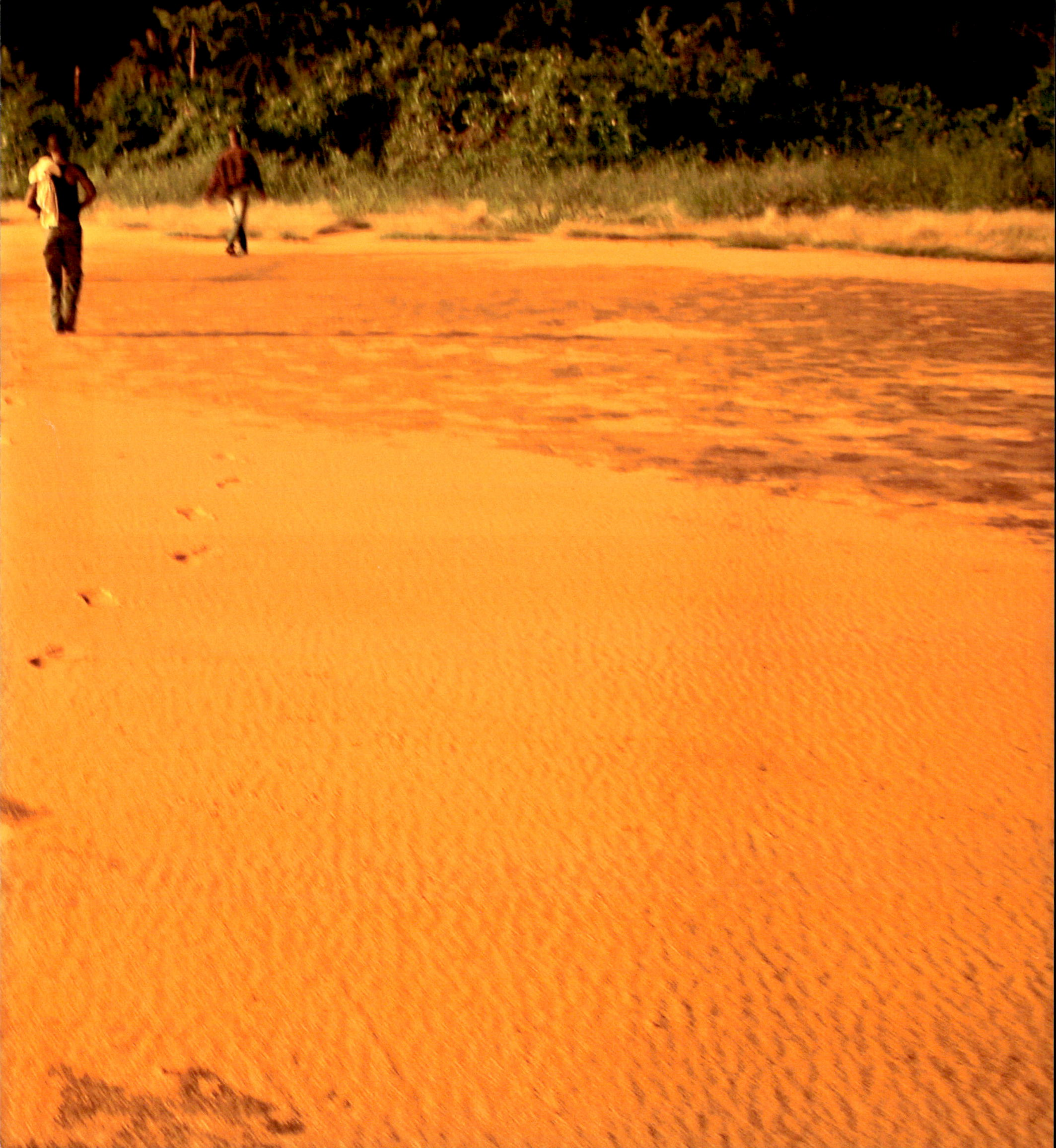

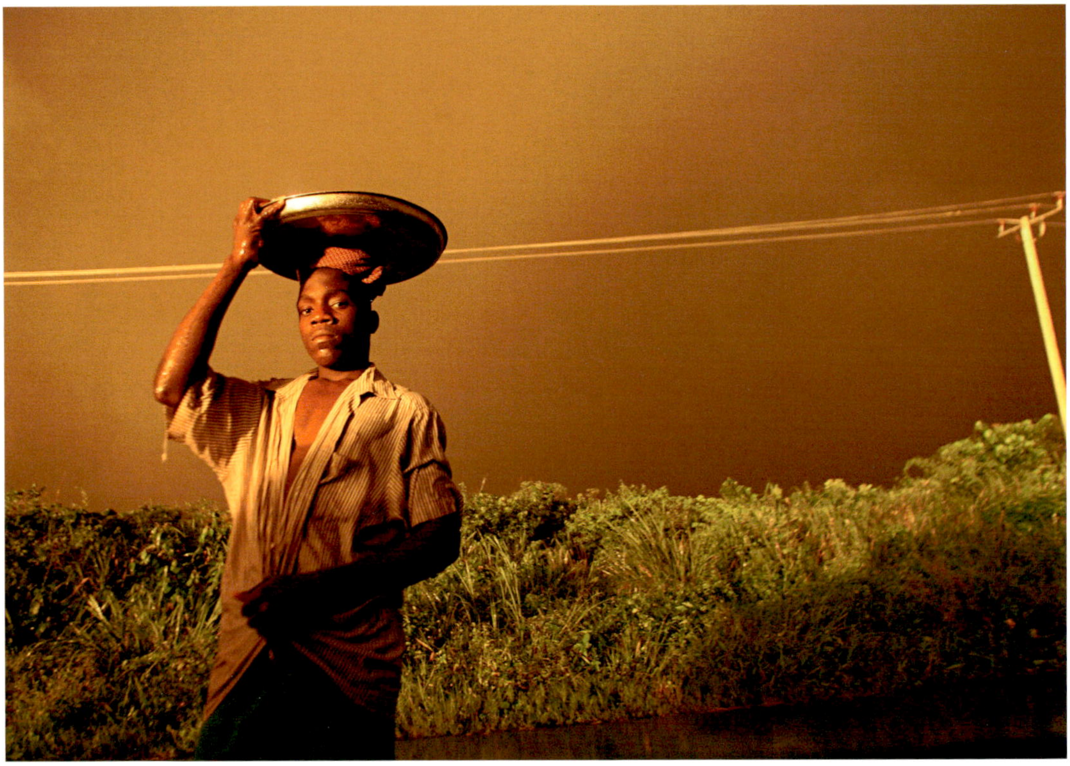

Above: Shell flare, Obigbo, near Port Harcourt, 2006.

Right: Agip flare, Ebocha, 2006.

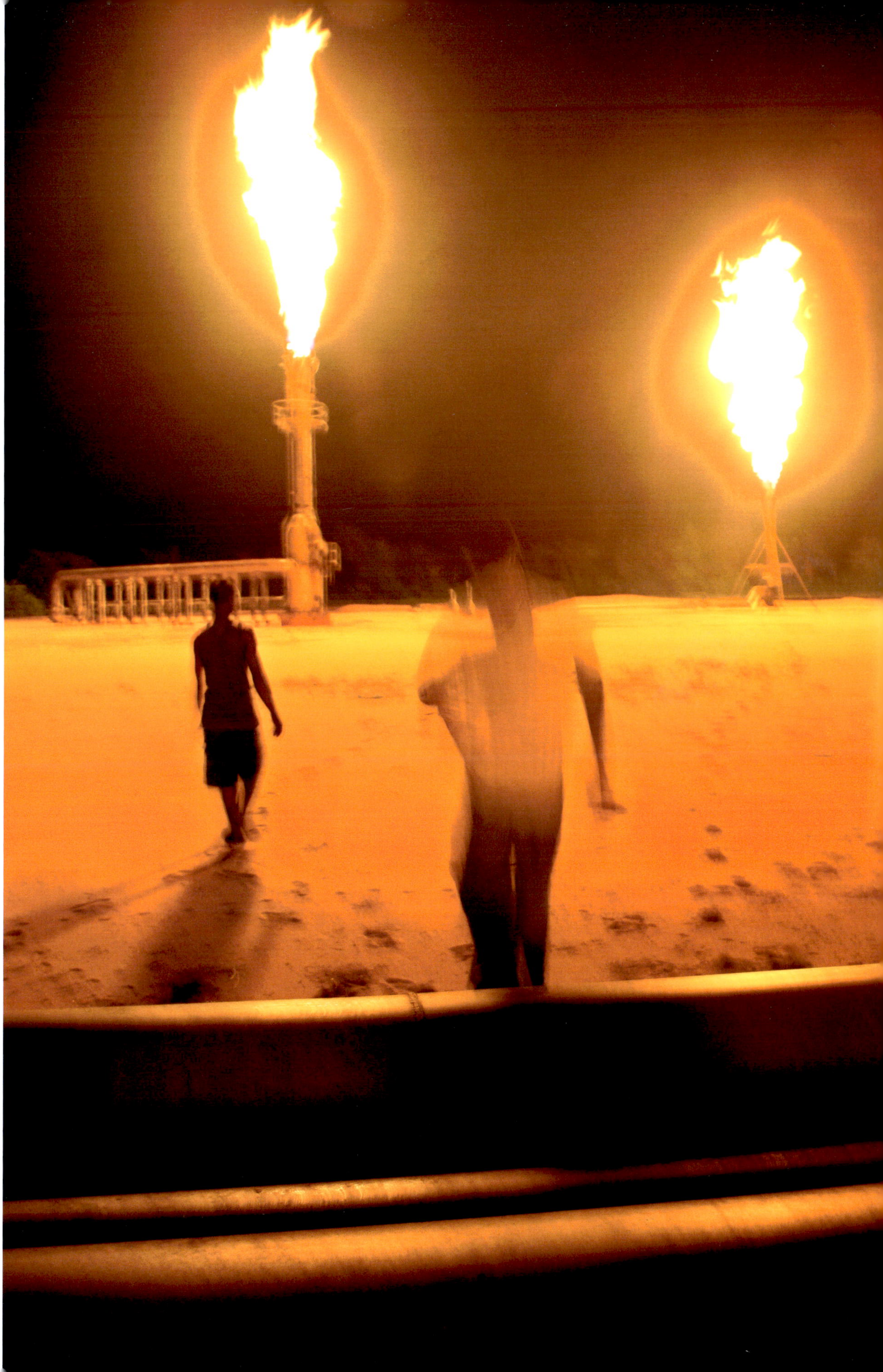

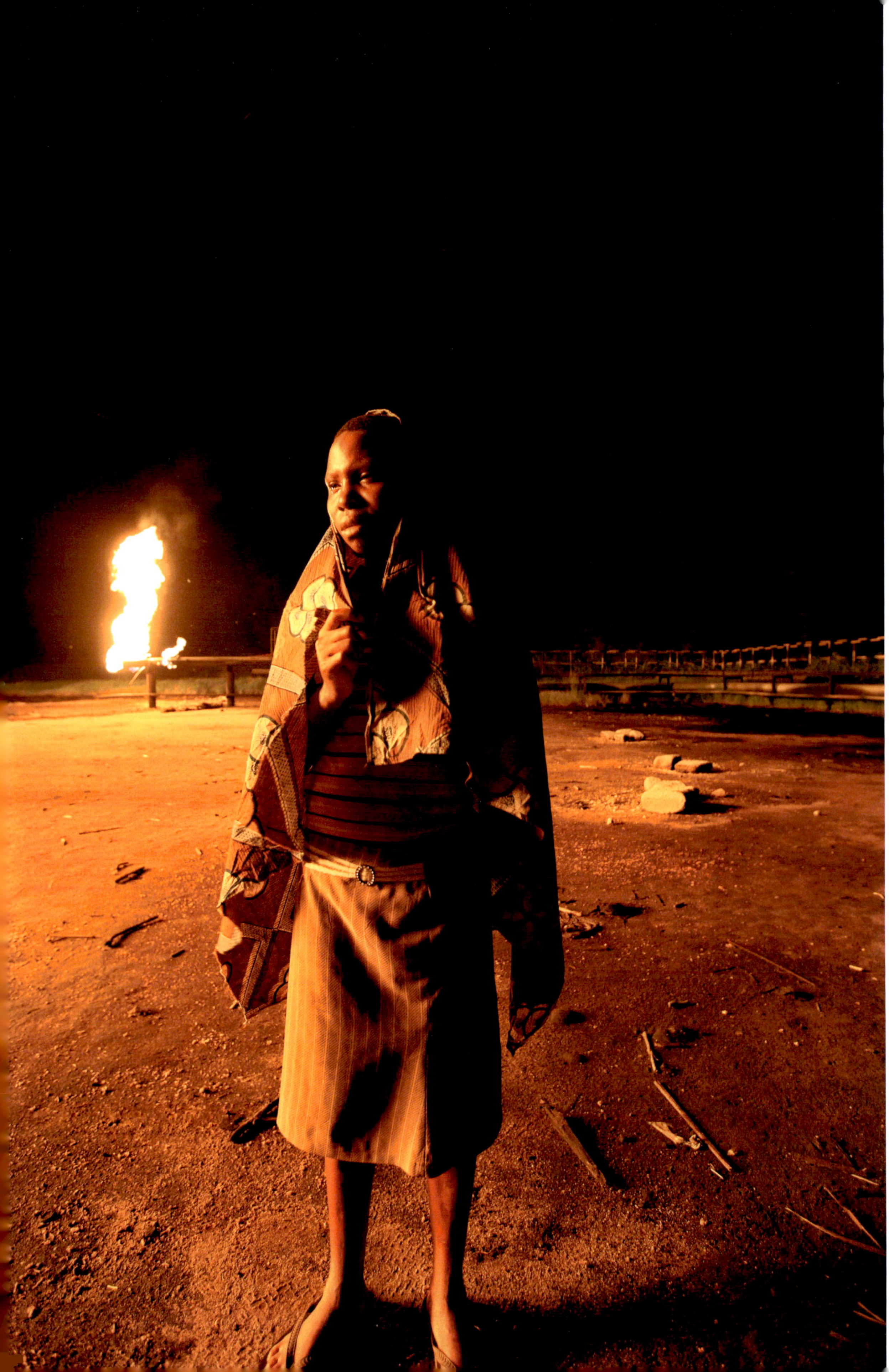

ORANGE LIGHT

Helen Ikri is a widow whose husband died of malaria more than a decade ago, leaving her alone with three children Samuel, Mavis and Joy. Helen lives with her children in a rented one room hut in Uzere, and to earn a living takes the risk of drying tapioca to make cassava cake by the gas flares day and night. Her children come to assist her after school and together they all spend almost half of their daily lives under the orange light from the gas flares, without knowing the danger associated with their actions. Helen and her children get home at midnight and are confronted with another orange light for their domestic needs - this time from the gas lantern due to lack of electricity. Helen is determined to pay their medical bills, to send her three children to school, and keep alive her desire for a better life. There are thousands of ordinary people like Helen all over the Delta region who against all odds try to make ends meet away from the oil politics and conflicts in the region, and they are THE REAL PEOPLE OF THE DELTA.

Mavis, daughter of Helen Ikri, stands near a Shell gas flare in Uzere, 2009.

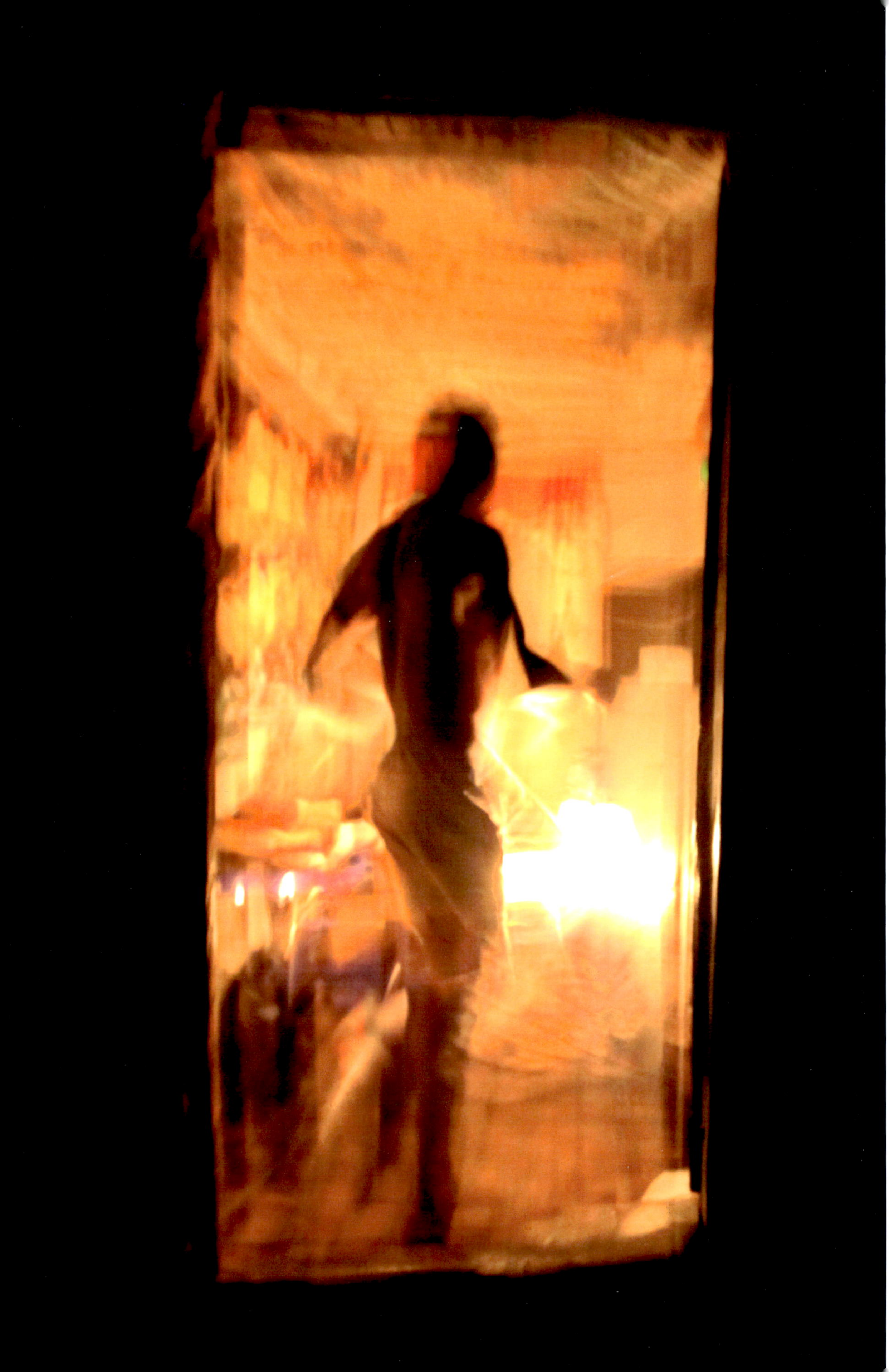

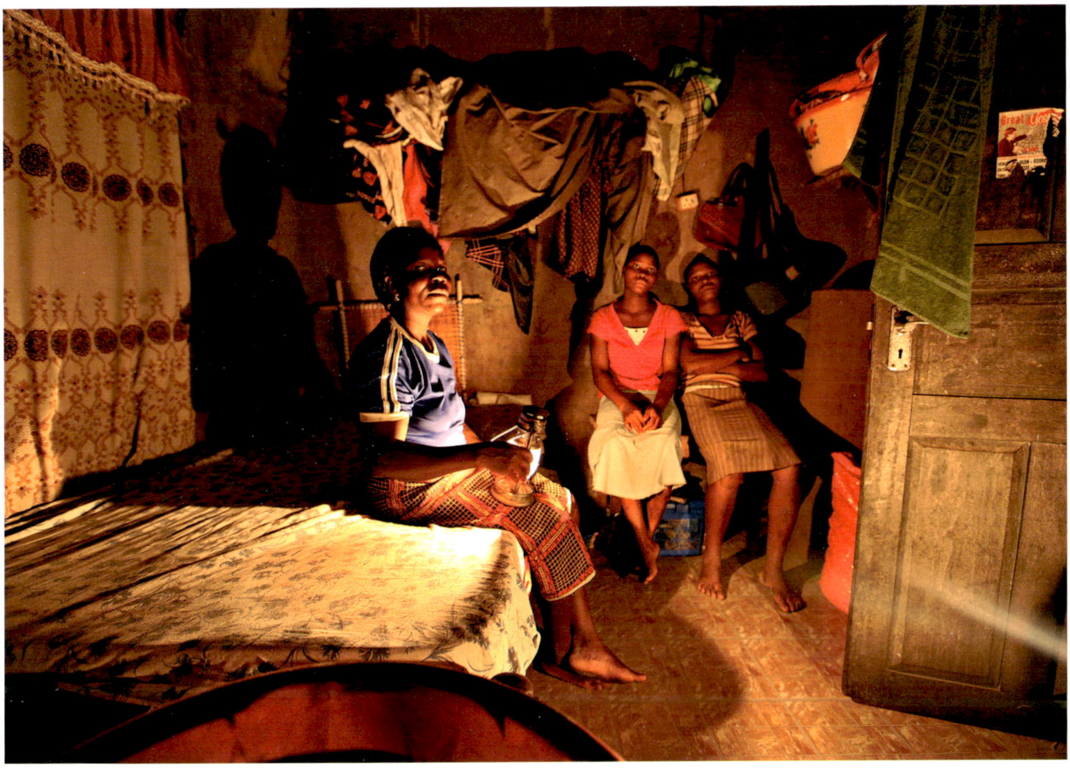

Left: Samuel Ikri, her son, walks in as light from a lantern illuminates the inside of their house.

Above: Helen Ikri sits holding a lantern, their major source of light, with her children after they arrive home from drying tapioca in Uzere, Delta State.

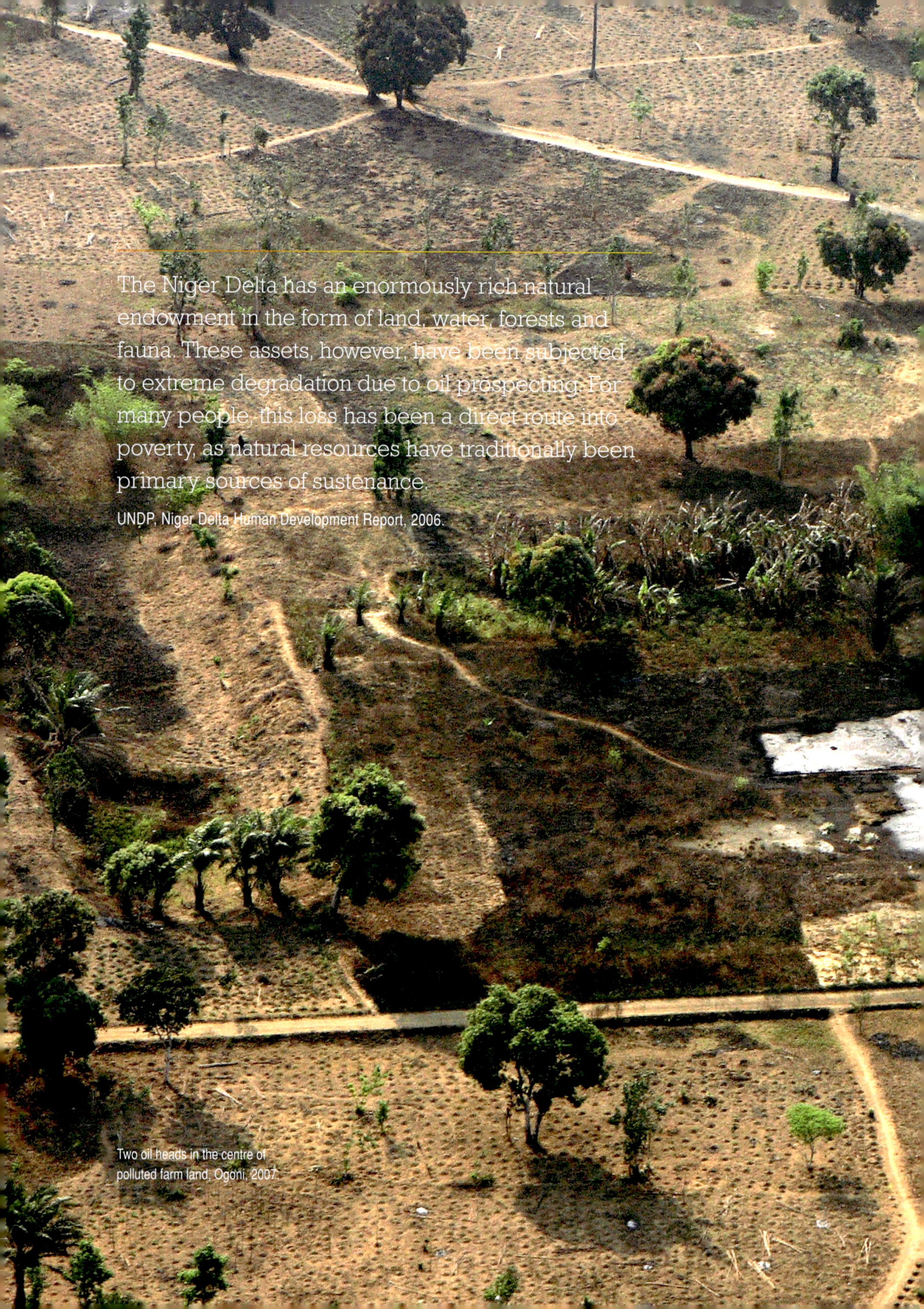

The Niger Delta has an enormously rich natural endowment in the form of land, water, forests and fauna. These assets, however, have been subjected to extreme degradation due to oil prospecting. For many people, this loss has been a direct route into poverty, as natural resources have traditionally been primary sources of sustenance.

UNDP, Niger Delta Human Development Report, 2006.

Two oil heads in the centre of polluted farm land, Ogoni, 2007.

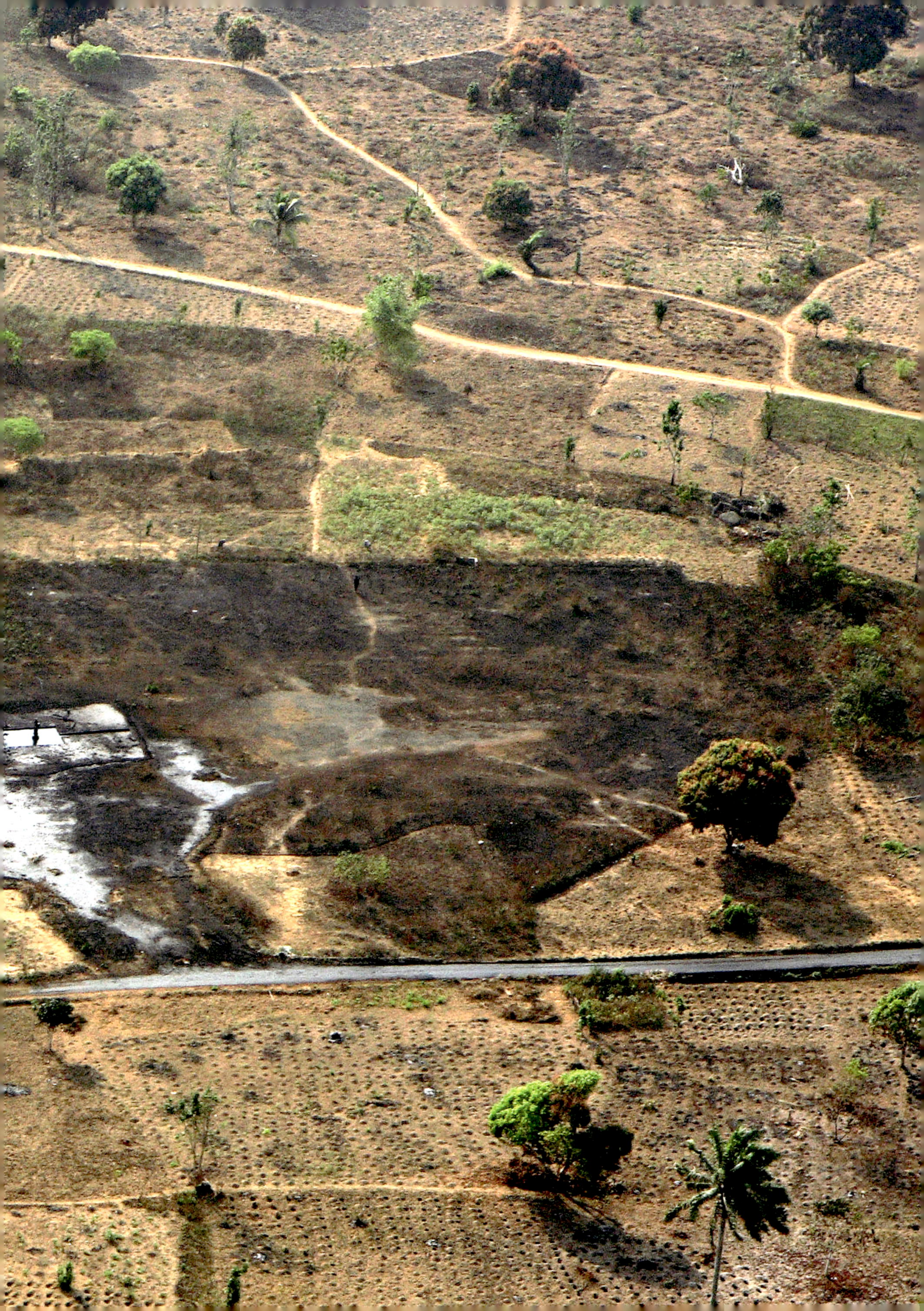

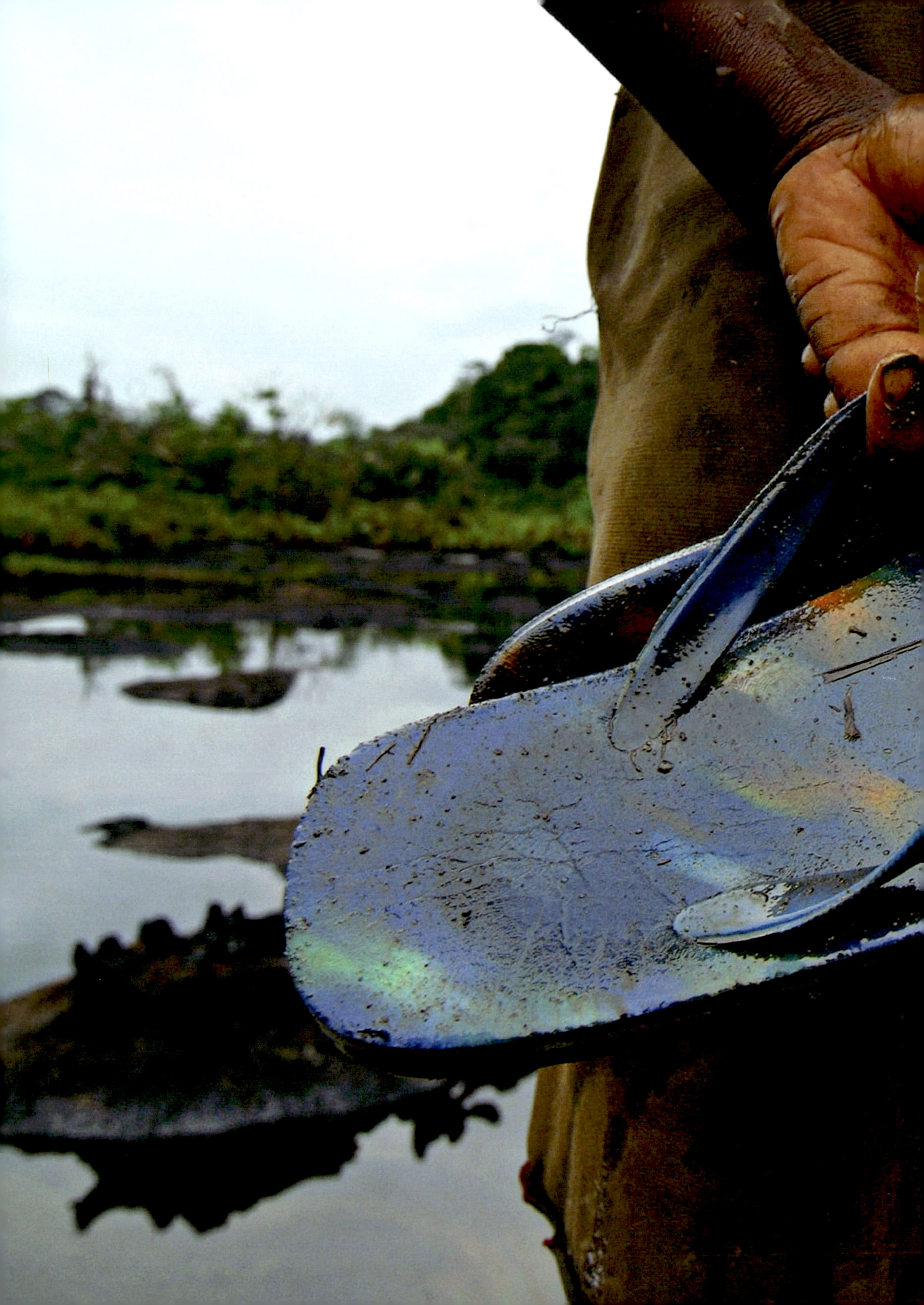

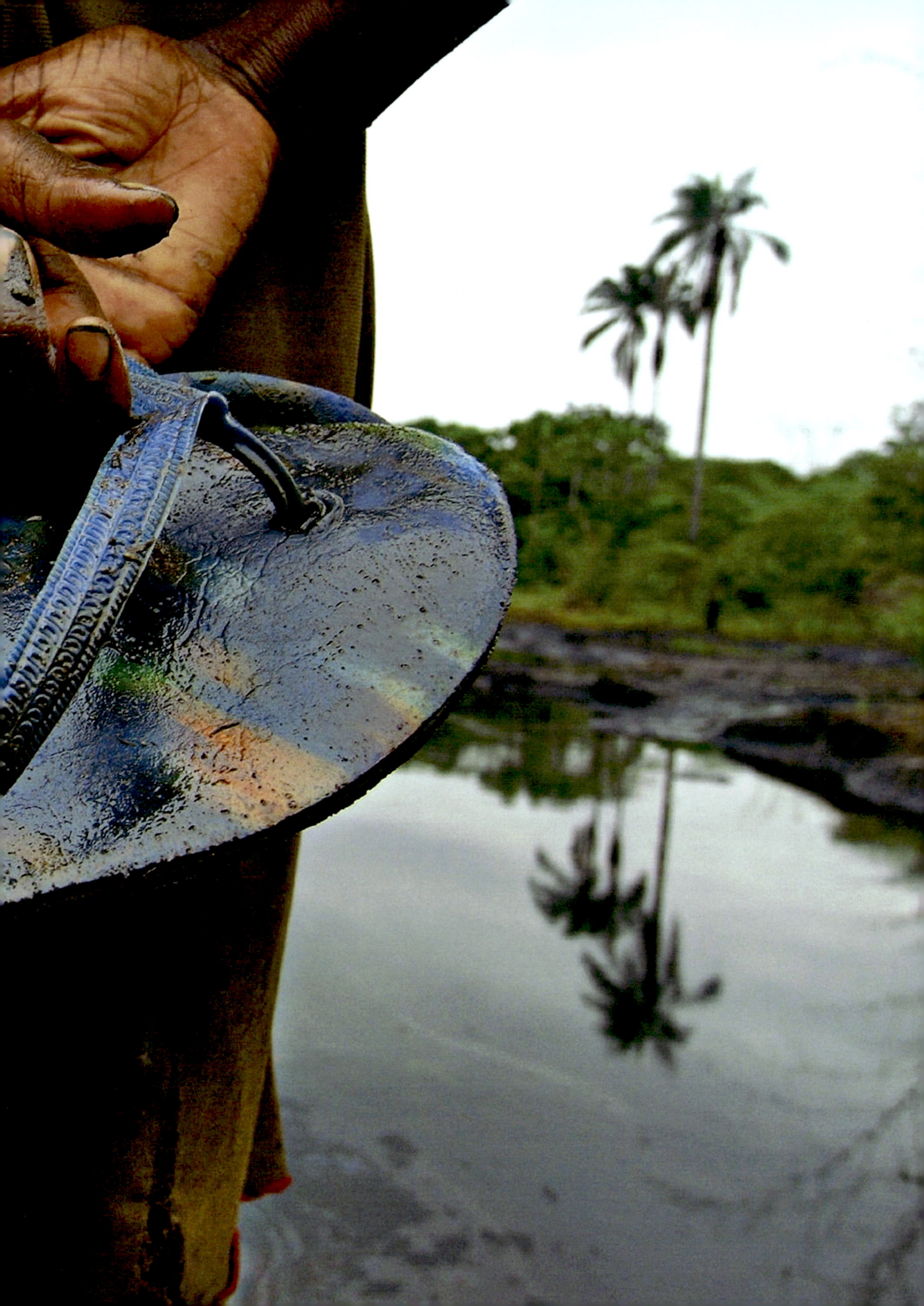

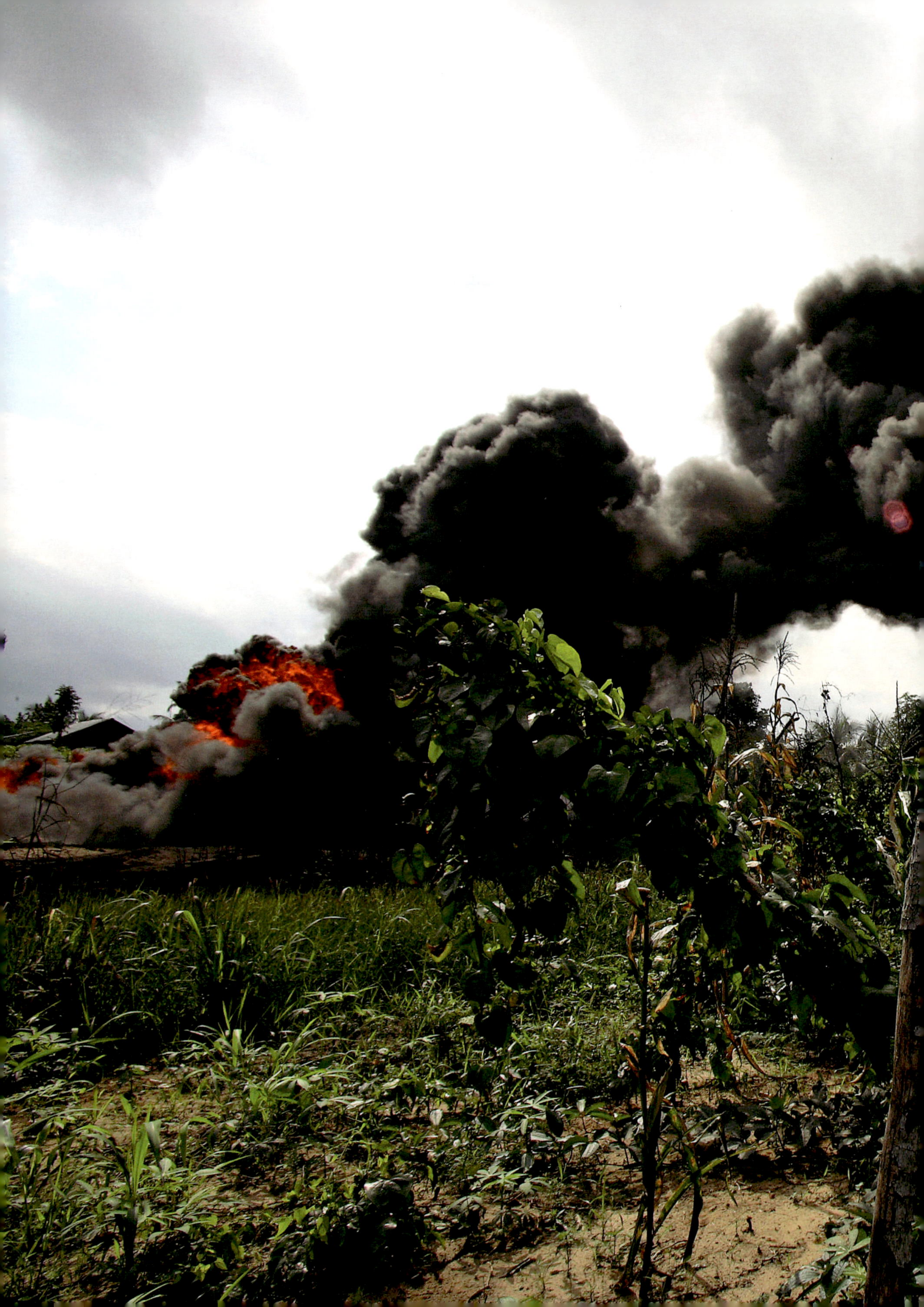

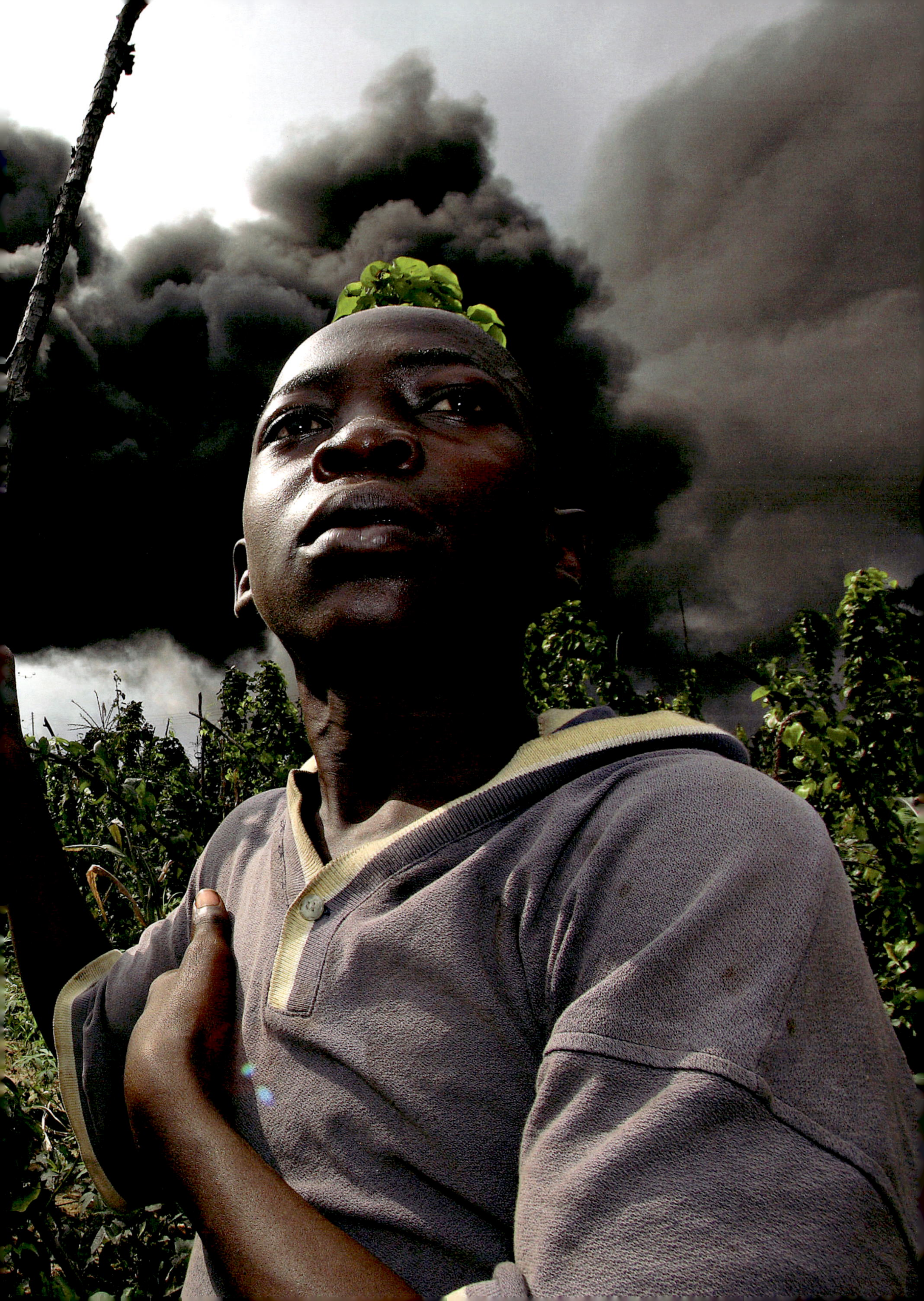

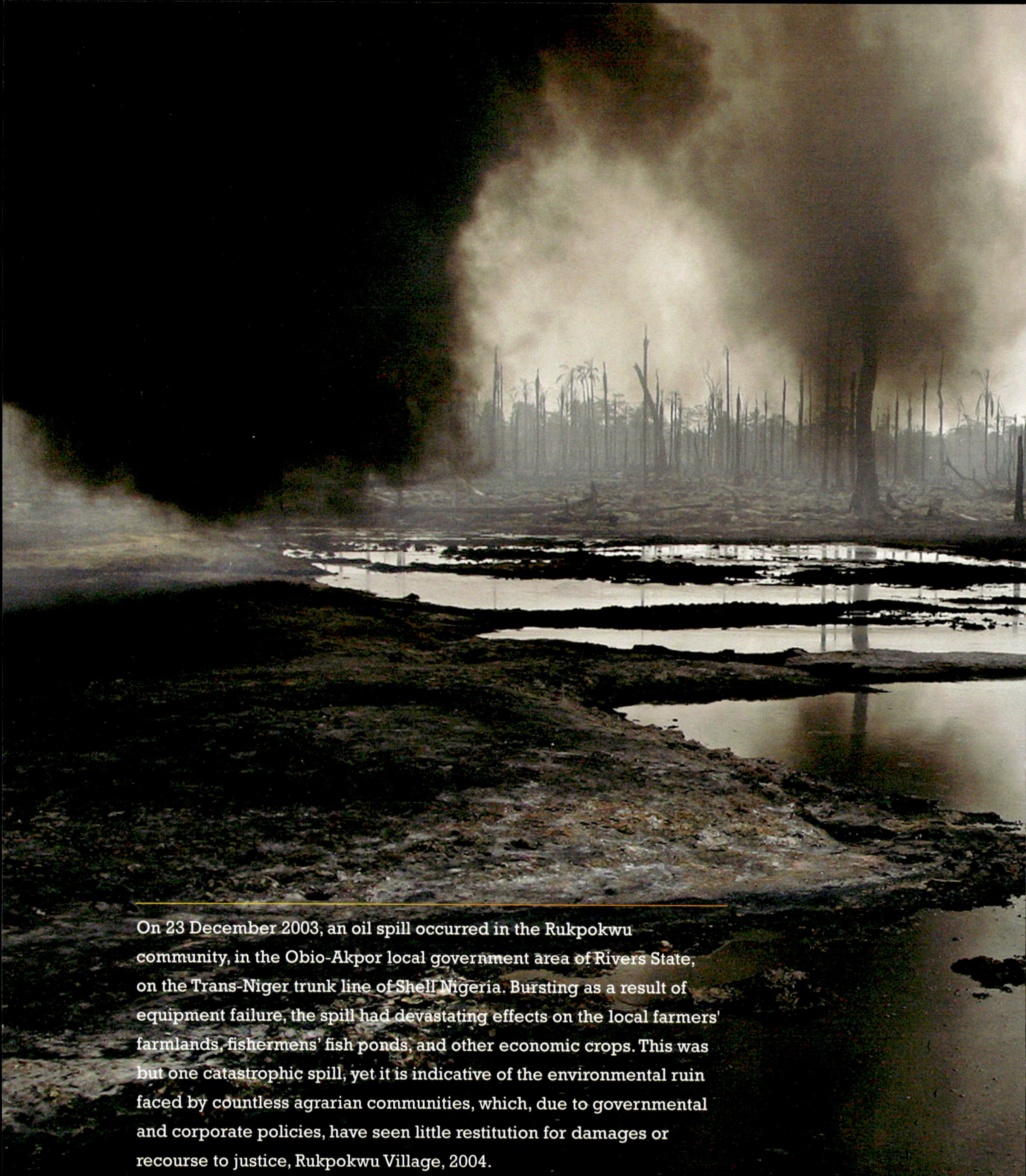

On 23 December 2003, an oil spill occurred in the Rukpokwu community, in the Obio-Akpor local government area of Rivers State, on the Trans-Niger trunk line of Shell Nigeria. Bursting as a result of equipment failure, the spill had devastating effects on the local farmers' farmlands, fishermens' fish ponds, and other economic crops. This was but one catastrophic spill, yet it is indicative of the environmental ruin faced by countless agrarian communities, which, due to governmental and corporate policies, have seen little restitution for damages or recourse to justice, Rukpokwu Village, 2004.

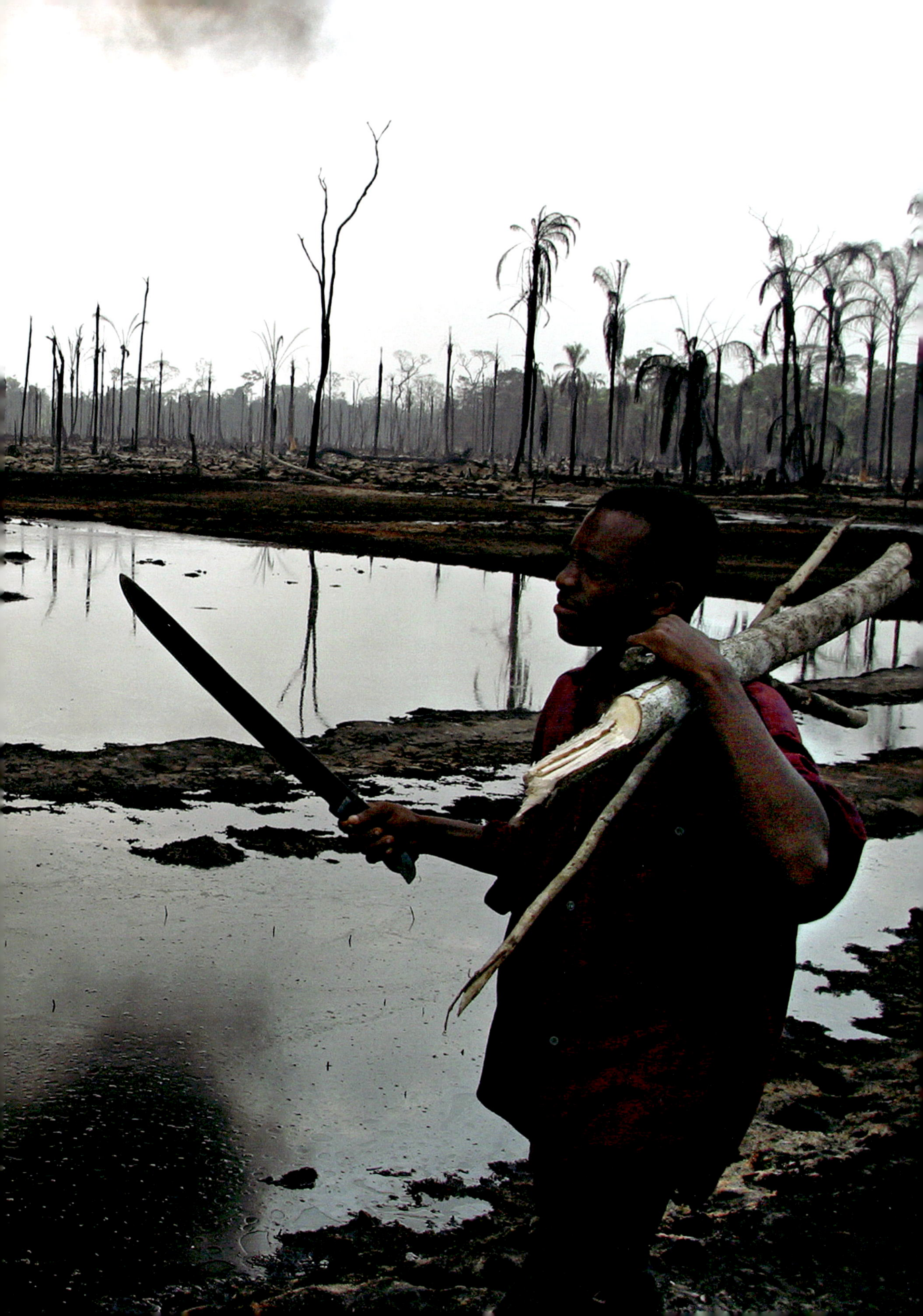

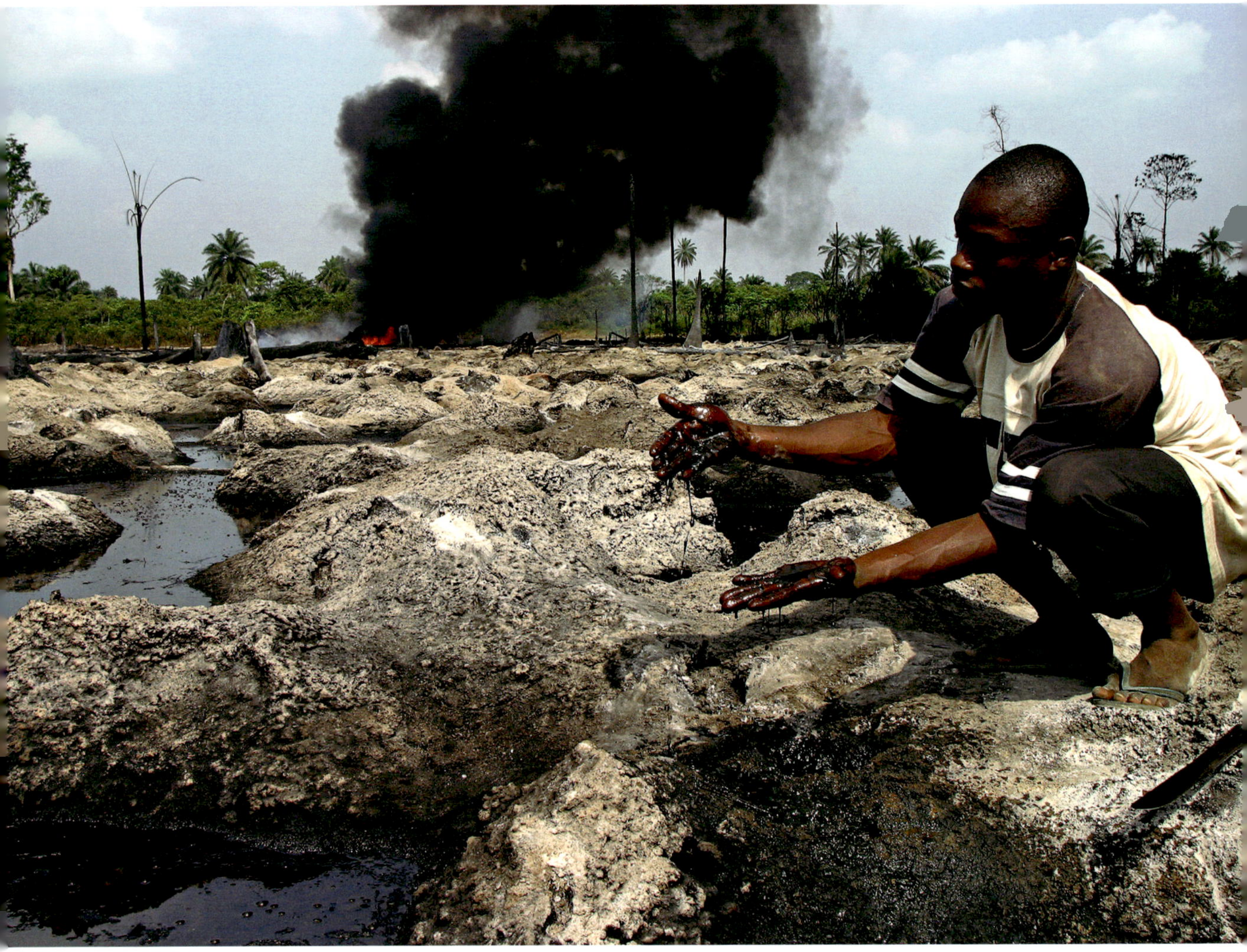

A local youth scoops crude oil from the ground as he assesses the damage done to their farm land by the oil spill in Rukpokwu, near the oil rich city of Port Harcourt, Nigeria, January 2004. The ruptured oil pipeline belonging to Shell Petroleum Development Company (SPDC) exploded damaging about 300 hectares of community farmlands.

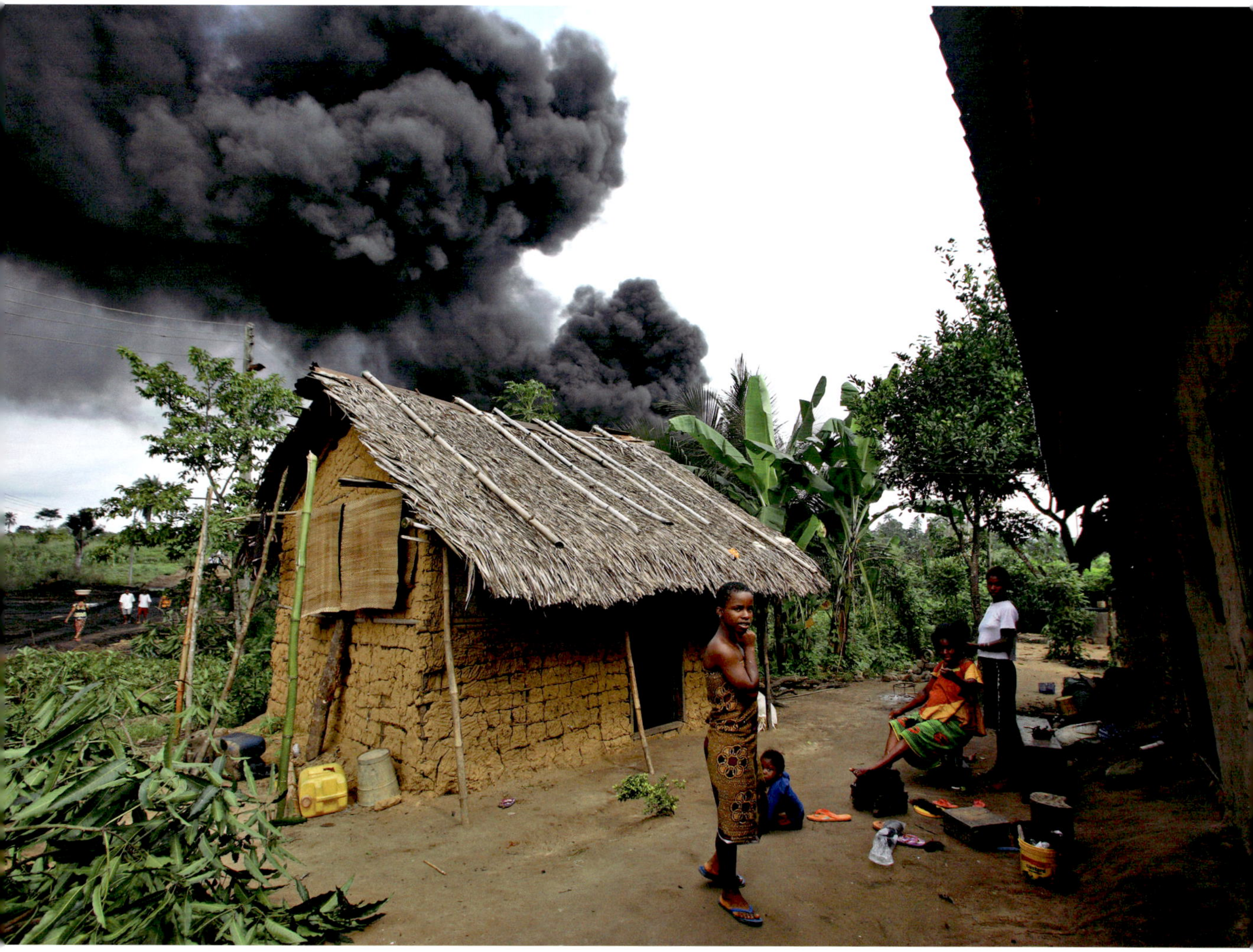

A Shell oil pipeline burns
behind houses in Kigbara-Dere.

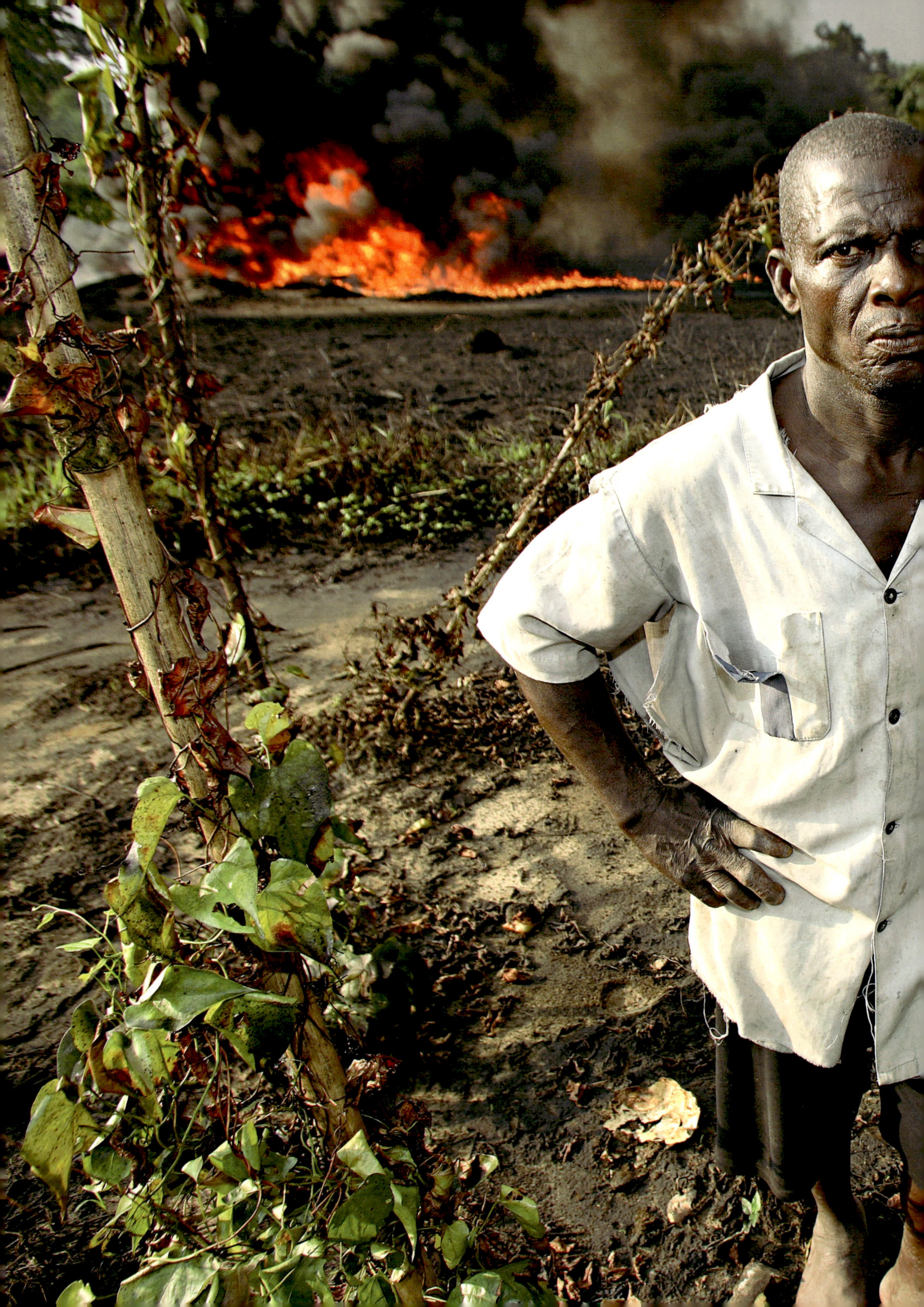

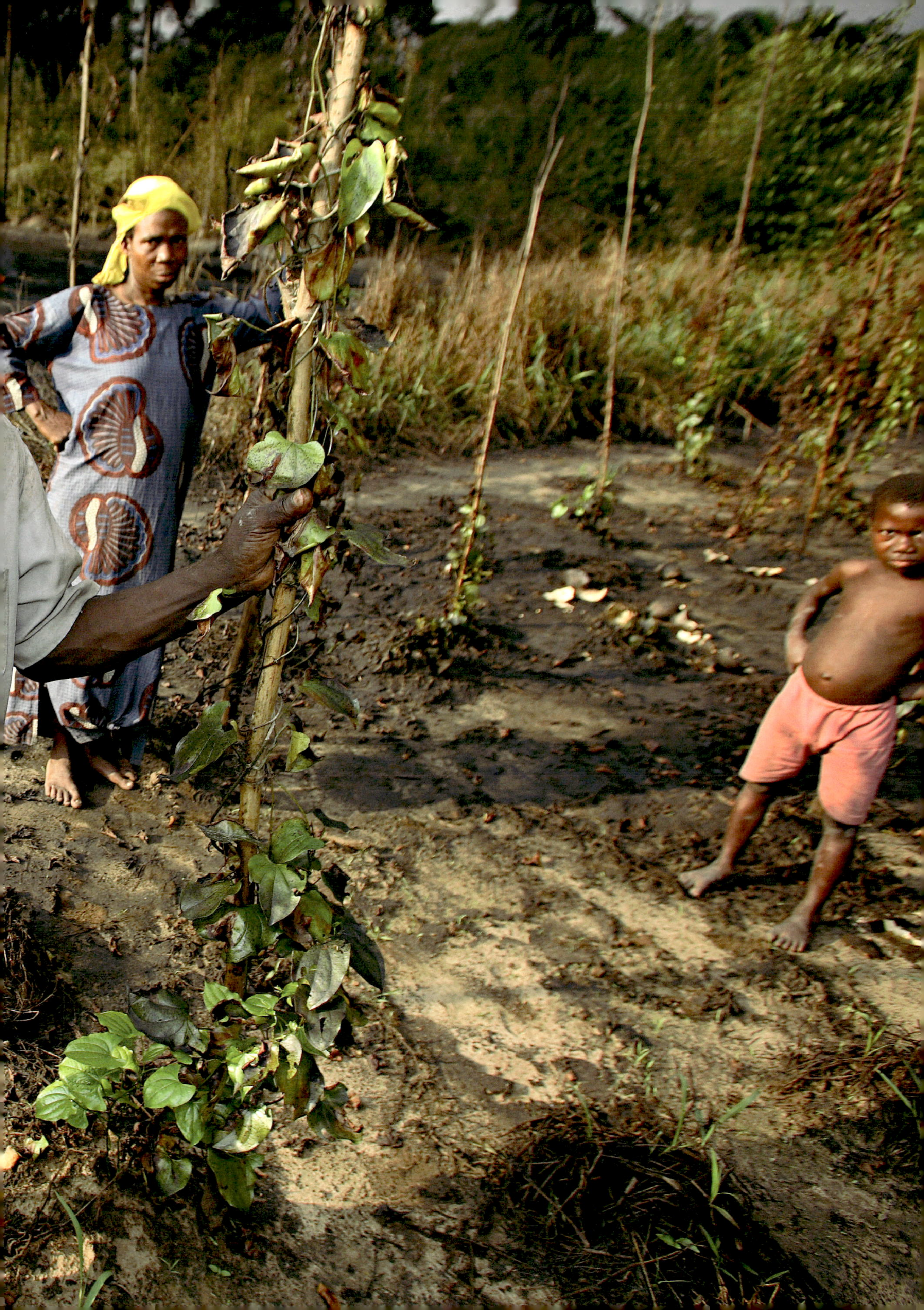

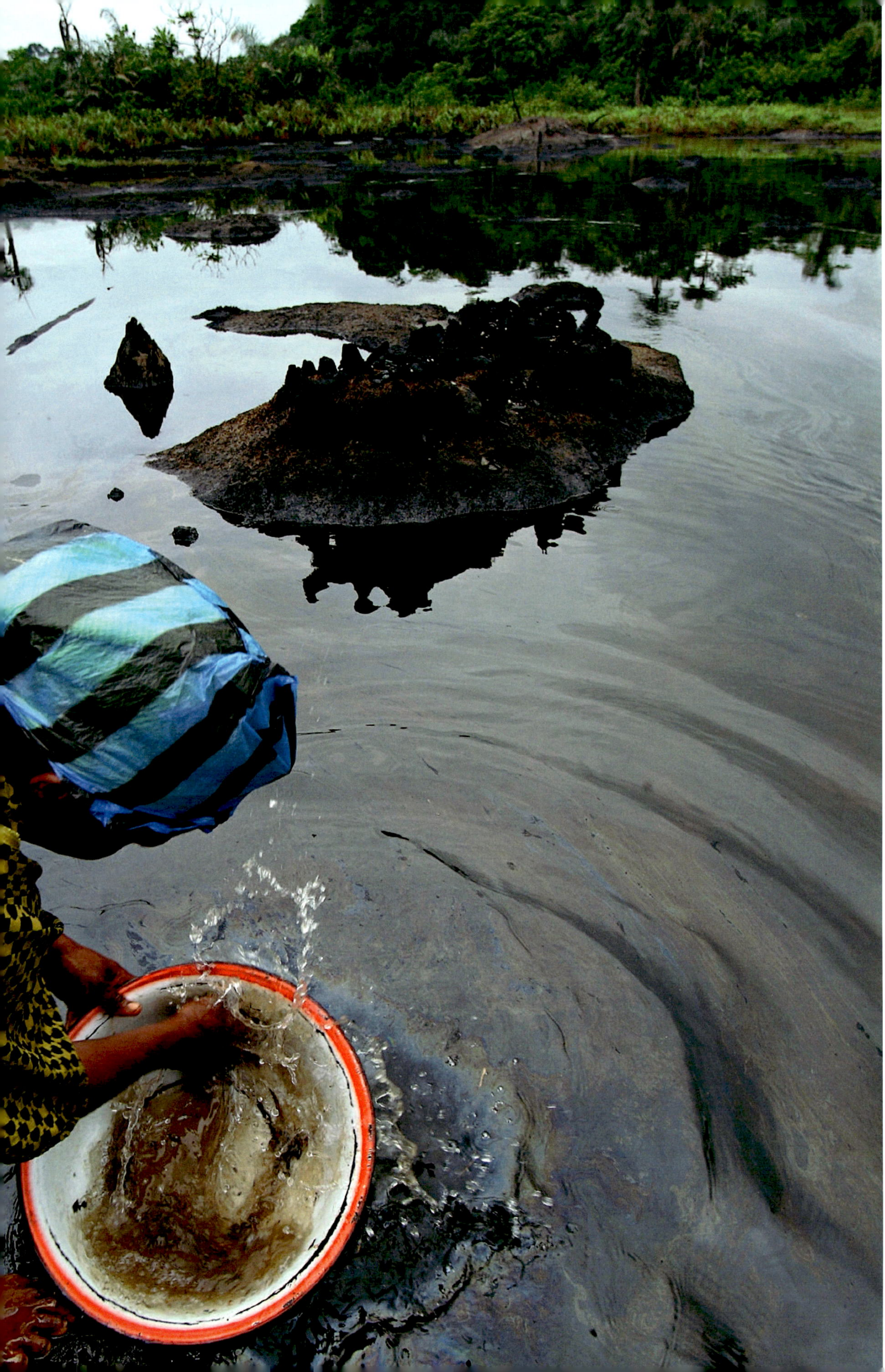

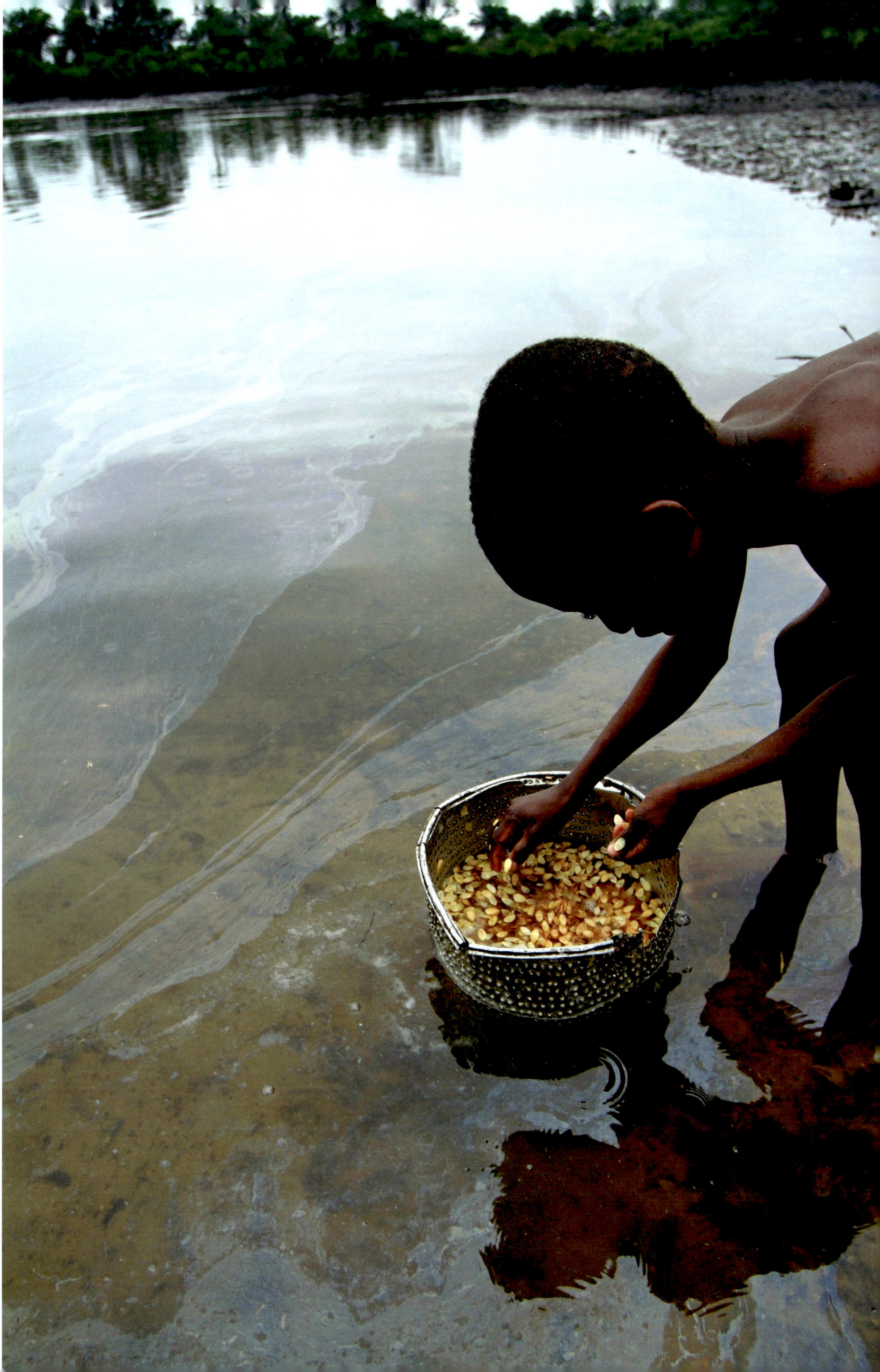

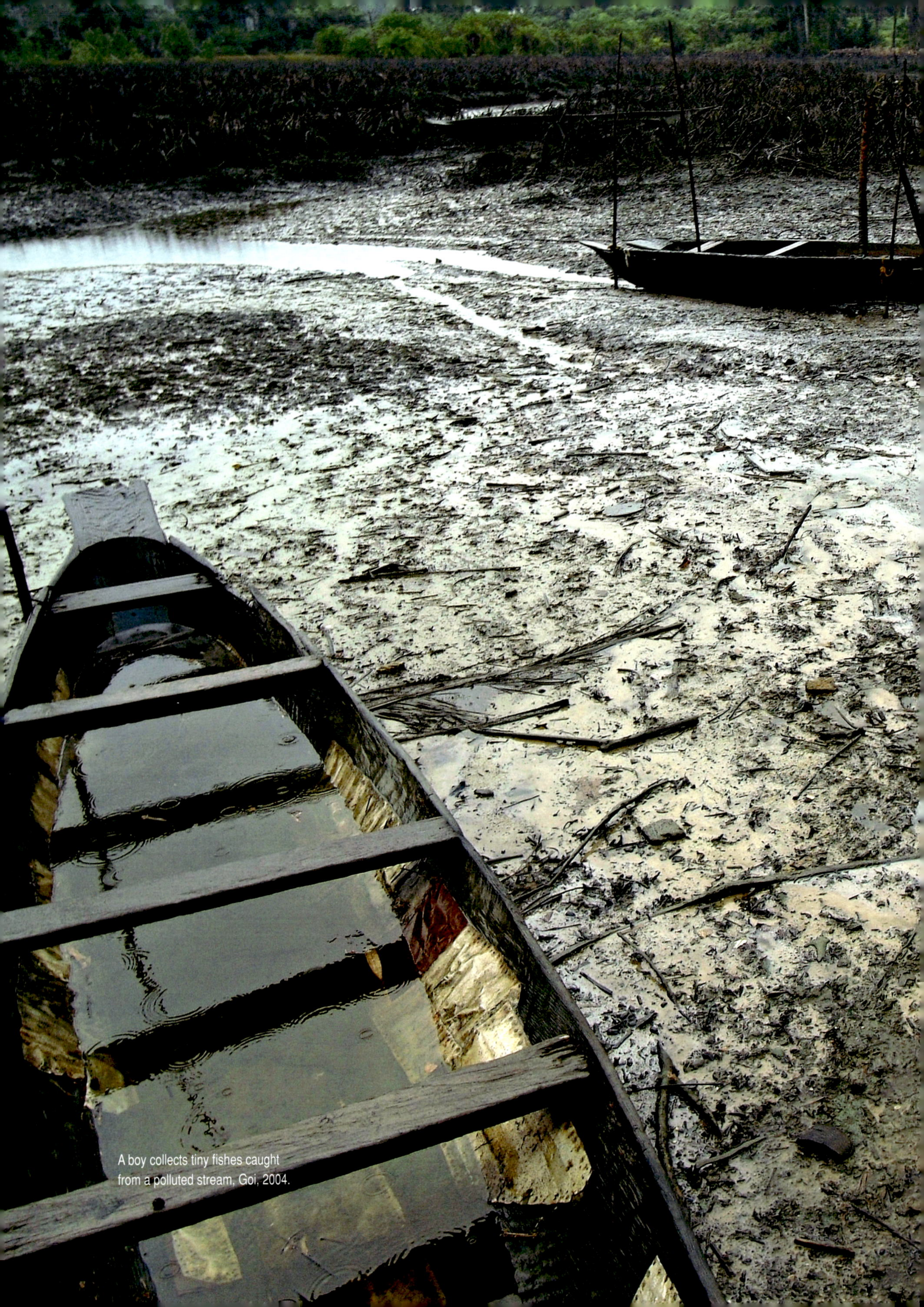

A boy collects tiny fishes caught
from a polluted stream, Goi, 2004.

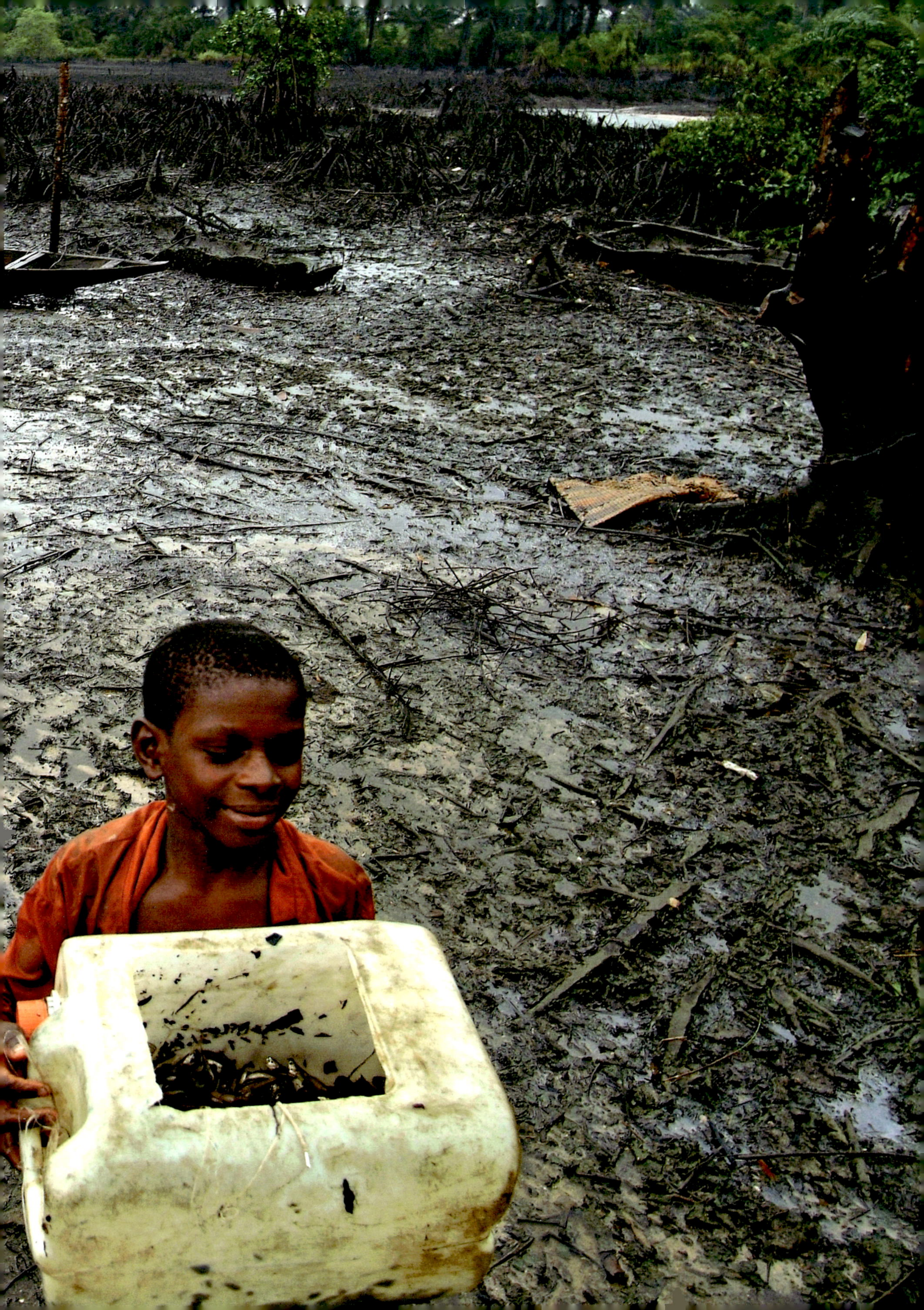

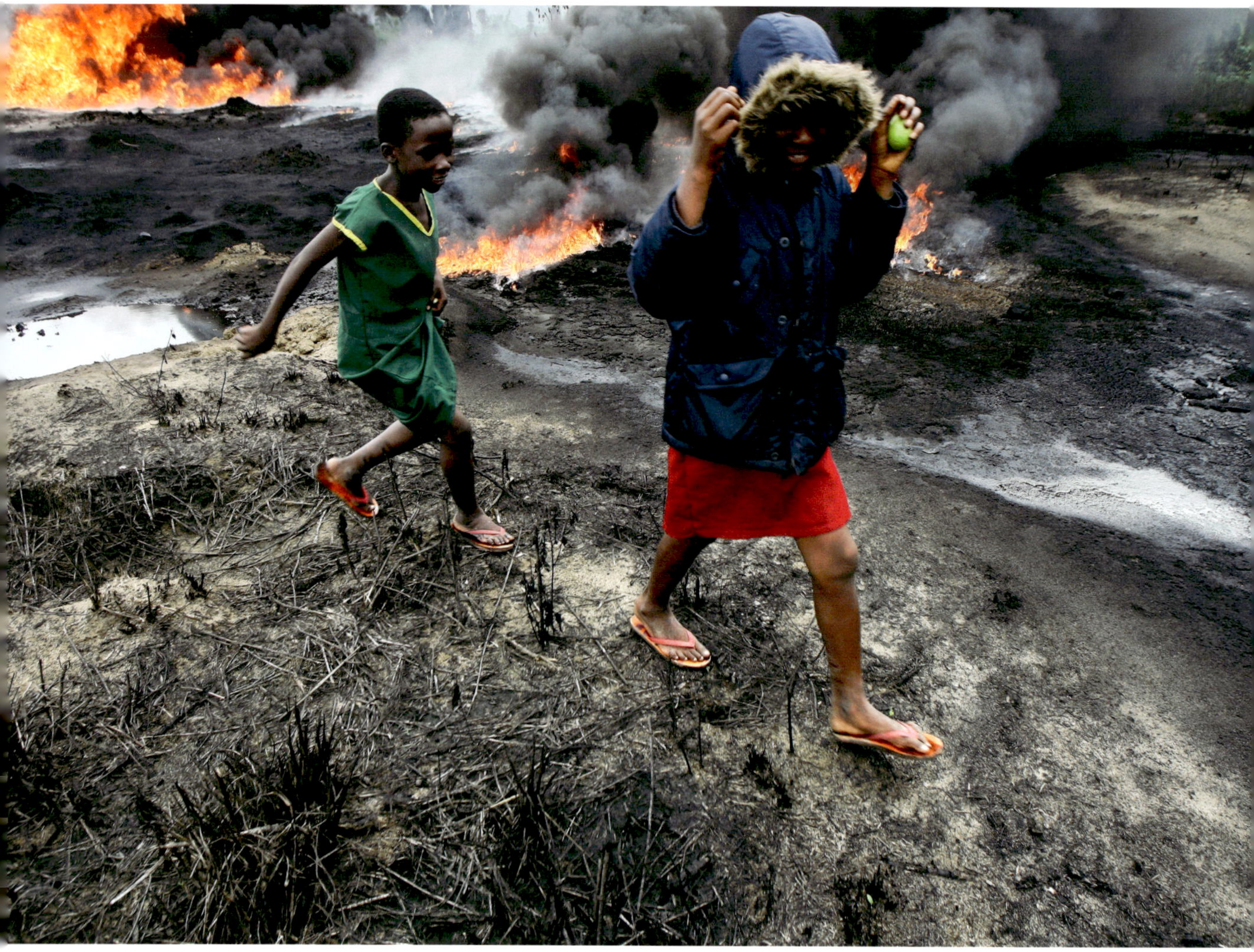

Above: Local kids walk pass an oil pipeline explosion in K-Dare, 2007.

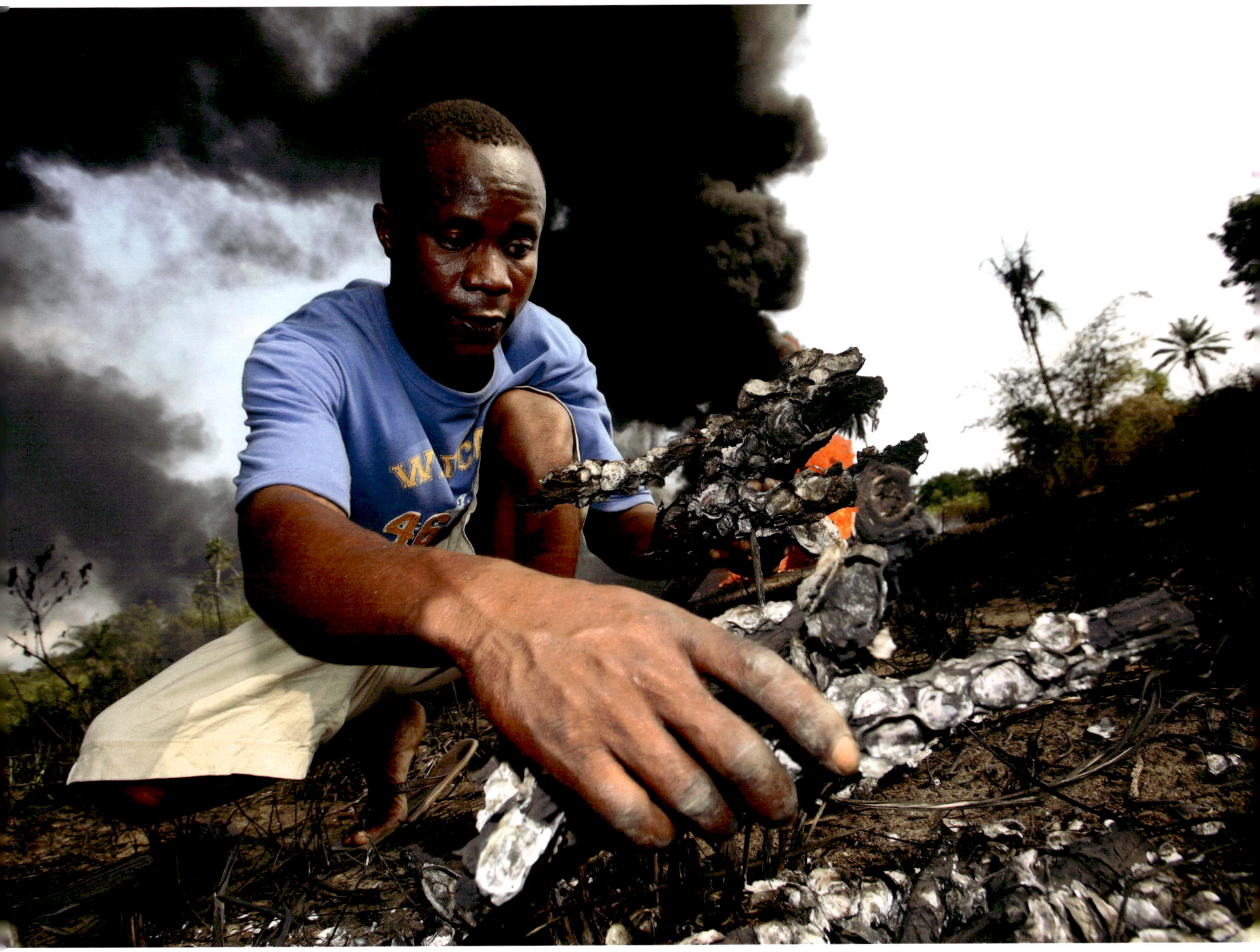

Above: A man sorts through pieces of burnt wood as a fire from an oil pipeline burns behind, Ogoniland, 2007.

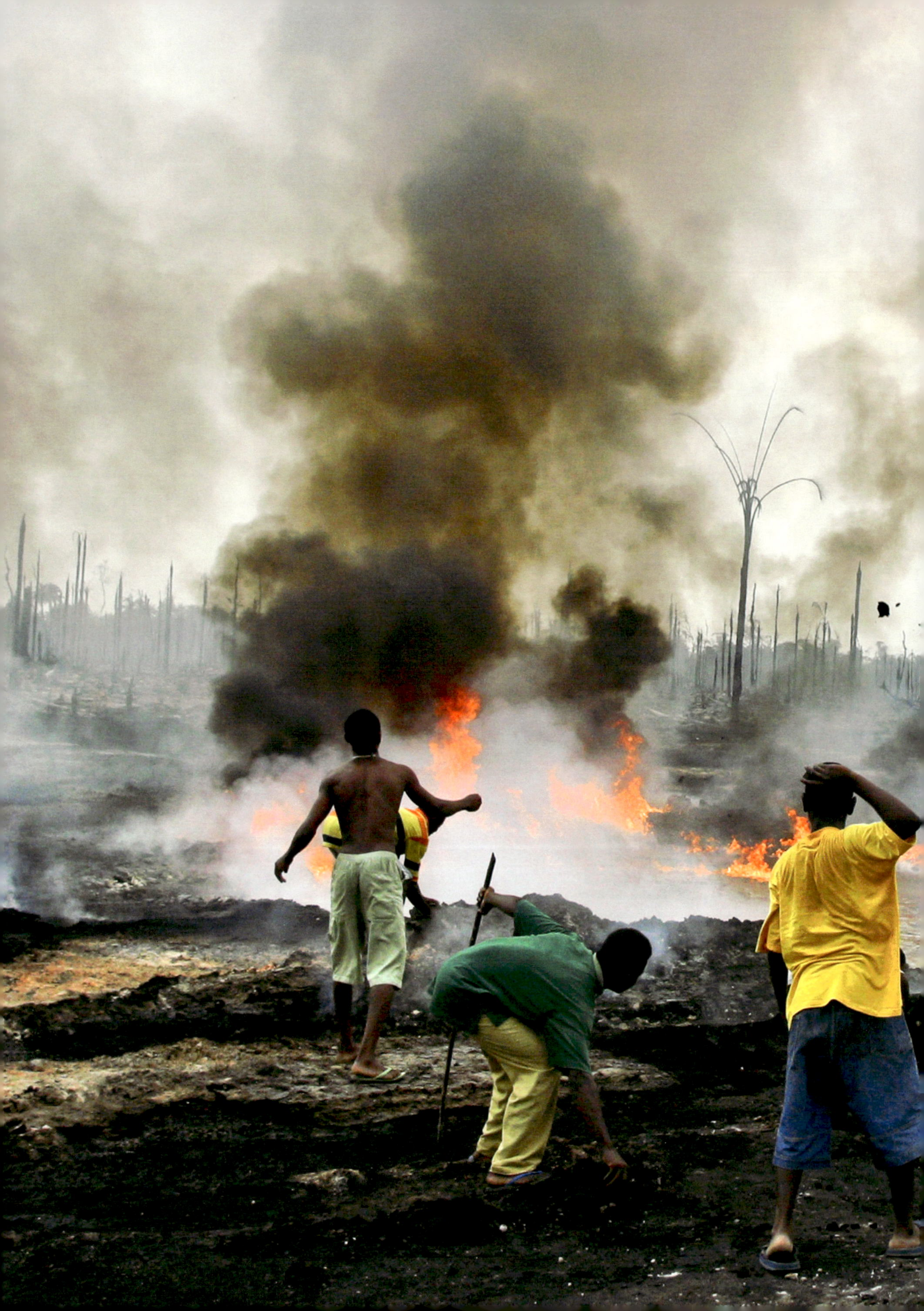

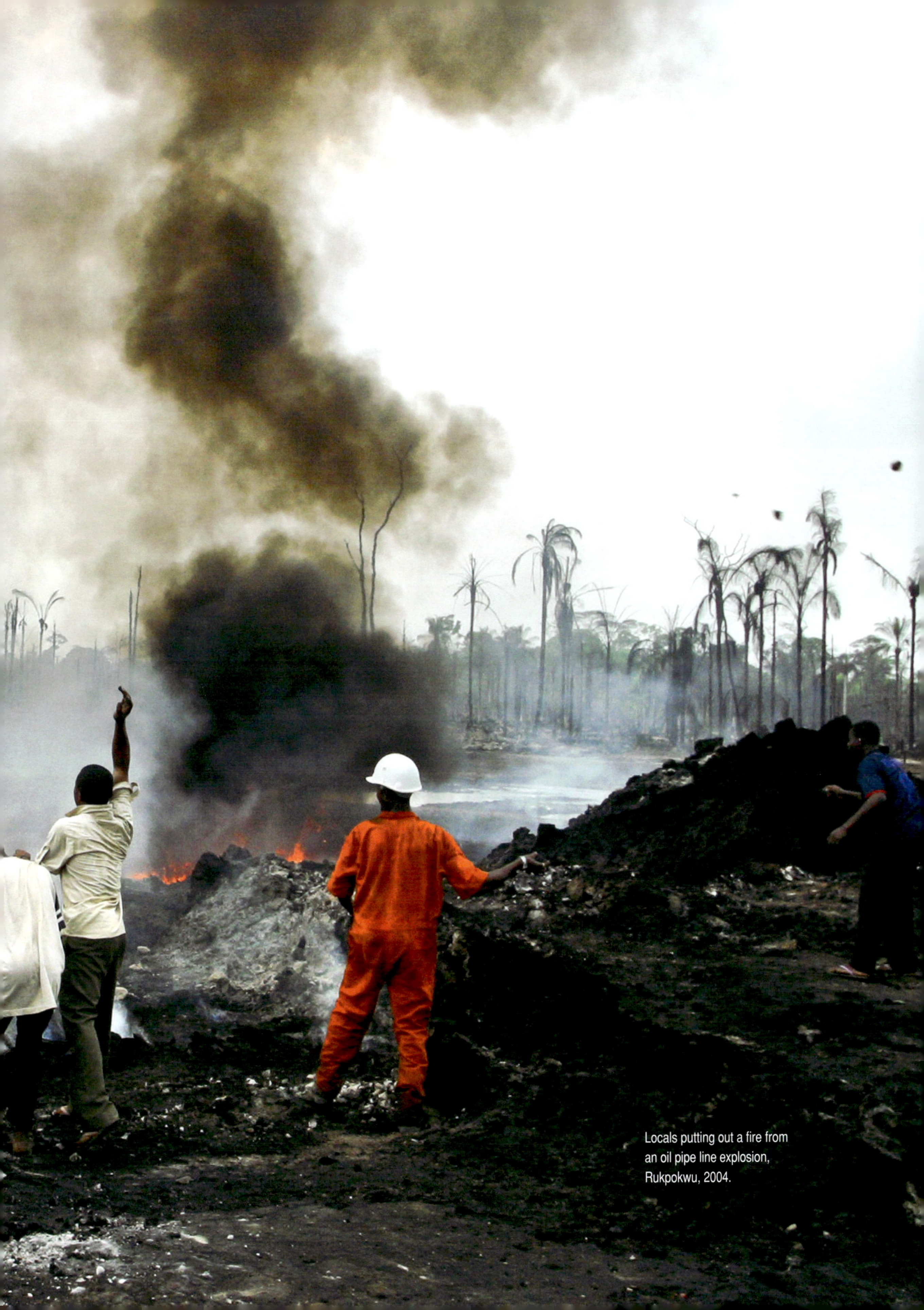

Locals putting out a fire from an oil pipe line explosion, Rukpokwu, 2004.

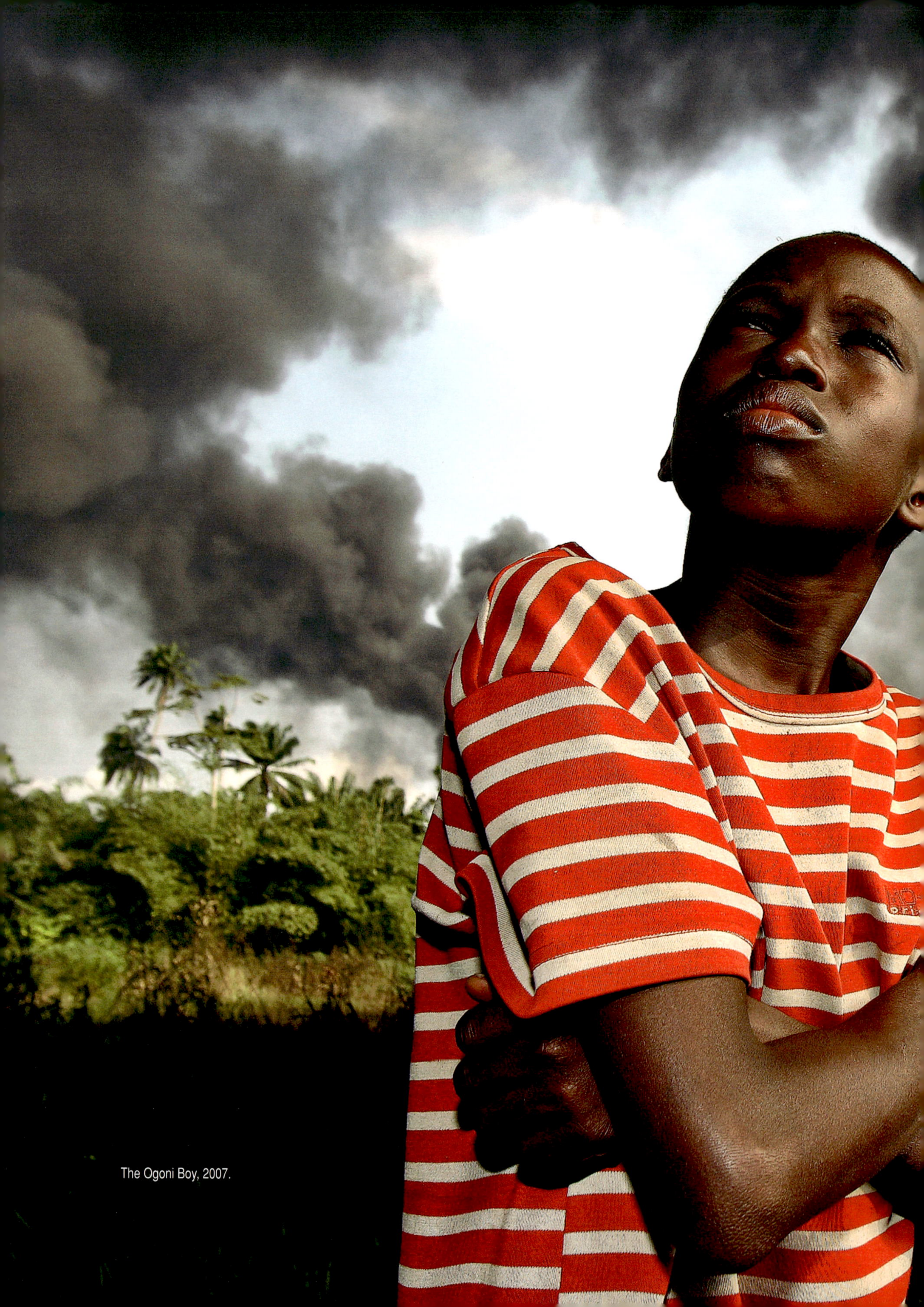

The Ogoni Boy, 2007.

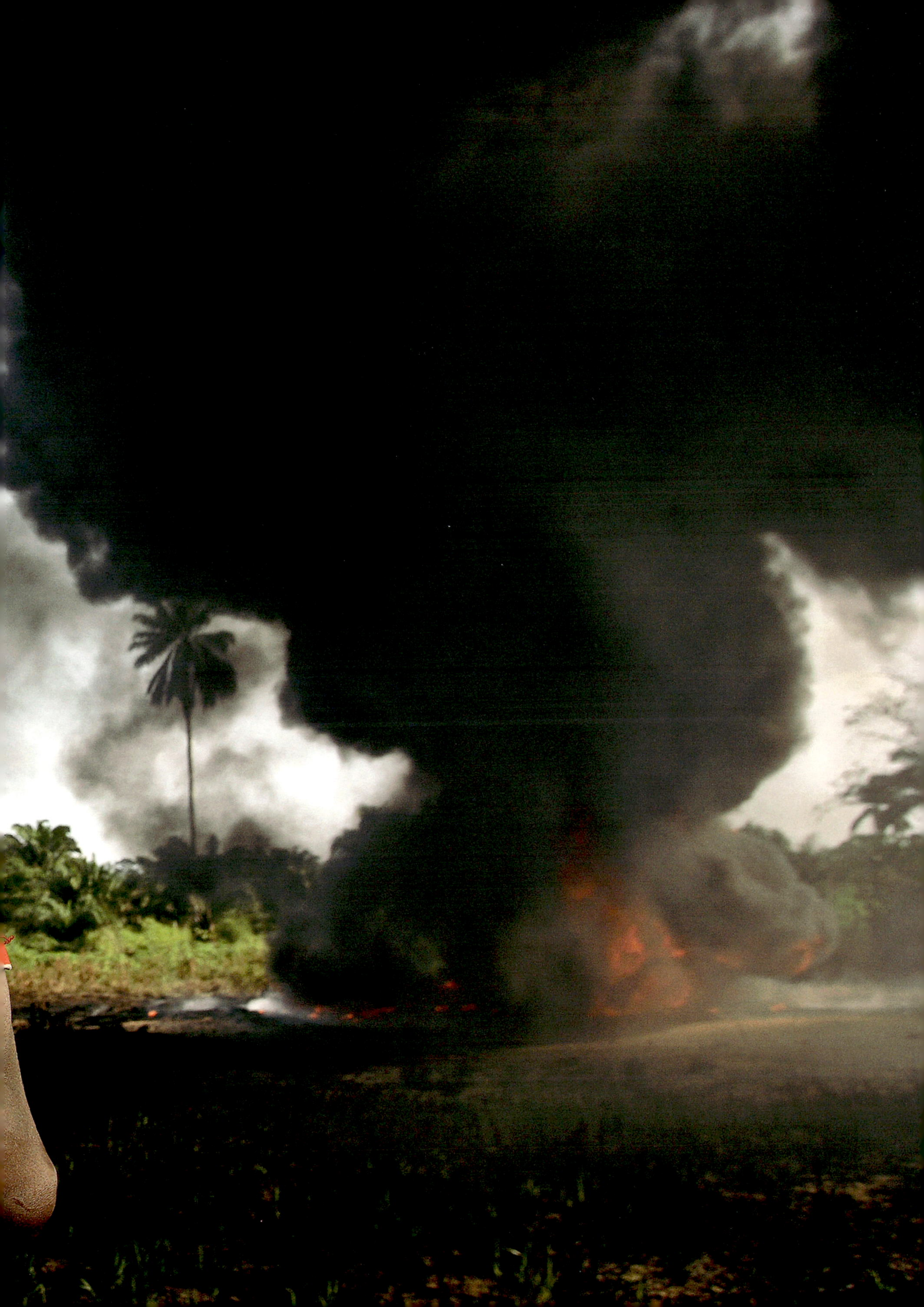

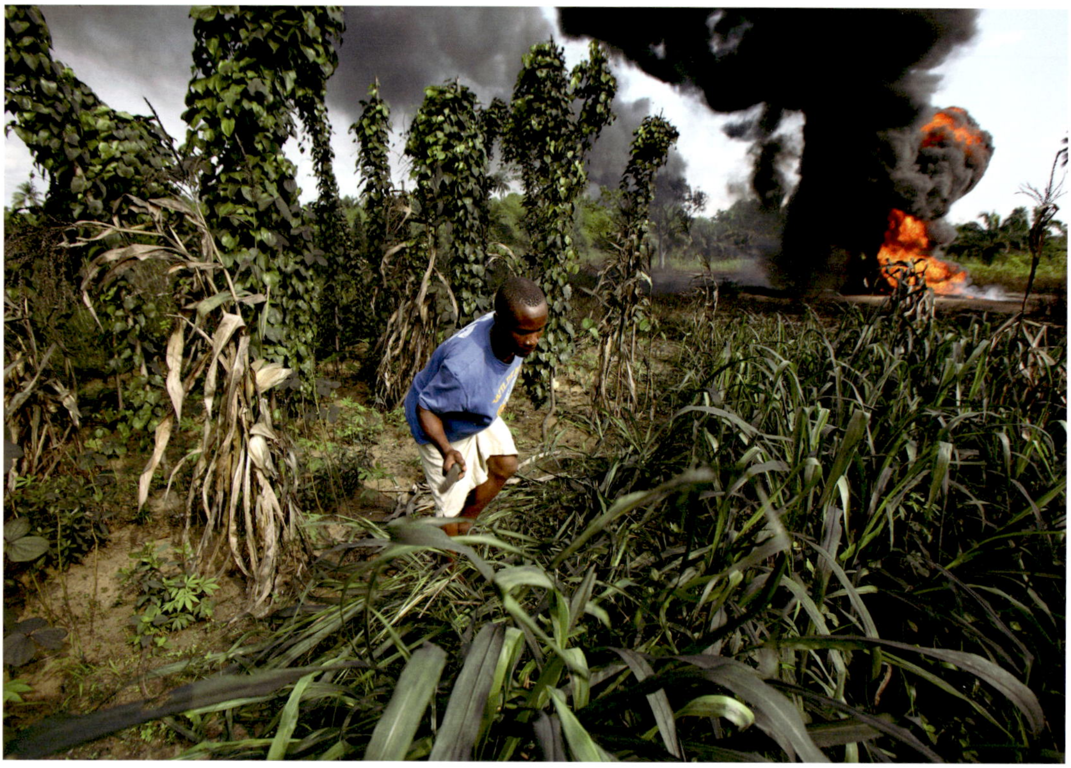

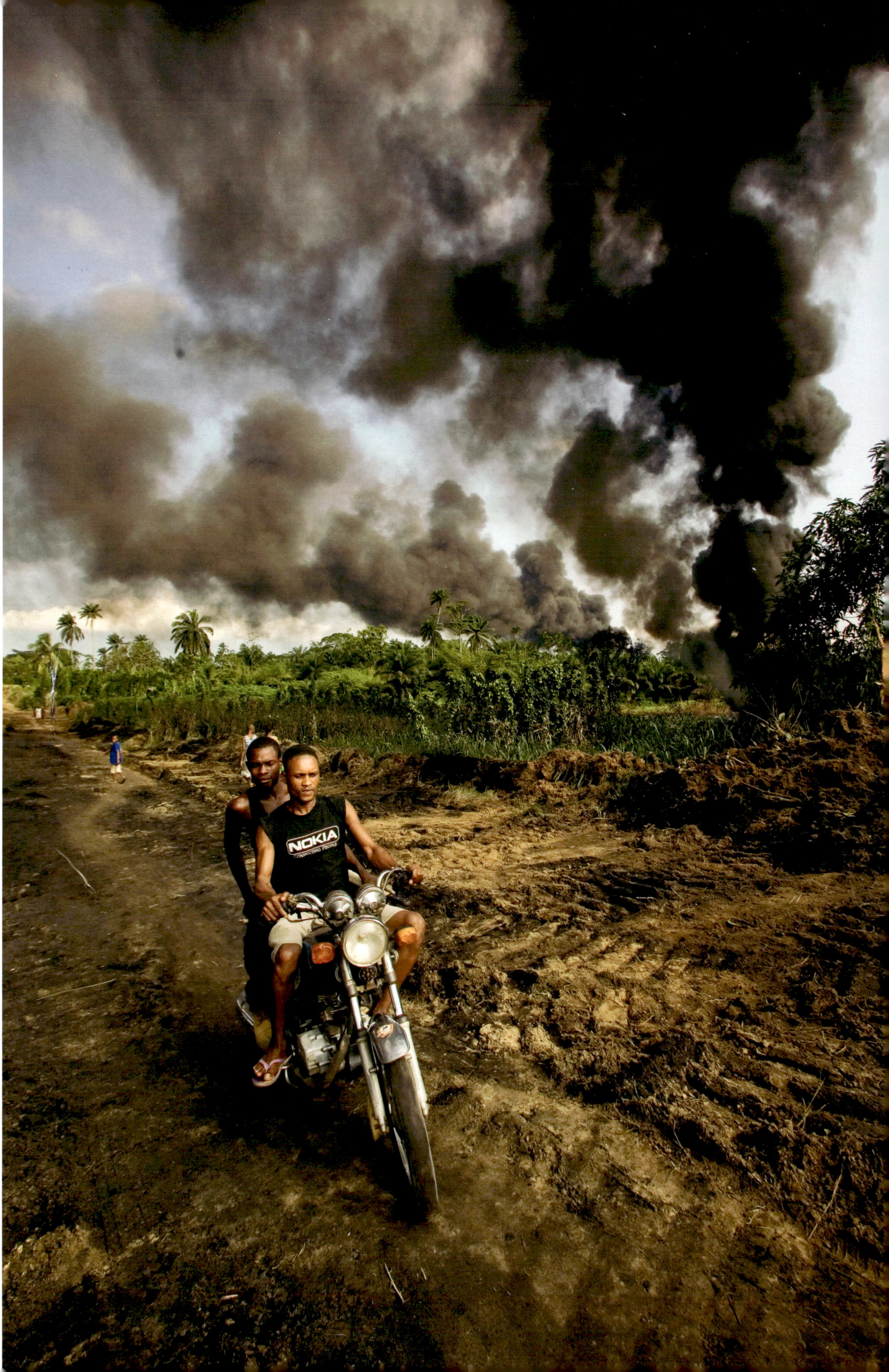
NOKIA

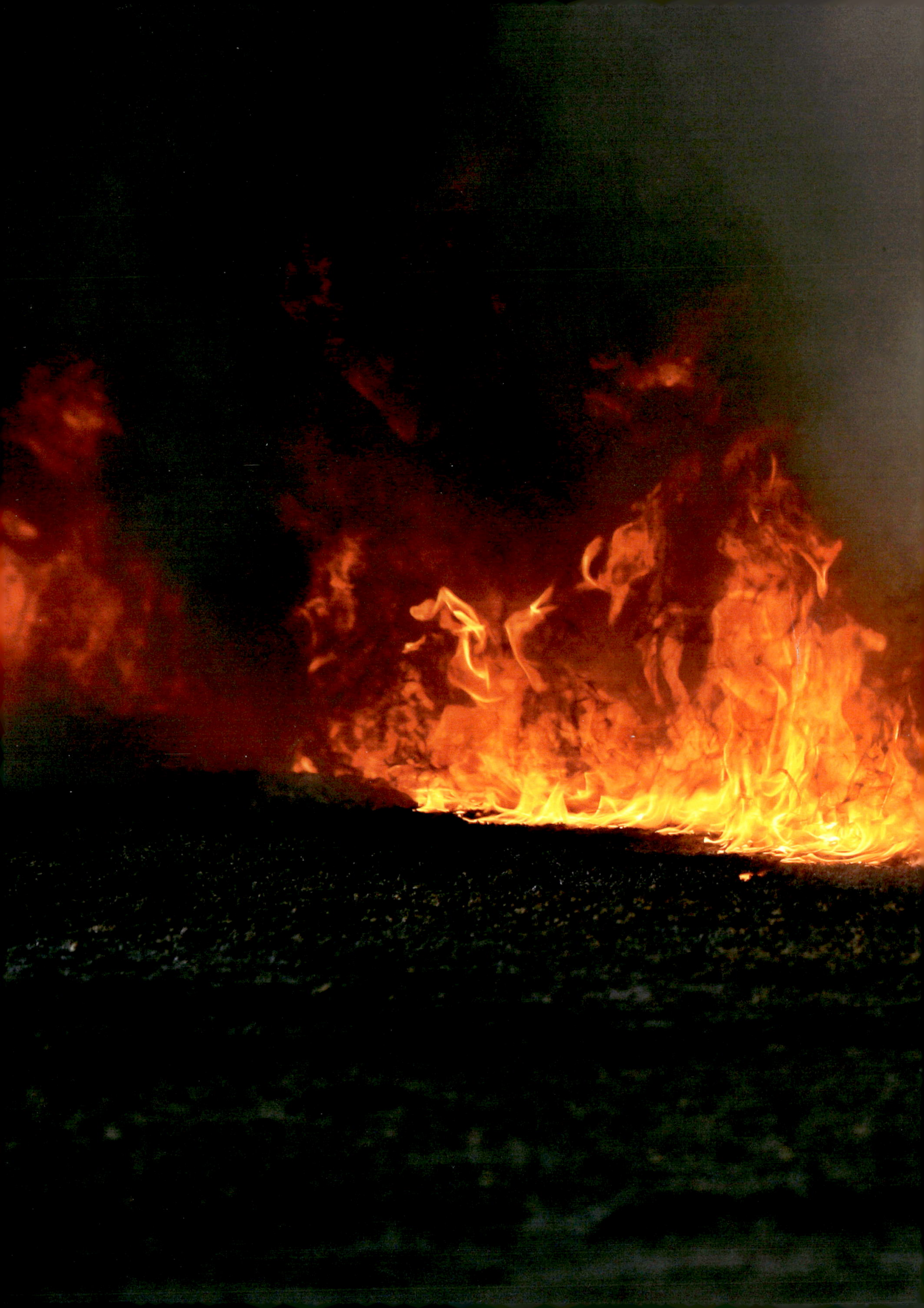

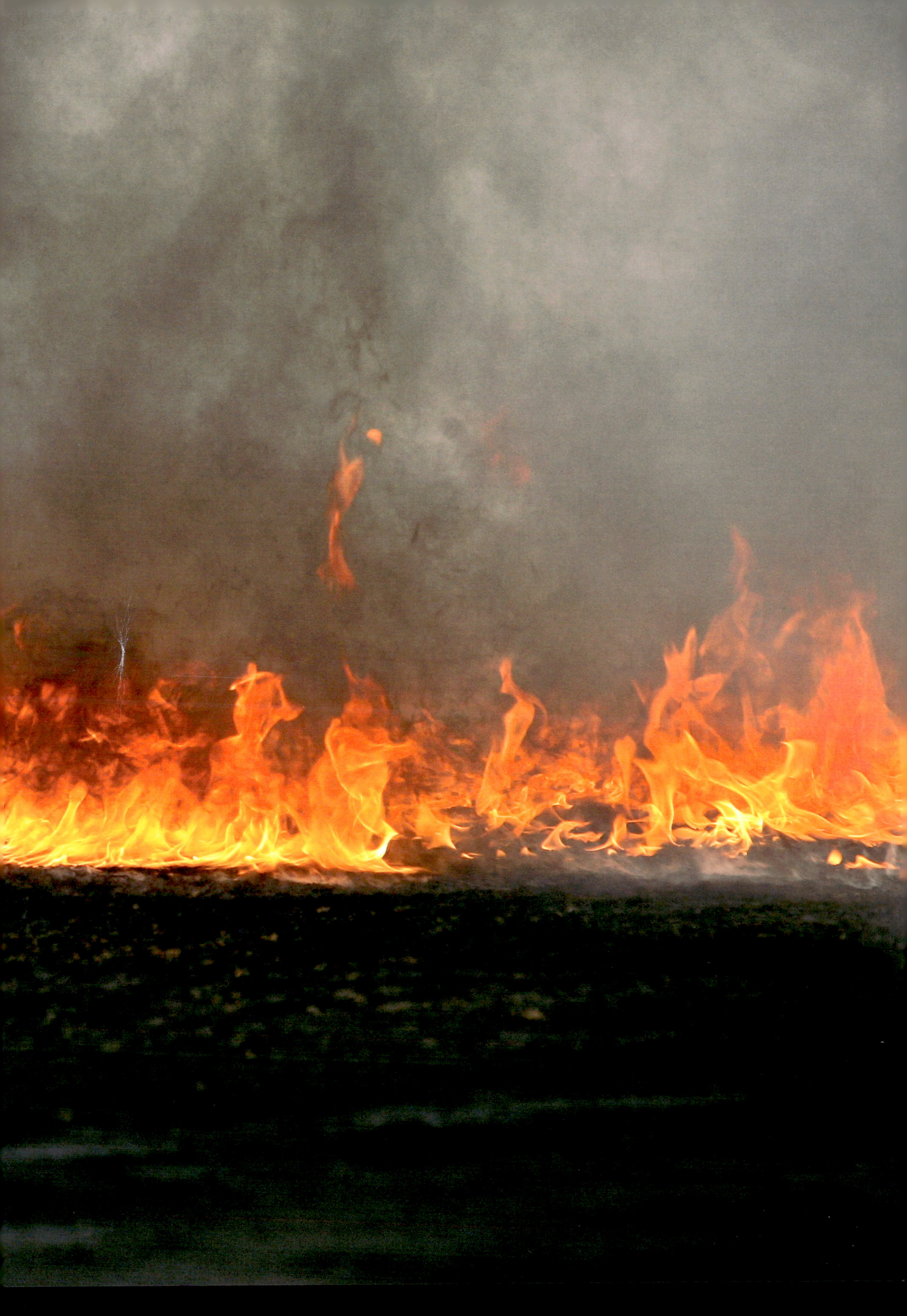

PETROLEUM, POLLUTION AND POVERTY IN THE NIGER DELTA

GOVERNMENT FAILURES

Systemic and serious government failures have left the people of the Niger Delta without adequate protection of their environment or their human rights. Despite recommendations by numerous bodies, including the African Commission on Human and Peoples' Rights, the Nigerian government has failed to ensure adequate oversight of the oil industry. The regulatory system in the Niger Delta is deeply flawed. Nigeria has laws and regulations that require companies to comply with internationally recognized standards of "good oil field practice", and laws and regulations to protect the environment but these laws and regulations are poorly enforced. The government agencies responsible for enforcement are ineffective and, in some cases, compromised by conflicts of interest.

The government of Nigeria has given the oil companies the authority to deal with matters that have a direct bearing on human rights, without adequate oversight – and often without any oversight at all. When communities suffer the consequences of environmental damage, they are frequently left to negotiate with the oil companies on action to address the problem and to obtain redress. The practice of allowing companies so much direct control over investigating oil spills and over the awarding of compensation represents a critical government failure to protect human rights and ensure an effective remedy. Nigeria's heavy dependence on oil and the fact that the Nigerian government is the majority partner in joint ventures are fundamental problems which underpin regulatory failures.

CORPORATE CULPABILITY

A company is not free to ignore the consequences of its actions because a government has failed to hold it to account. There are international standards that relate to oil industry operations, and accepted standards on social and environmental impacts, which oil companies in the Niger Delta should be well aware of. In addition, companies have in place policies which commit them to good practice in terms of environment and social impacts. The vast majority of cases reported to, and investigated by, Amnesty International relate to the Shell Petroleum Development Company (SPDC). However, other companies are not blameless – there is evidence of poor practice in other areas and by other companies. SPDC has operated in the Niger Delta for decades and has engaged in practices known to be damaging to the environment and to people. After a brief period of openness during the 1990s, when SPDC recognized that its operations were causing serious problems, SPDC and its parent company Royal Dutch Shell today appear unwilling to acknowledge many of the problems associated with their operations in the Niger Delta. Although SPDC operates within a delta system and oil infrastructure is frequently located close to farmland and waterways, few, if any, adequate protective measures have been taken by SPDC – pipelines have not been properly maintained and waste products have been released into the environment without adequate monitoring of the impact on people's lives and livelihoods. Despite legal requirements to clean-up and remediate land and water swiftly and adequately, these actions frequently do not happen. In several cases, SPDC's failure to adequately address oil pollution has had devastating consequences for human rights.

HUMAN RIGHTS

The people of the Niger Delta have seen their human rights undermined by oil companies that their government cannot or will not hold to account. They have been systematically denied access to information about how oil exploration and production will affect them, and are repeatedly denied access to justice.

The Niger Delta provides a stark case study of the lack of accountability of a government to the people and of multinational companies' almost total lack of accountability when it comes to the impact of their operations on human rights.

Amnesty International June 2009

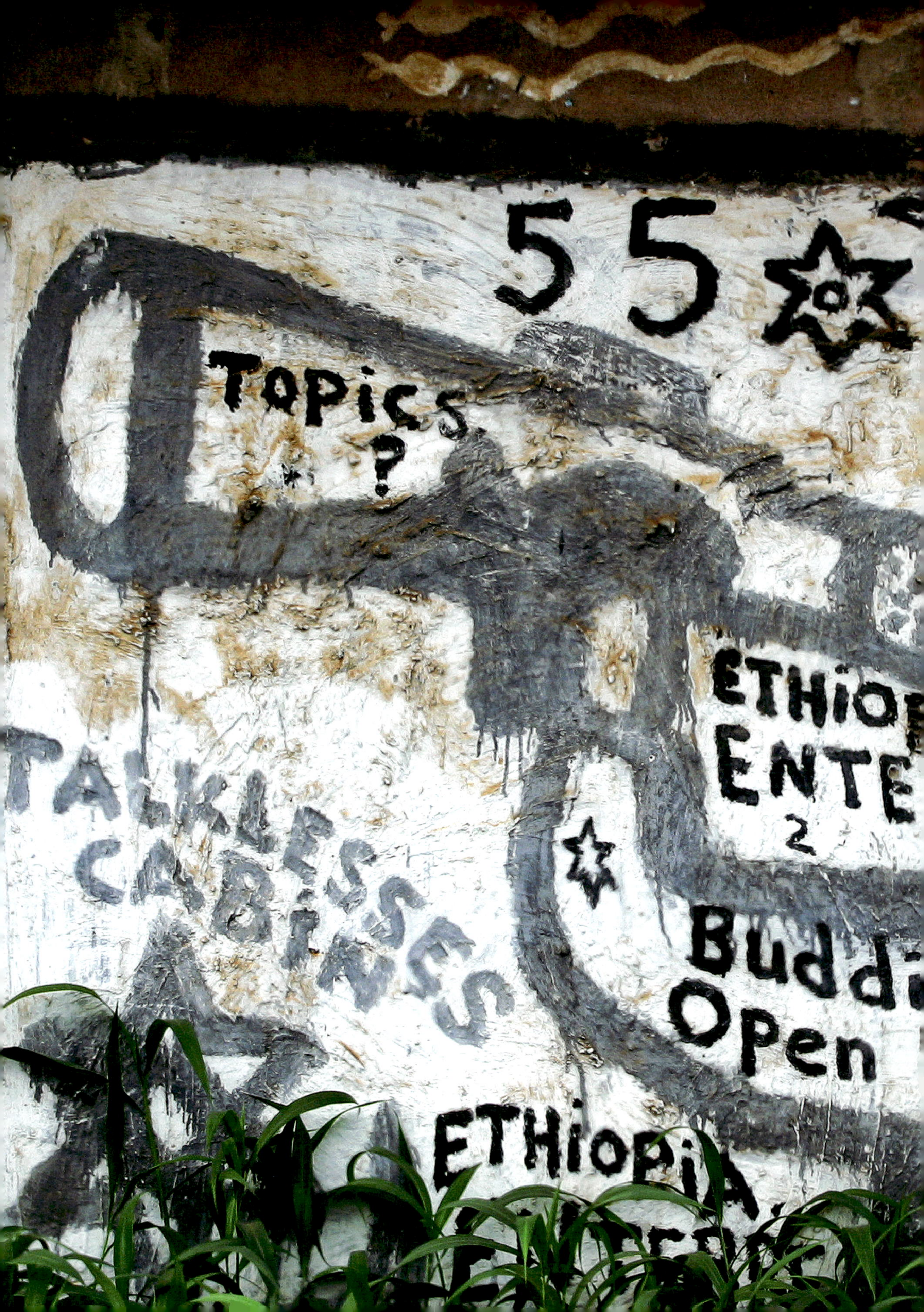
55
TOPICS
ETHIOPIA
Open

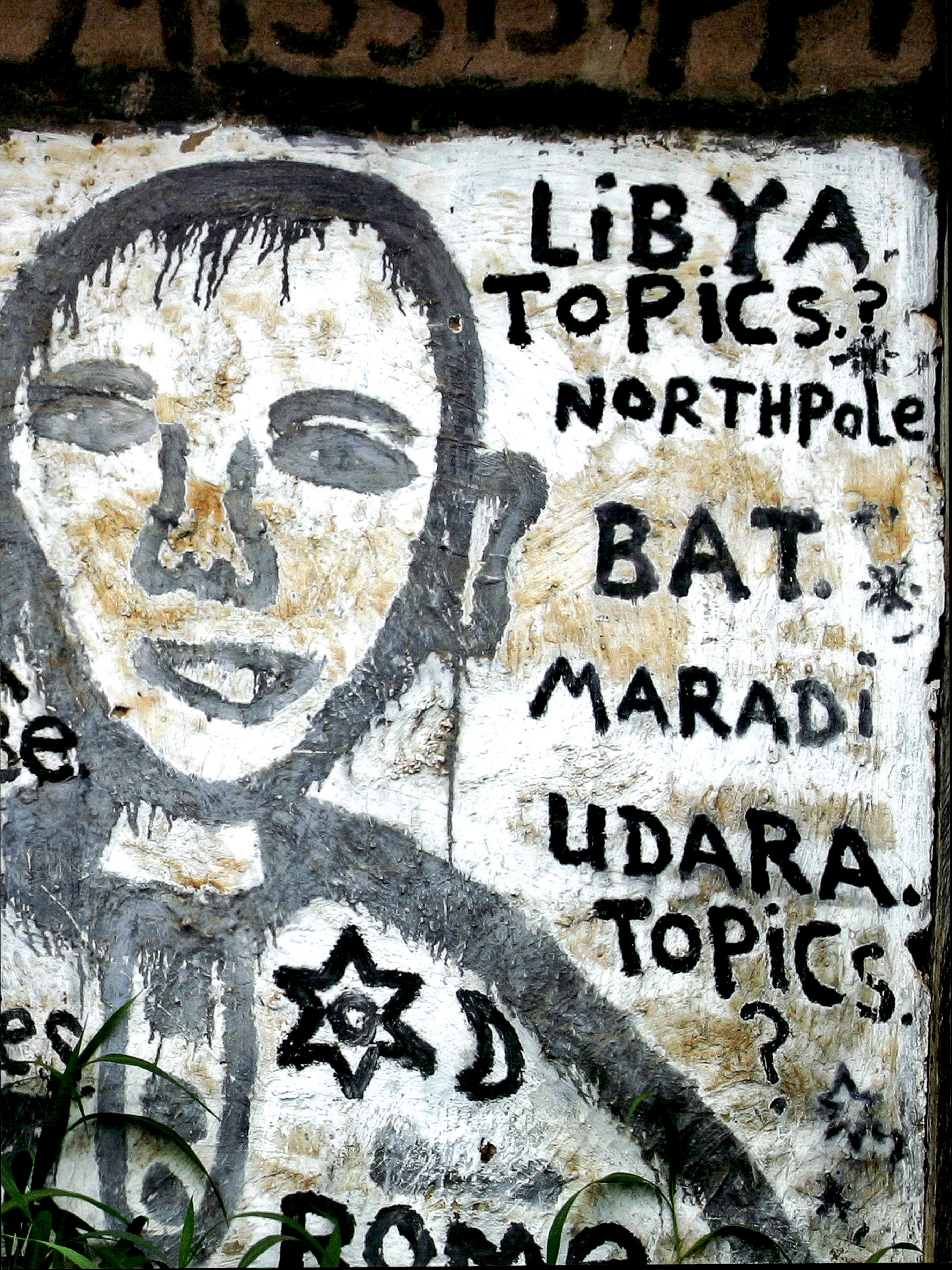
LiBYA
TOPiCS.?
NORTHPoLe
BAT.
MARADi
UDARA.
TOPiCS.
?

DEMOCRACY!!!
DEMONSTRATION OF CRAZE
CRAZY DEMONSTRATION

Fela Anikulapo Kuti

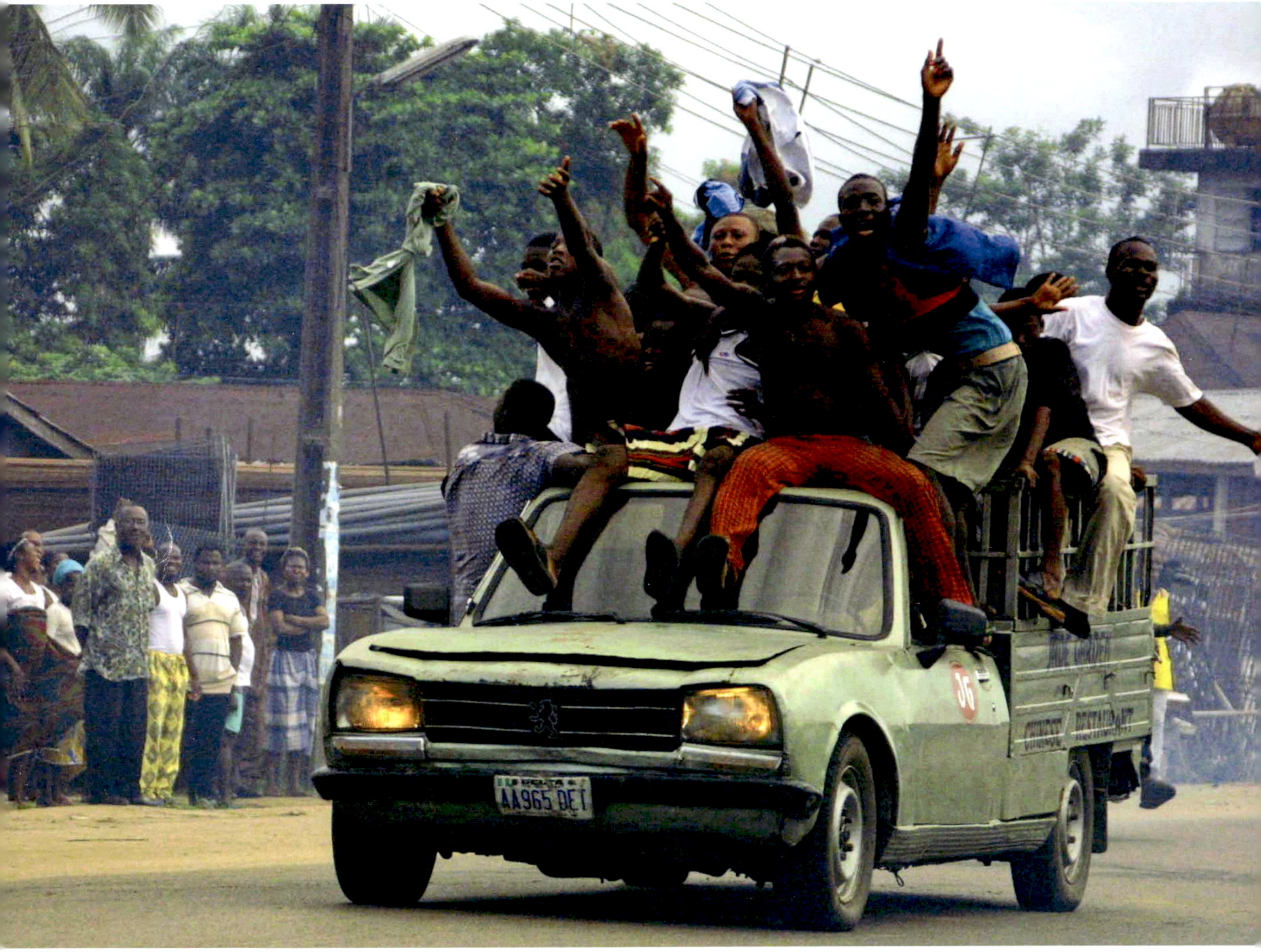

Opposition party supporters celebrate following uncomfirmed voting results in Warri, Niger Delta, 2003.

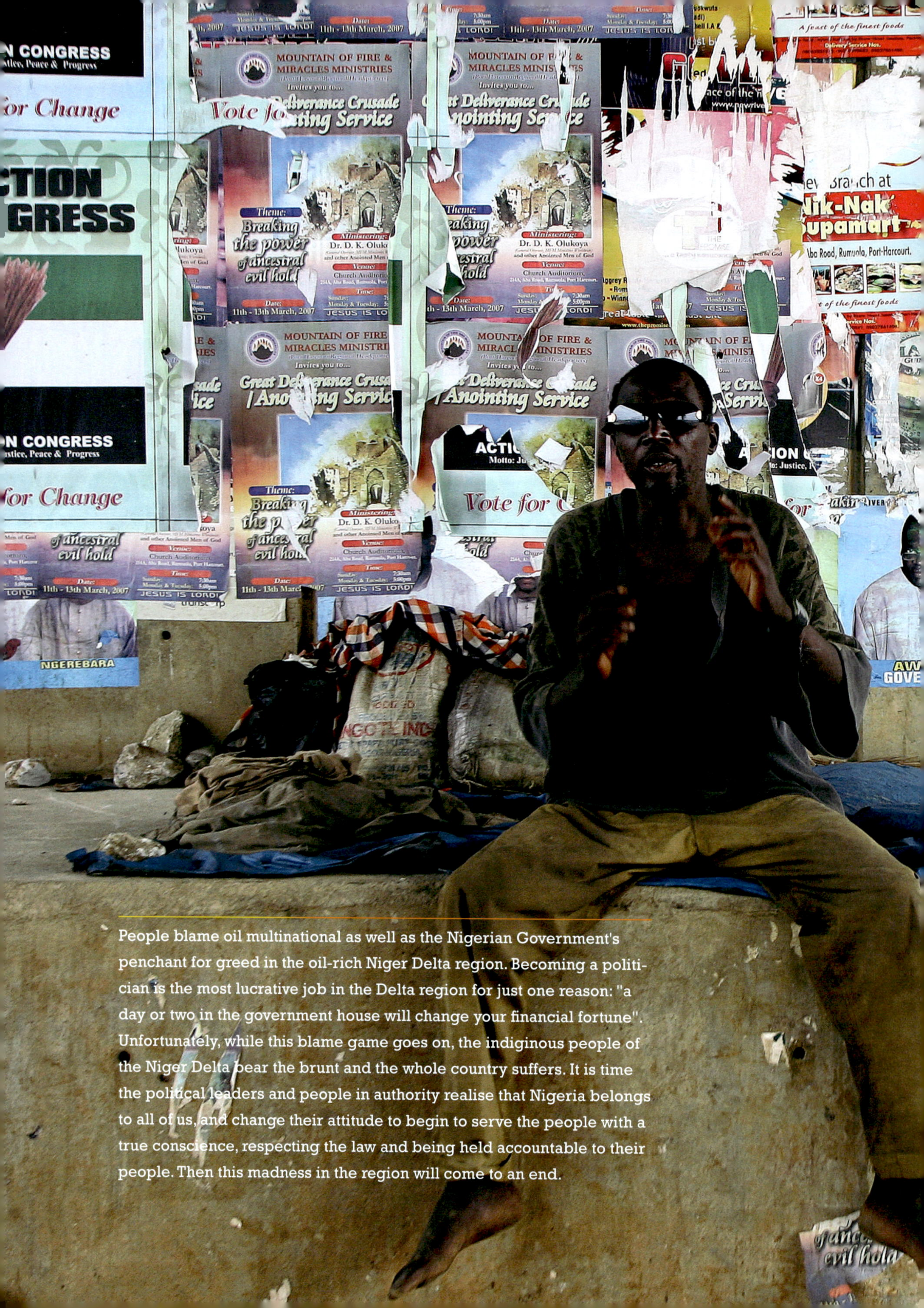

People blame oil multinational as well as the Nigerian Government's penchant for greed in the oil-rich Niger Delta region. Becoming a politician is the most lucrative job in the Delta region for just one reason: "a day or two in the government house will change your financial fortune". Unfortunately, while this blame game goes on, the indiginous people of the Niger Delta bear the brunt and the whole country suffers. It is time the political leaders and people in authority realise that Nigeria belongs to all of us, and change their attitude to begin to serve the people with a true conscience, respecting the law and being held accountable to their people. Then this madness in the region will come to an end.

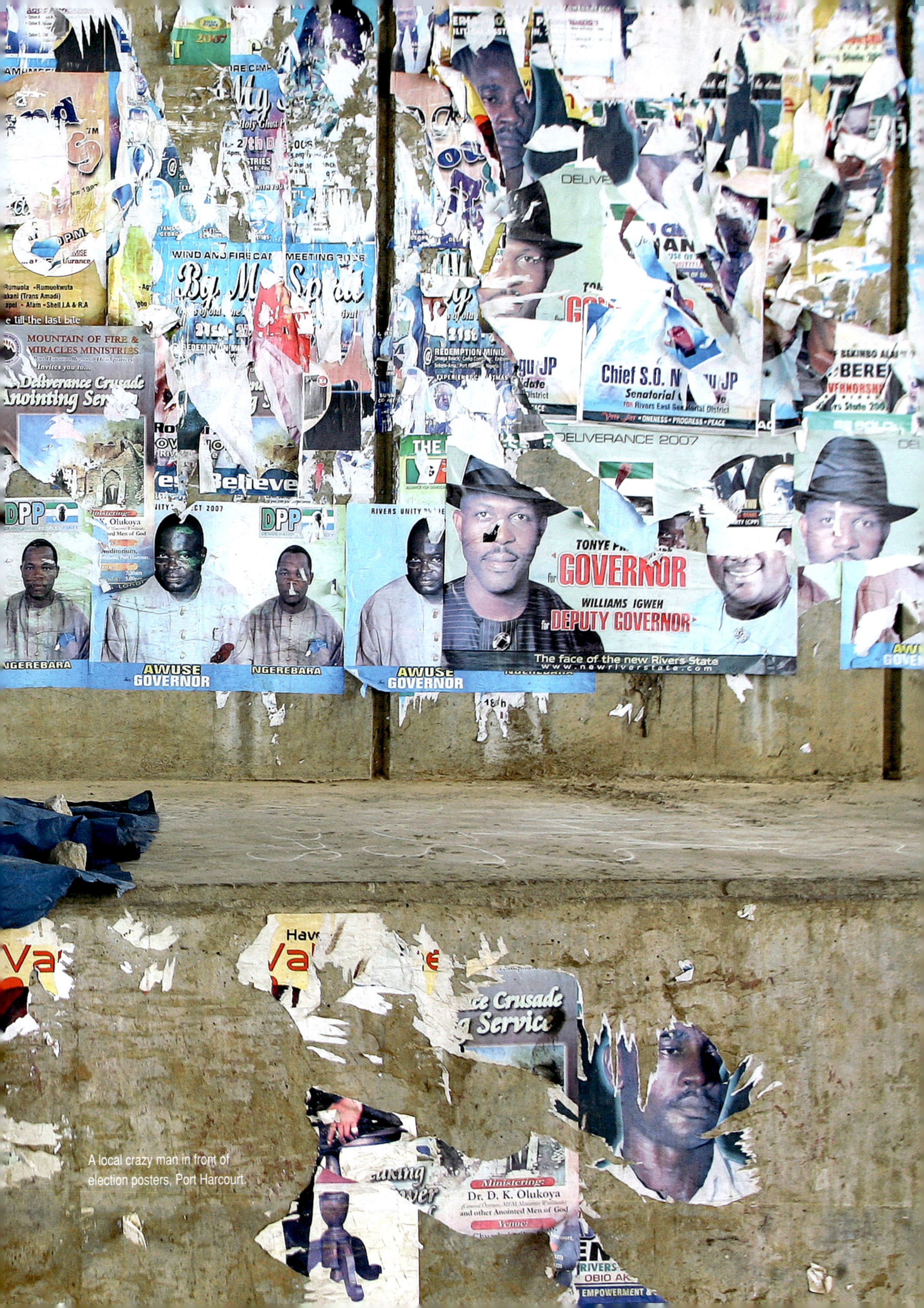

A local crazy man in front of election posters, Port Harcourt.

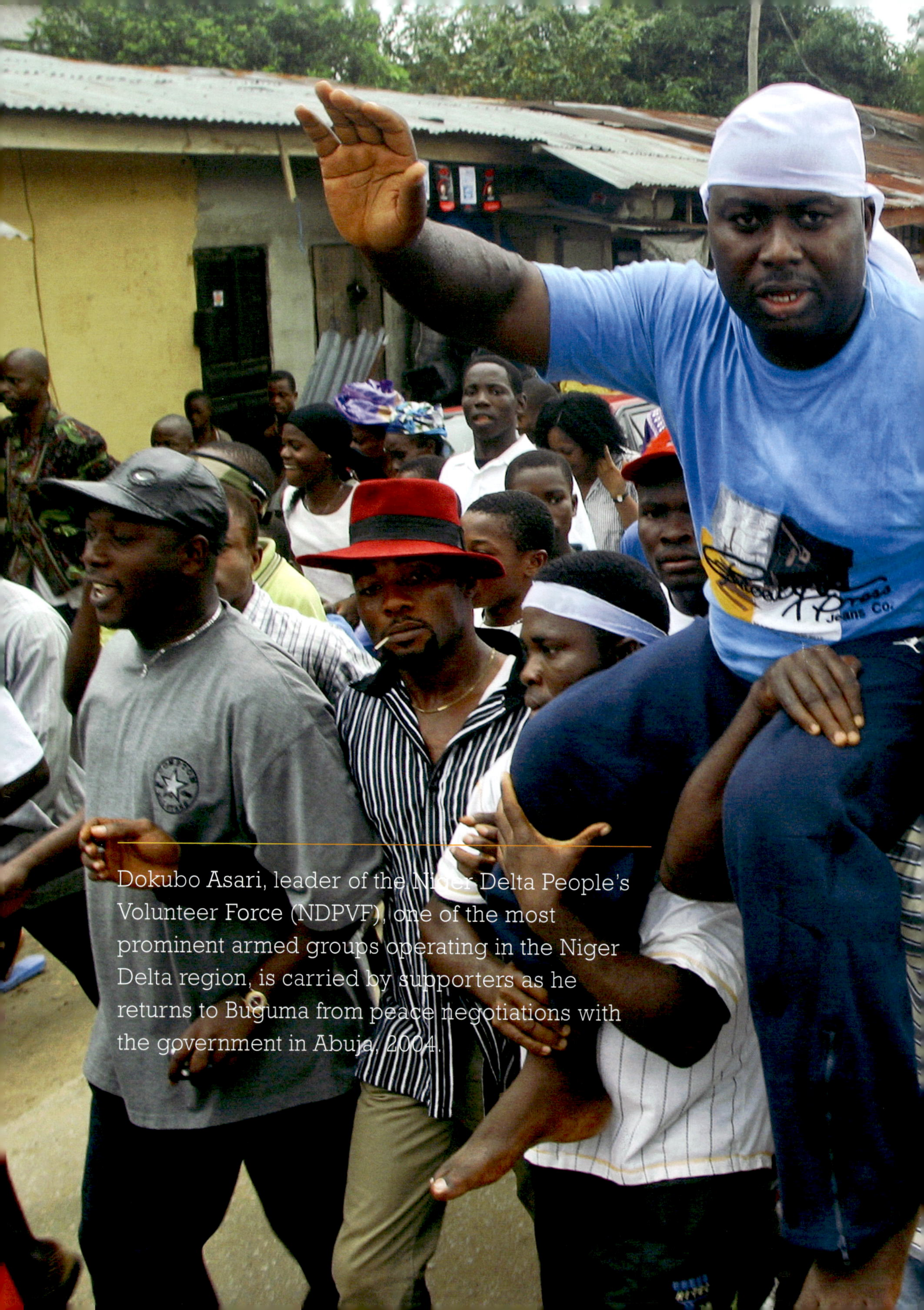

Dokubo Asari, leader of the Niger Delta People's Volunteer Force (NDPVF), one of the most prominent armed groups operating in the Niger Delta region, is carried by supporters as he returns to Buguma from peace negotiations with the government in Abuja, 2004.

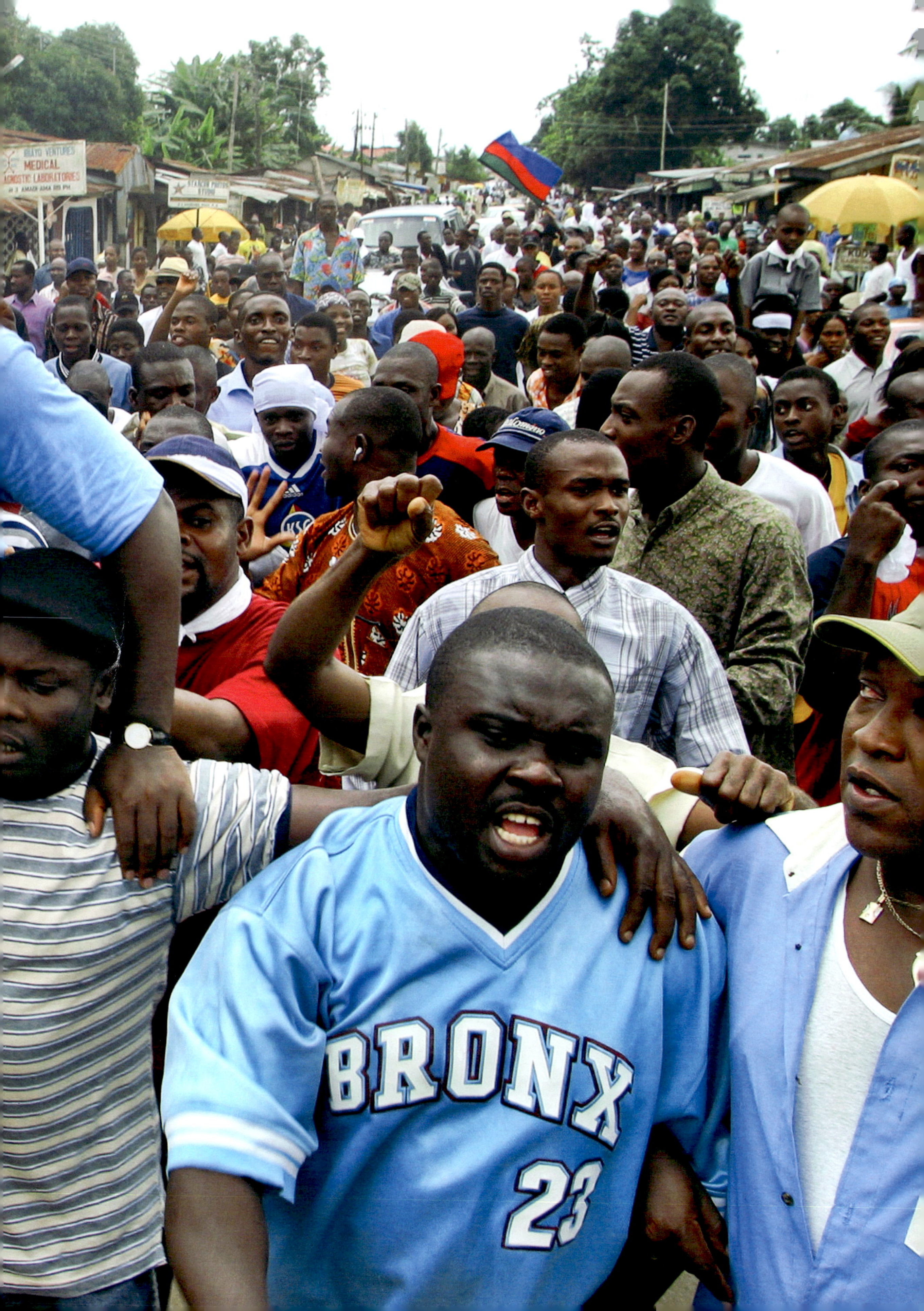
MEDICAL
BRONX
23

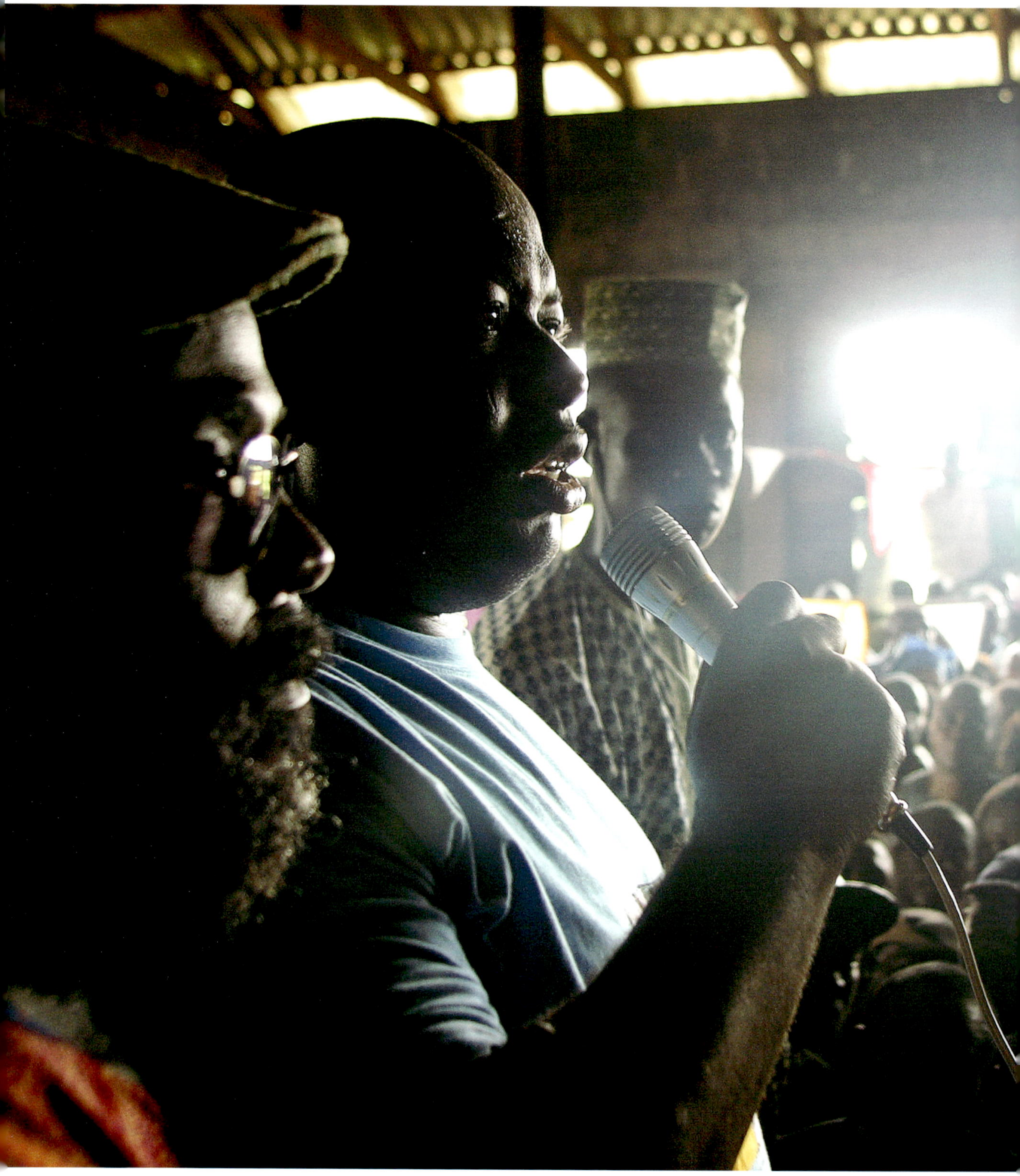

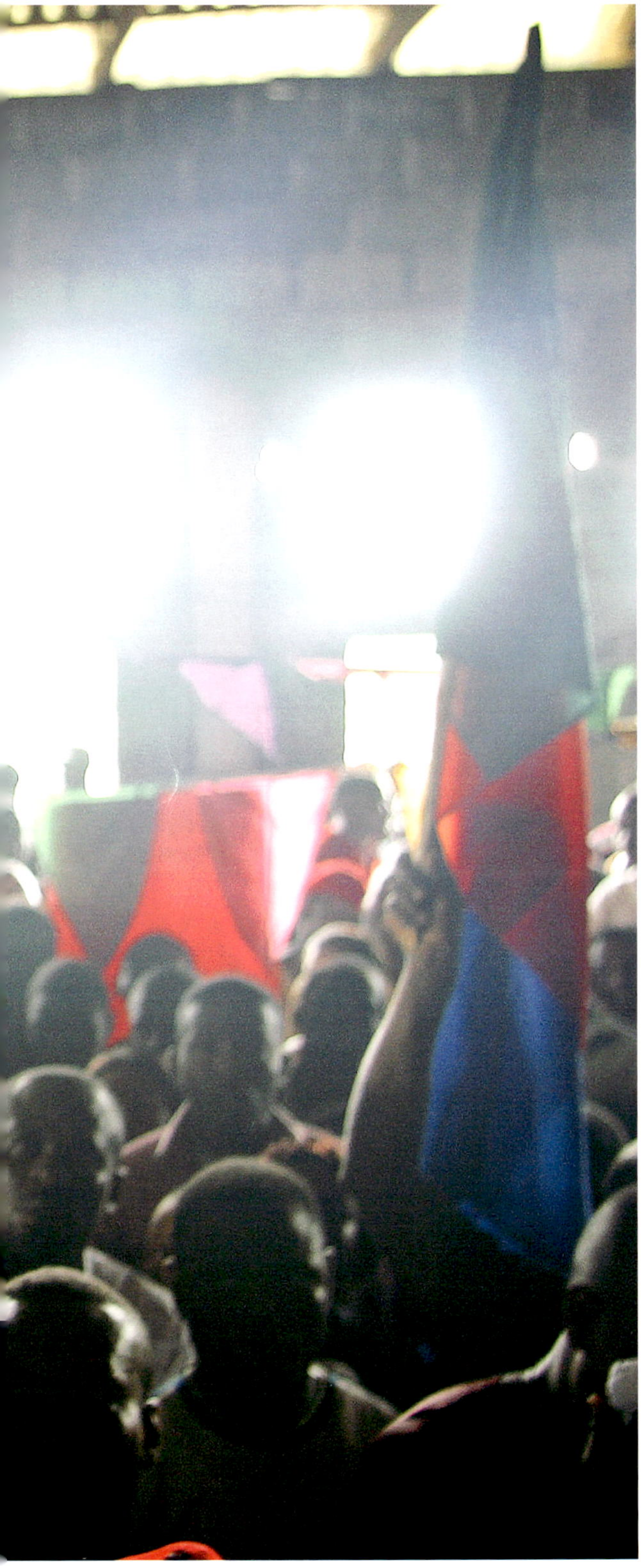

Dokubo Asari, leader of The Niger Delta People's Volunteer Force (NDPVF) speaks in Port Harcourt to supporters. The previous year in 2003, he had opposed the presidential election and formed the NDPVF armed group in retaliation, falling out with his political godfathers and heading to the creeks to begin grooming members of his armed group. His aim was to confront the Nigerian state in the quest for resource control, asking for self determination for the Ijaws. His subsequent threat to blow up oil installations in the Delta pushed up oil prices to 50 dollars per barrel for the first time in history.

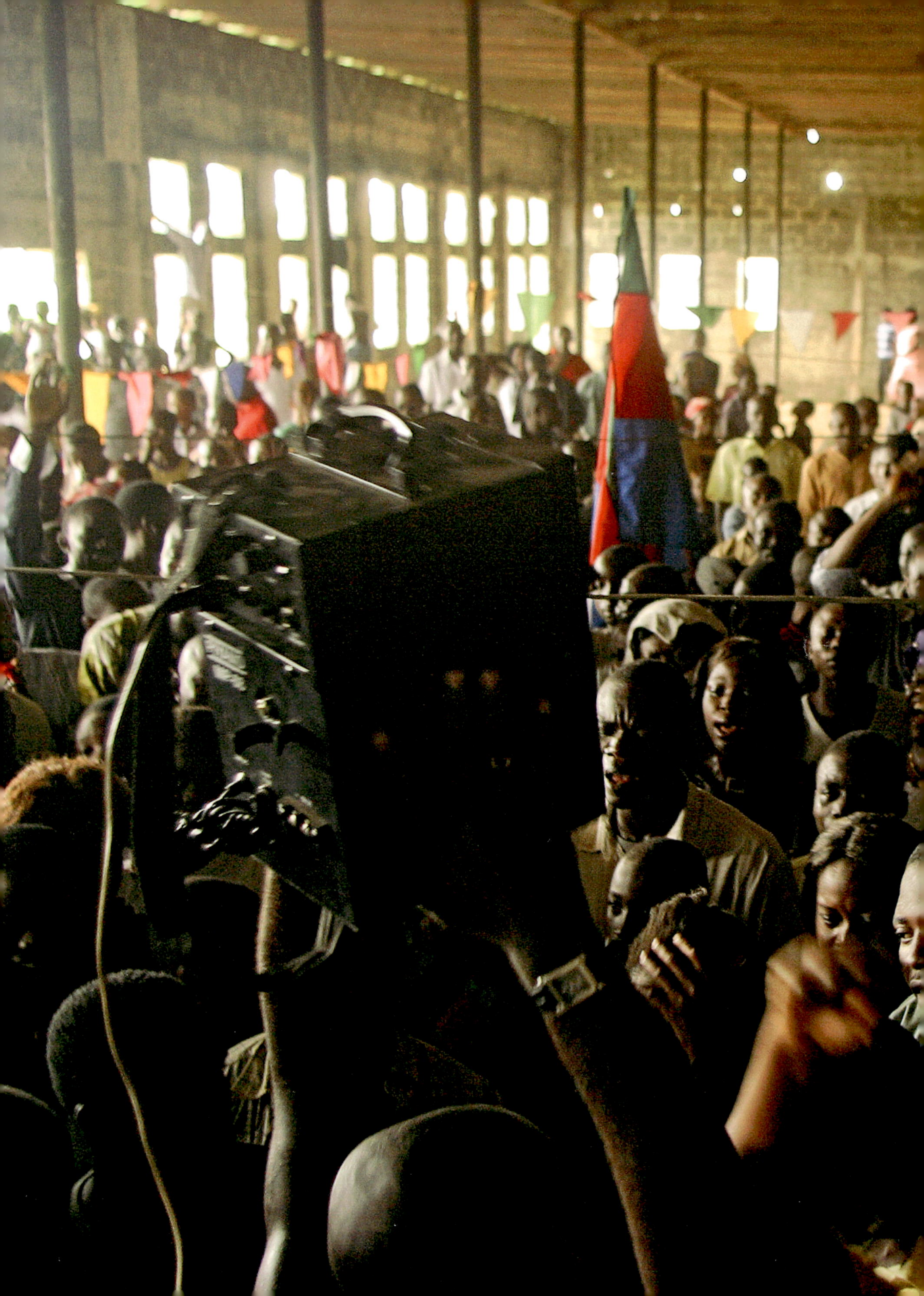

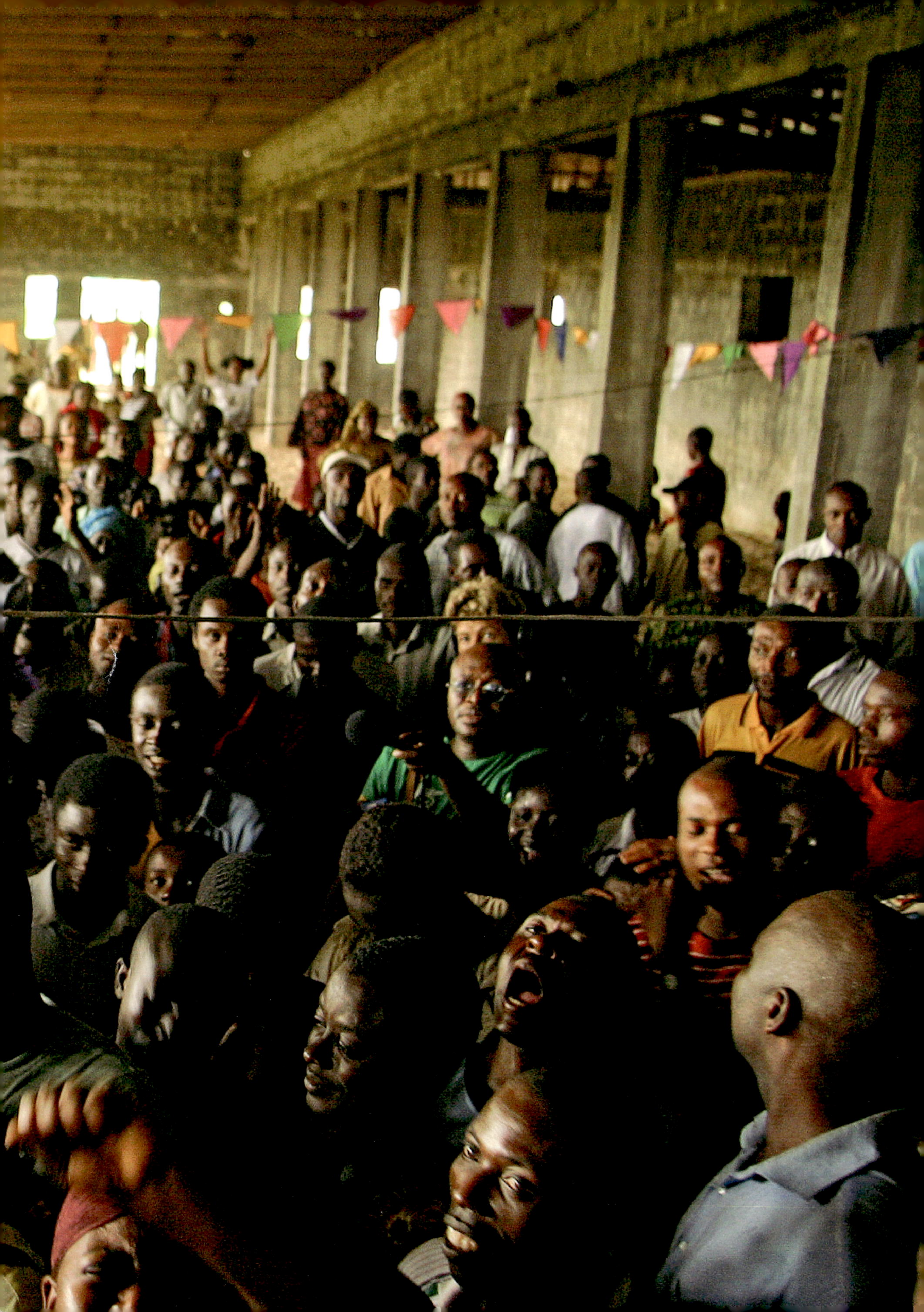

DOKUBO ASARI BECOMES A PUBLIC ENEMY

The then state government were threatened by Asari's creation of the NDPVF. By September 2004, the NDPVF had announced an "all-out war" against the Nigerian state, turning the former collaborators into enemies. In addition to directing the Nigerian security services at the NDPVF, the government allegedly employed another Delta militant, Ateke Tom, to use his private militia as part of the fight. Ateke Tom created the Niger Delta Vigilantes (NDV) - which largely drew men from his cult gang known as the Icelanders - and waged war against Asari's NDPVF. The two groups fought over territory and access to lucrative energy bunkering routes. By late 2004, instability in the Delta was rising rapidly, and, at the end of September, Asari's NDPVF issued an ultimatum to multinational energy companies to shut down their operations by October 1, 2004, otherwise their employees would become targets in a major guerrilla operation called Operation Locust Feast. The announcement sparked panic among energy companies and caused the price of oil to shoot to a new record of $50 per barrel. As a result, the government of Olusegun Obasanjo was forced to recognize Asari's influence in the Delta by inviting him and his rival, Ateke Tom, to Abuja in order to reach an agreement

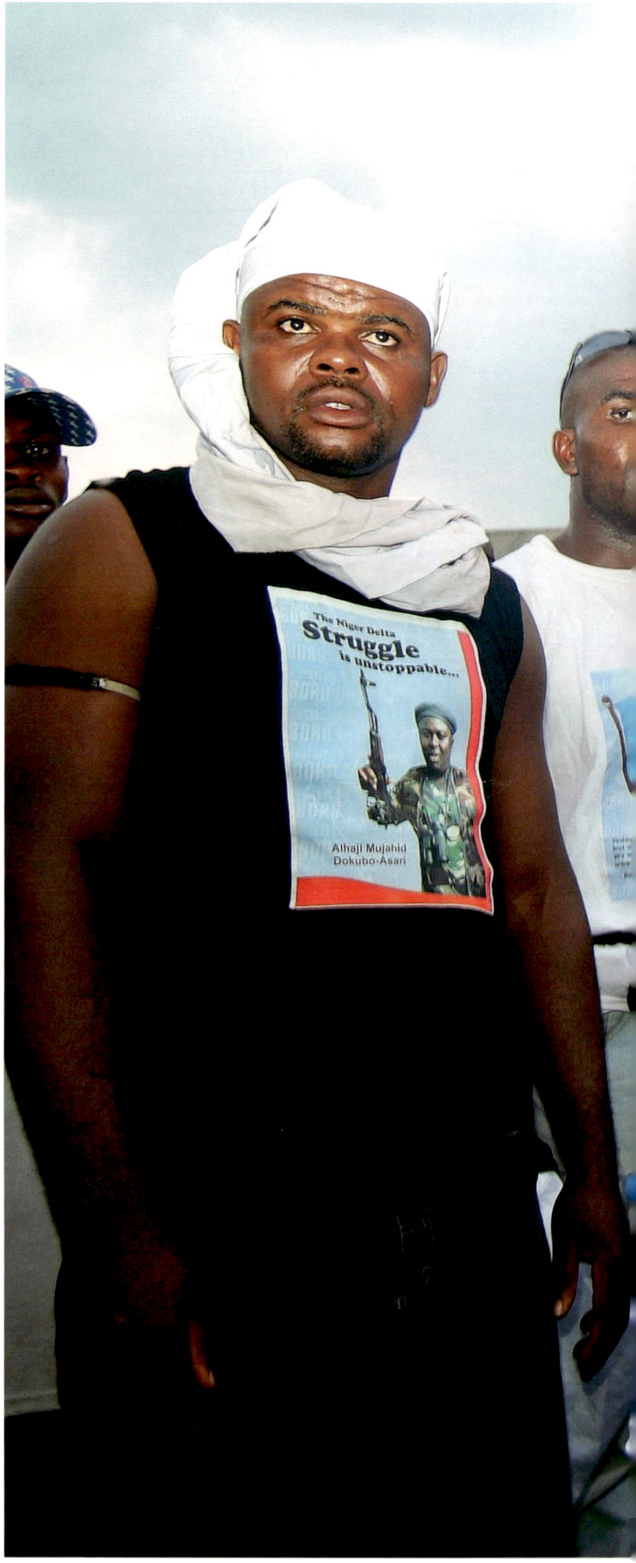

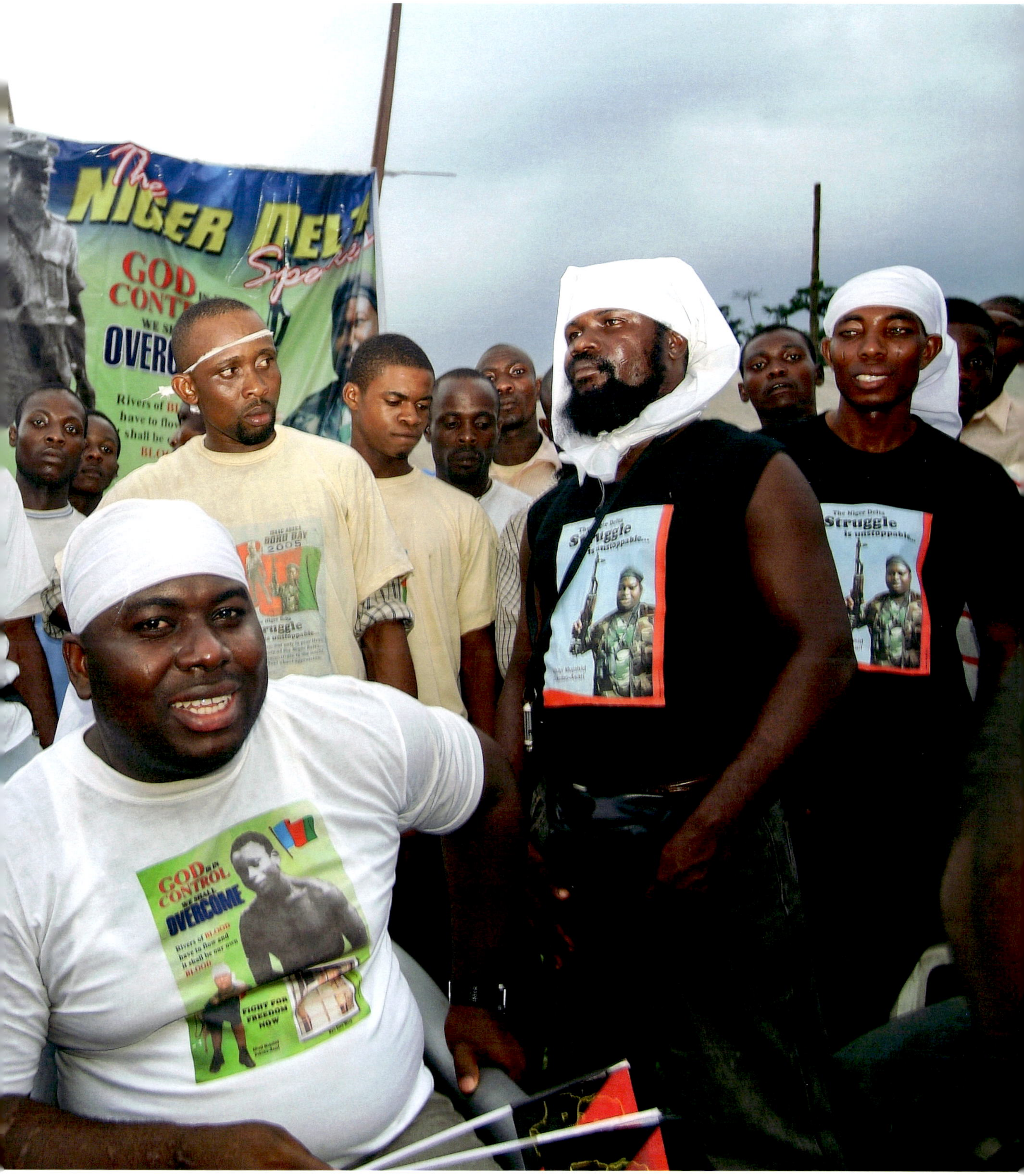
GOD is in CONTROL
WE SHALL
OVERCOME
Rivers of BLOOD have to flow and it shall be our own BLOOD
FIGHT FOR FREEDOM NOW
The Niger Delta
Struggle
is unstoppable...

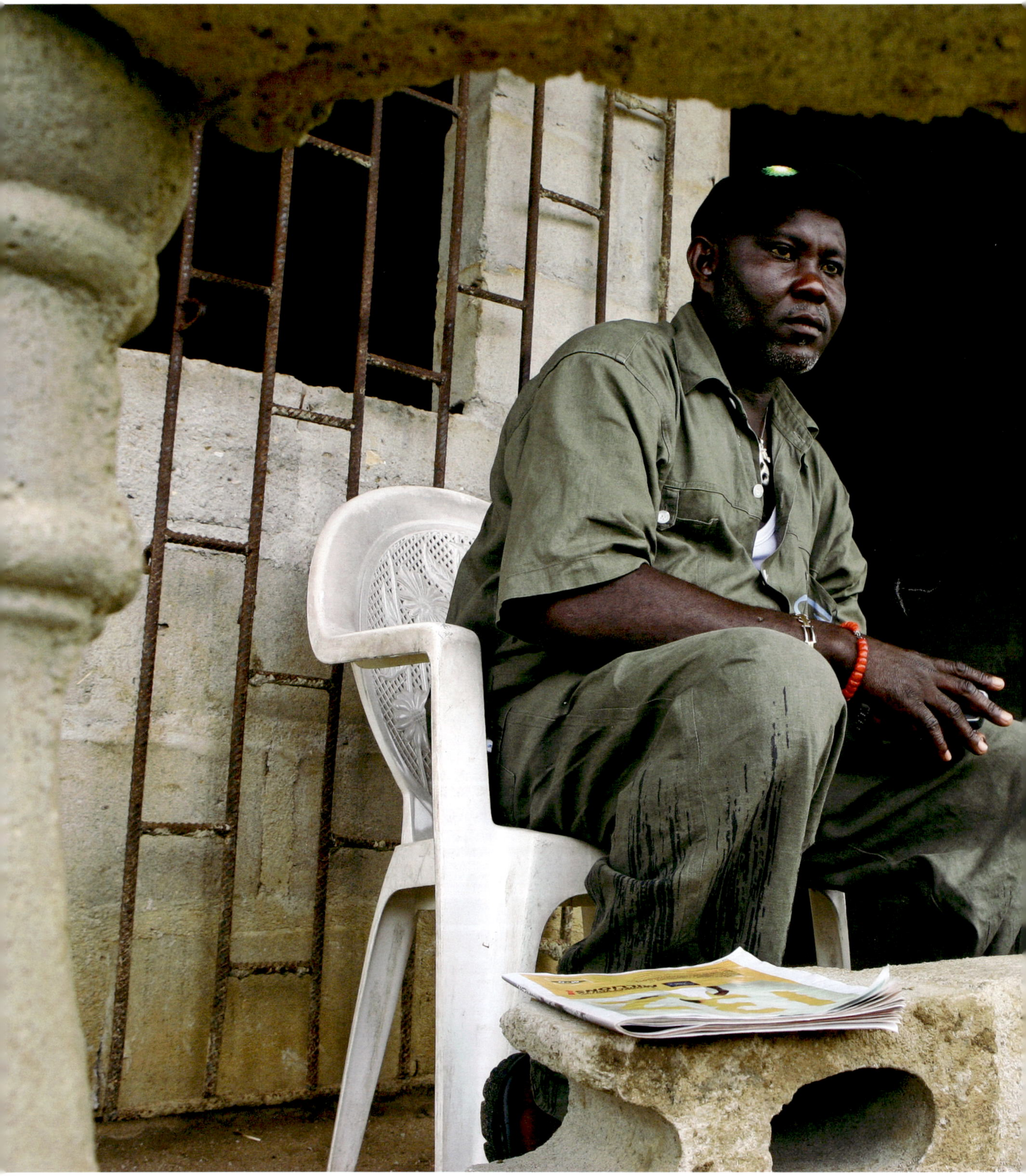

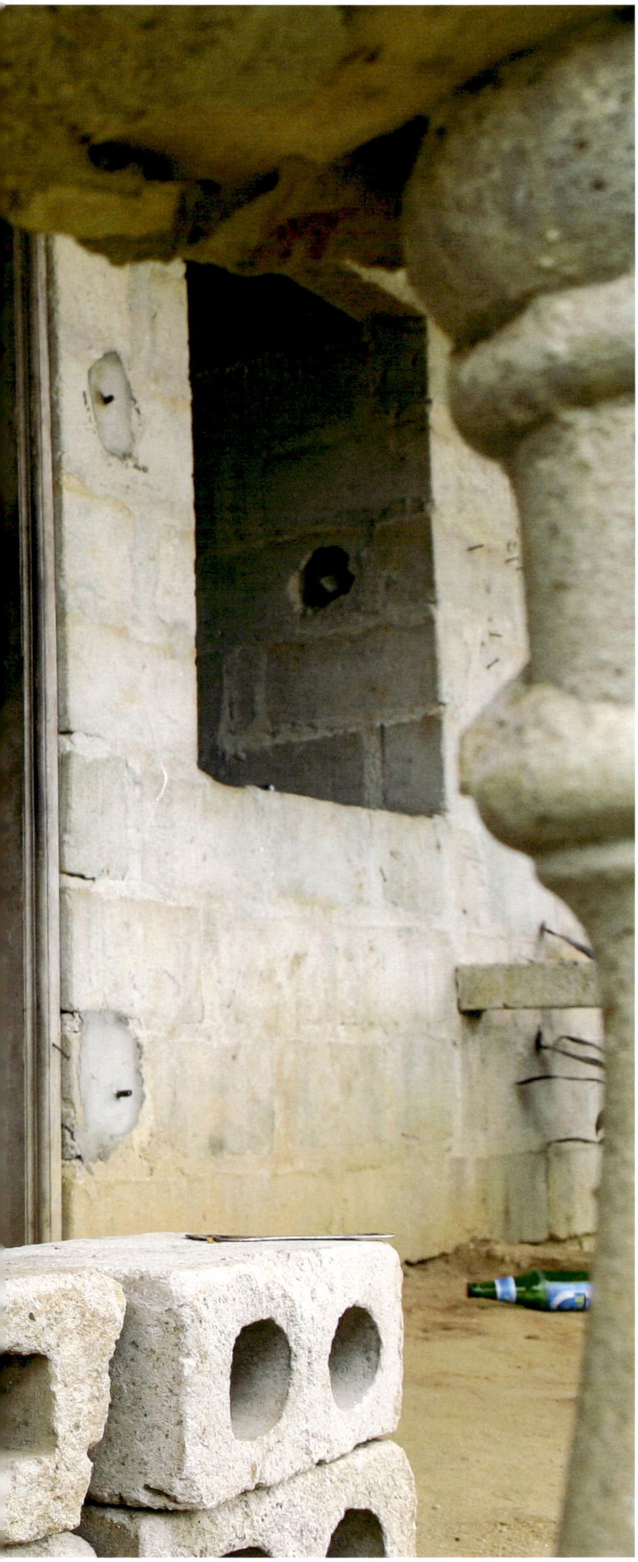

on the cessation of hostilities. A deal was reached on October 1 that was supposed to lead to the disbanding of their respective militia groups and their total disarmament as well as keeping them within the democratic process. Although some of the deal was implemented, it soon fell apart. On September 20, 2005, Nigerian authorities arrested Asari on charges of treason and imprisoned him at the federal security services headquarters in Abuja. His arrest came after he issued persistent separatist rhetoric. Throughout his detention, Asari was rebellious, especially in public venues, such as during his various court hearings. In early 2007, for example, Asari threatened to kill Federal High Court Judge Peter Olayiwola. Furthermore, in the wake of his detention, violence in the Delta actually escalated. So it happened that the NDPVF fractured, and former fighters now conducted attacks in the name of the Movement for the Emancipation of the Niger Delta (MEND) and several other armed groups. There followed this a rise in criminality, and the political ideology that previously characterised the struggle, lost its prominence.

Ateke Tom, leader of the Niger Delta Vigilante (NDV) an armed militia group, sits outside one of his uncompleted buildings in Okrika, 2006.

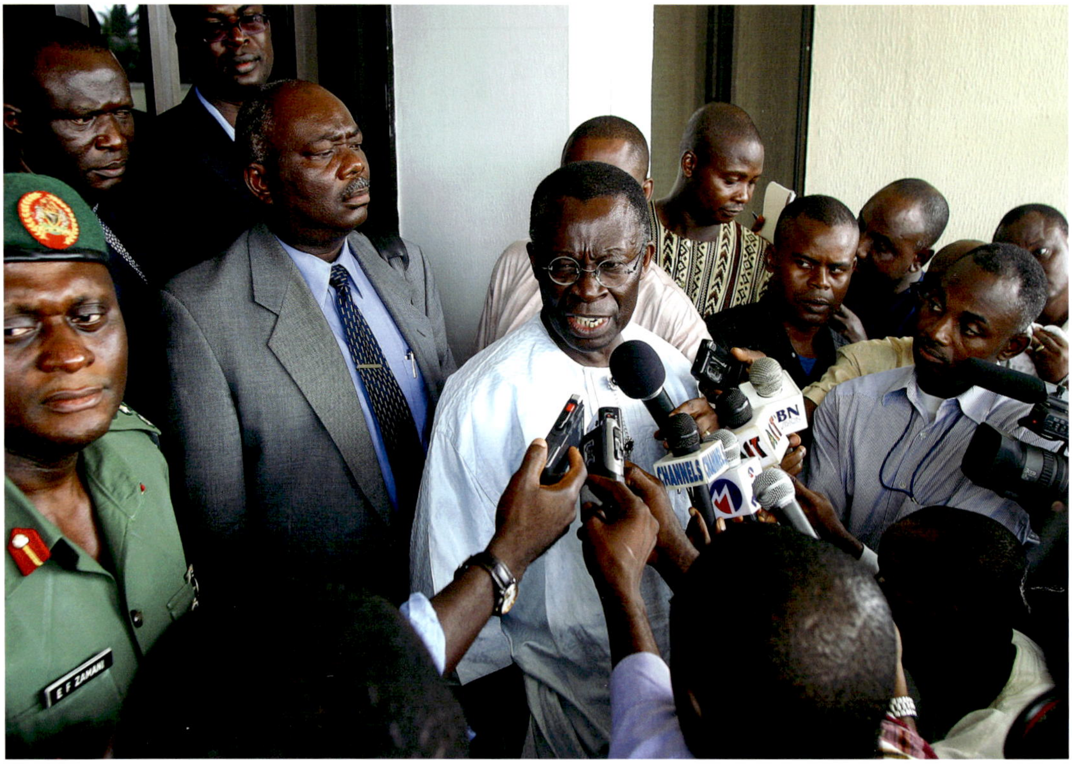

Above: Dr. Edmund Daukoru, Chairman of Nigerian National Petroleum Corporation (NNPC) (centre) speaks to journalists during a visit to Warri, Delta State, 2005.

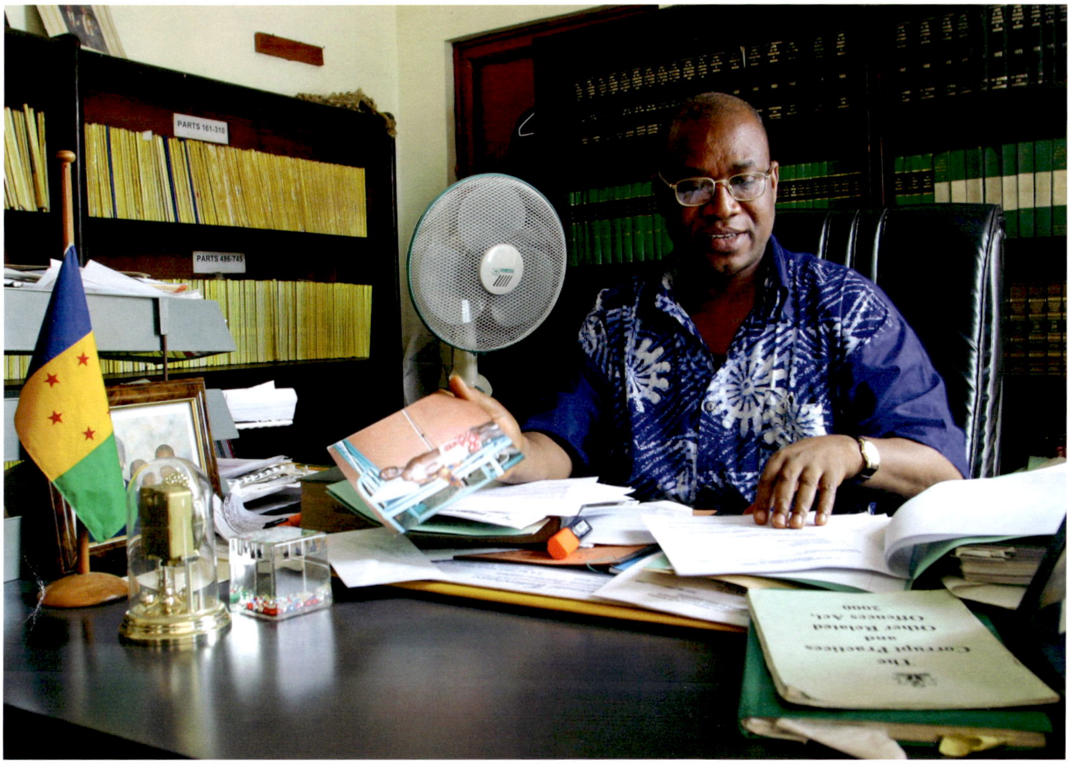

Above: Ledum Mitte, President of the Movement for the Survival of Ogoni People (MOSOP) holds a picture of an injured youth from conflict while he looks through documents in his office, Port Harcourt, 2005.

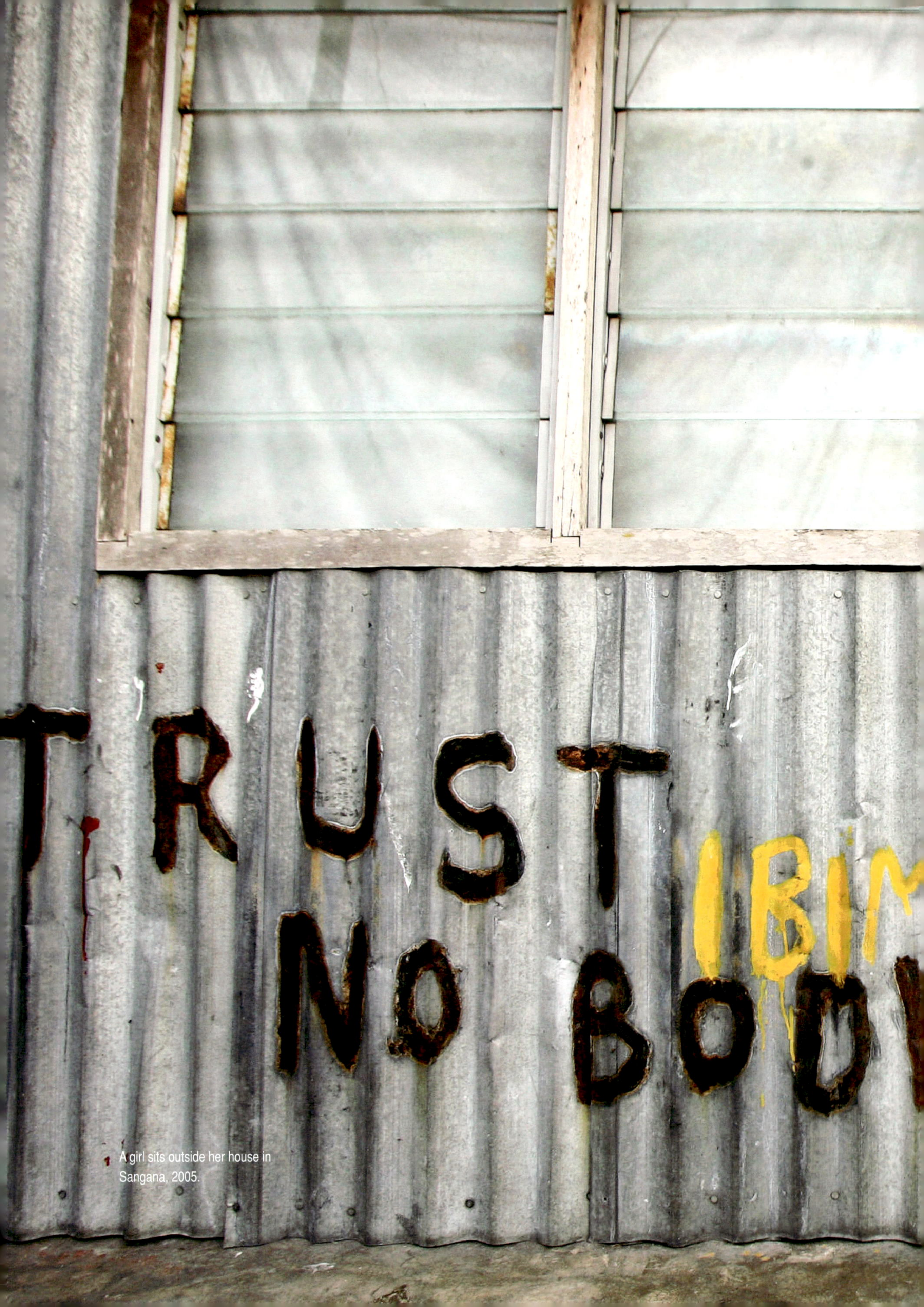

A girl sits outside her house in Sangana, 2005.

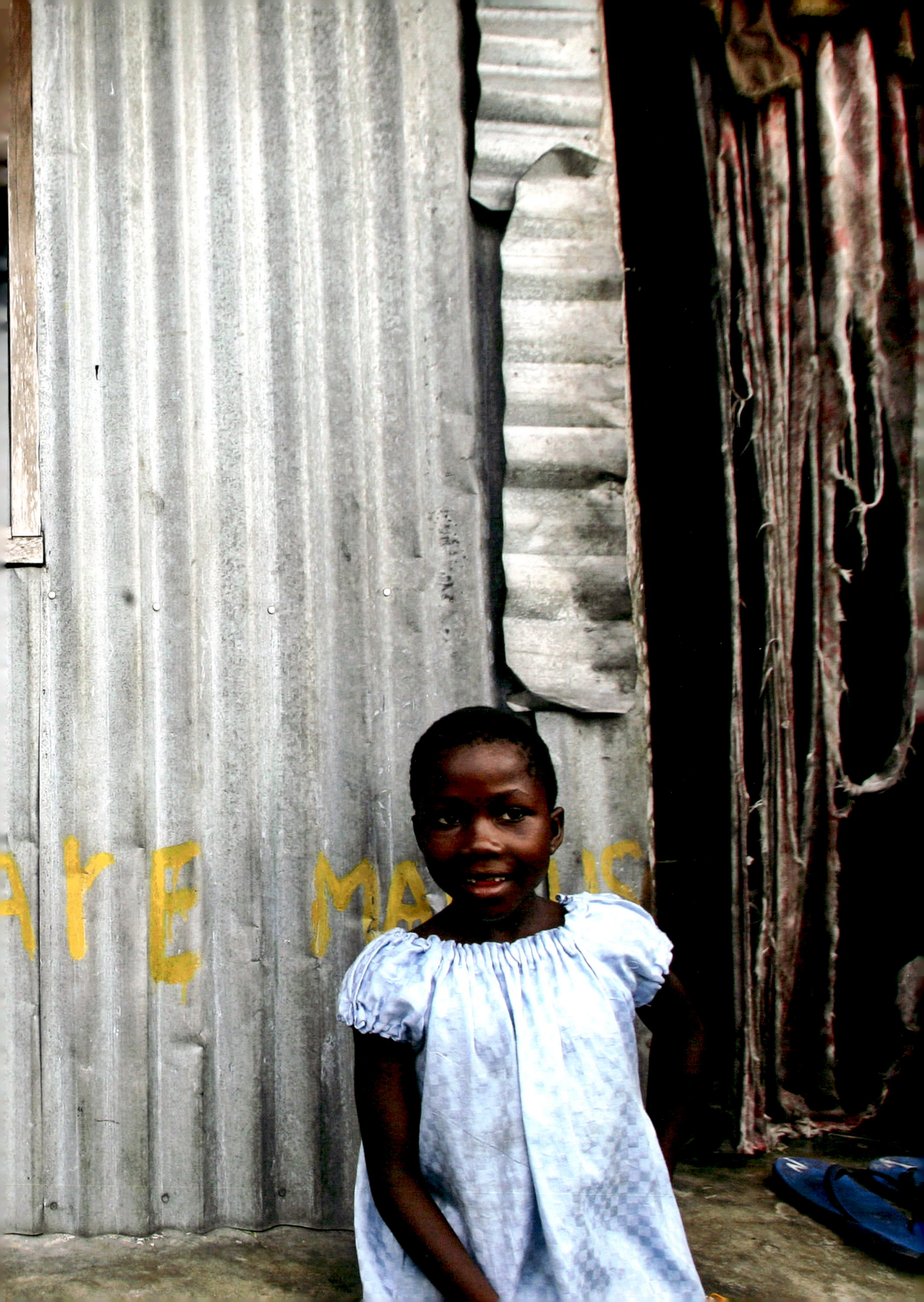

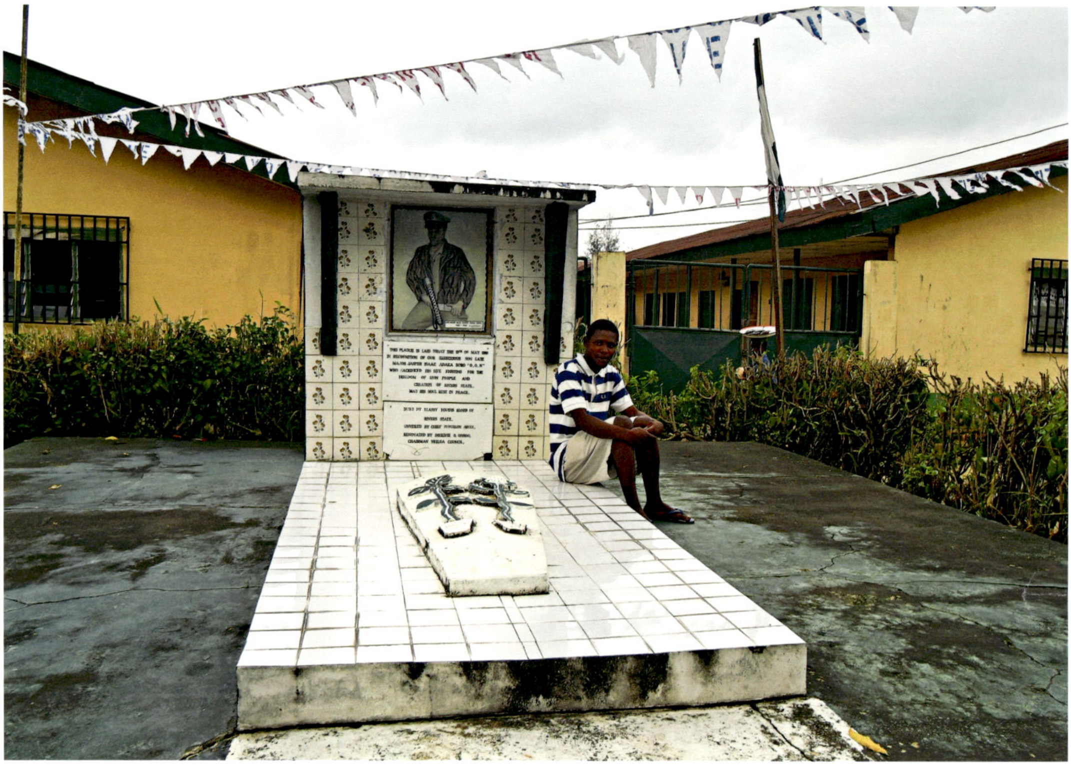

The late Major Isaac Adaka Boro, struggled for the socio-economic liberation of the Ijaws in the Niger Delta. A young Ijaw man, he left school in the early 1960's to lead a protest group against the exploration of oil in the region, arguing that the oil belonged to the people of the Niger Delta as a natural heritage and it should be for their exclusive benefit. He armed his group and declared the secession of the Niger Delta. He is considered the first ever rebel from the Niger Delta, and many major militant groups in the Delta look up to him as their HERO.

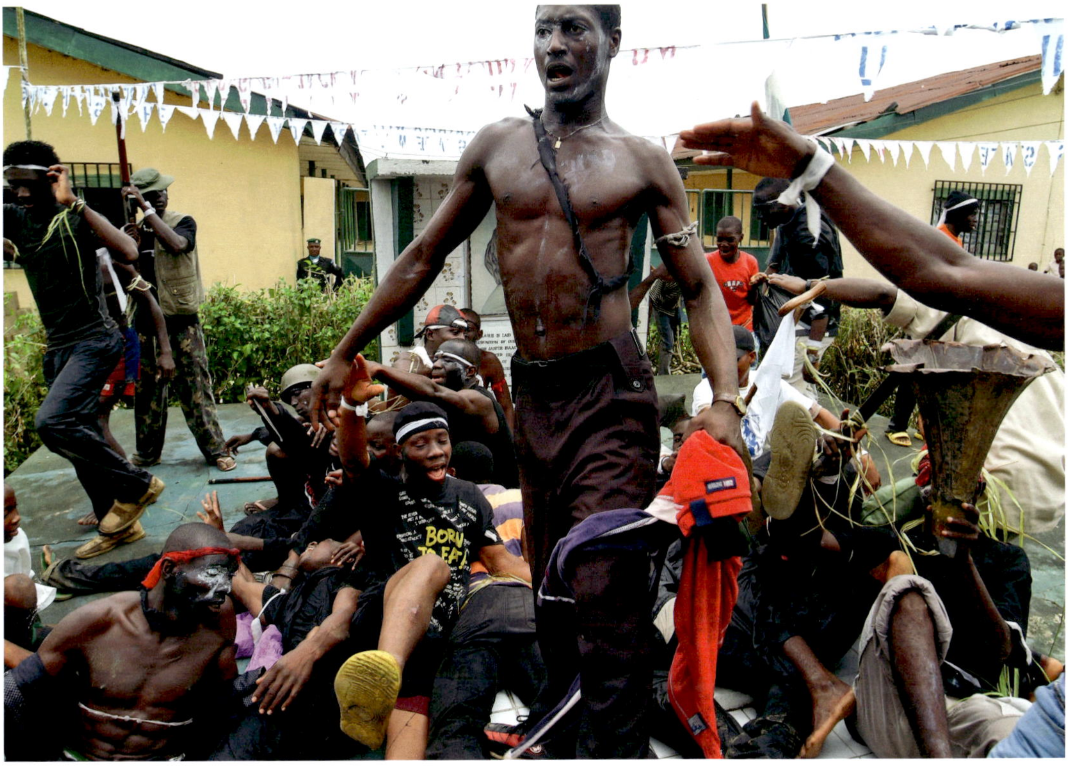

Above and left: The grave of Isaac Adaka Boro, Kaiama, Bayelsa state. Ijaw youths celebrate Isaac Adaka Boro's day, 2005.

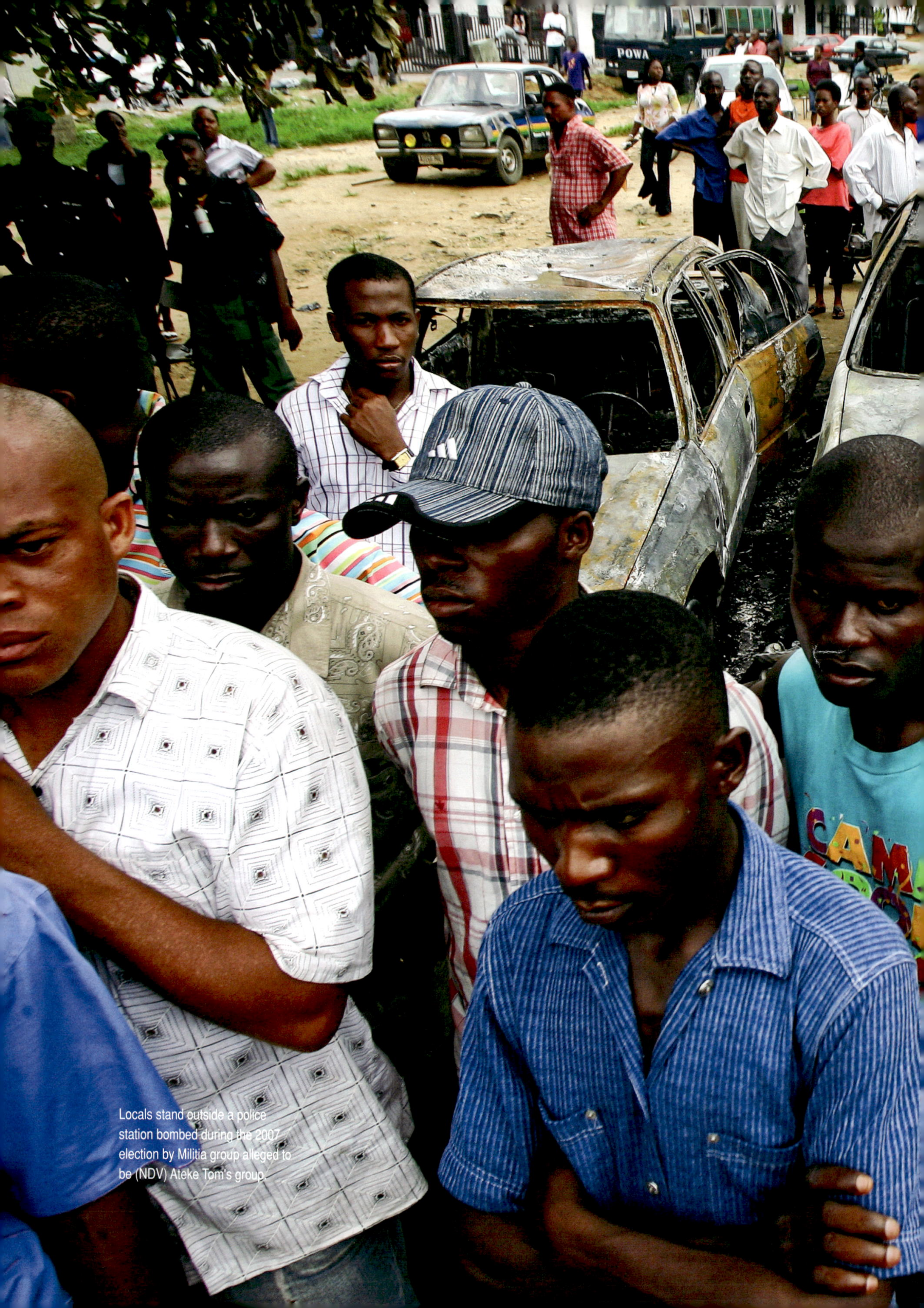

Locals stand outside a police station bombed during the 2007 election by Militia group alleged to be (NDV) Ateke Tom's group.

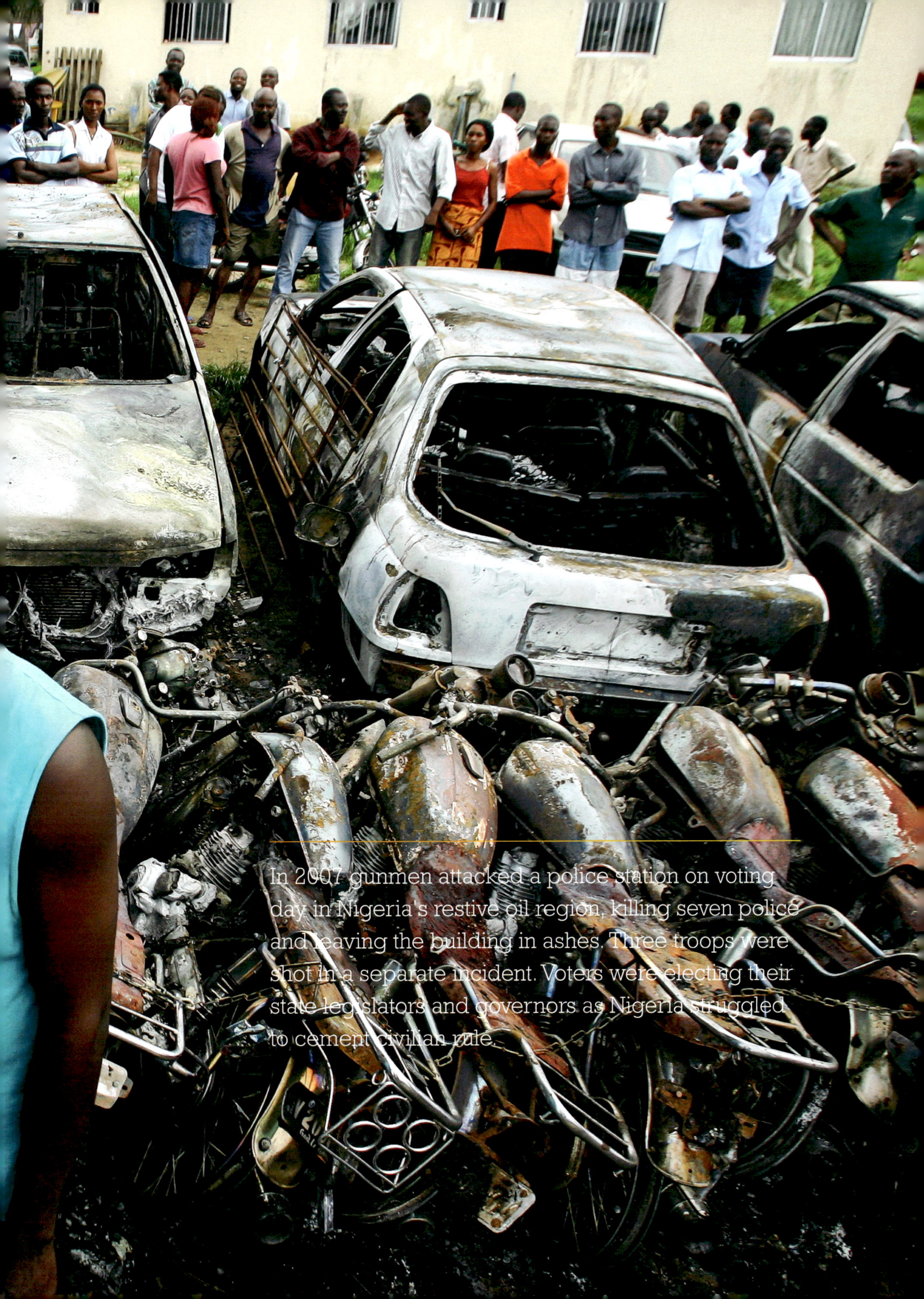

In 2007 gunmen attacked a police station on voting day in Nigeria's restive oil region, killing seven police and leaving the building in ashes. Three troops were shot in a separate incident. Voters were electing their state legislators and governors as Nigeria struggled to cement civilian rule.

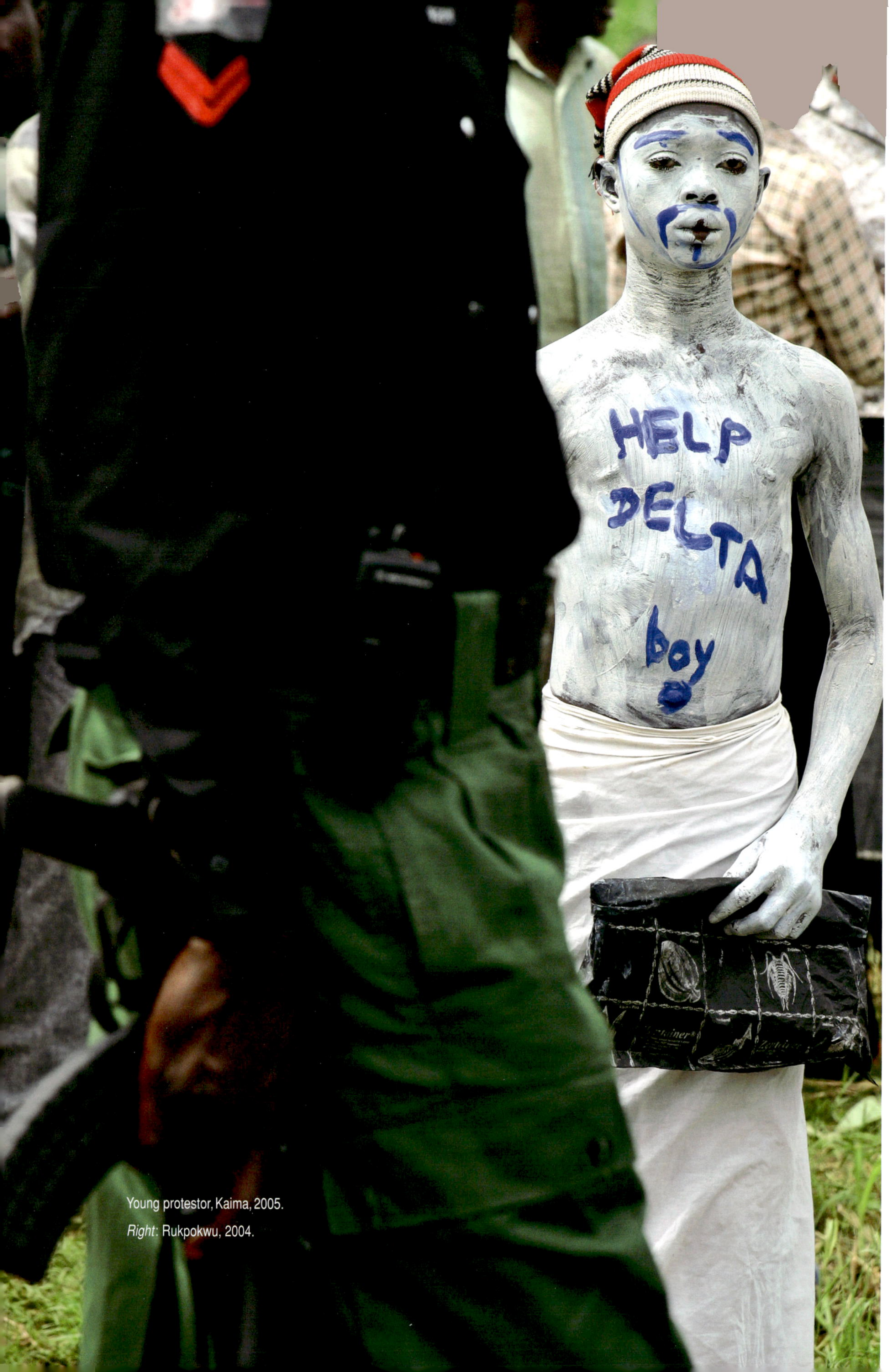

Young protestor, Kaima, 2005.

Right: Rukpokwu, 2004.

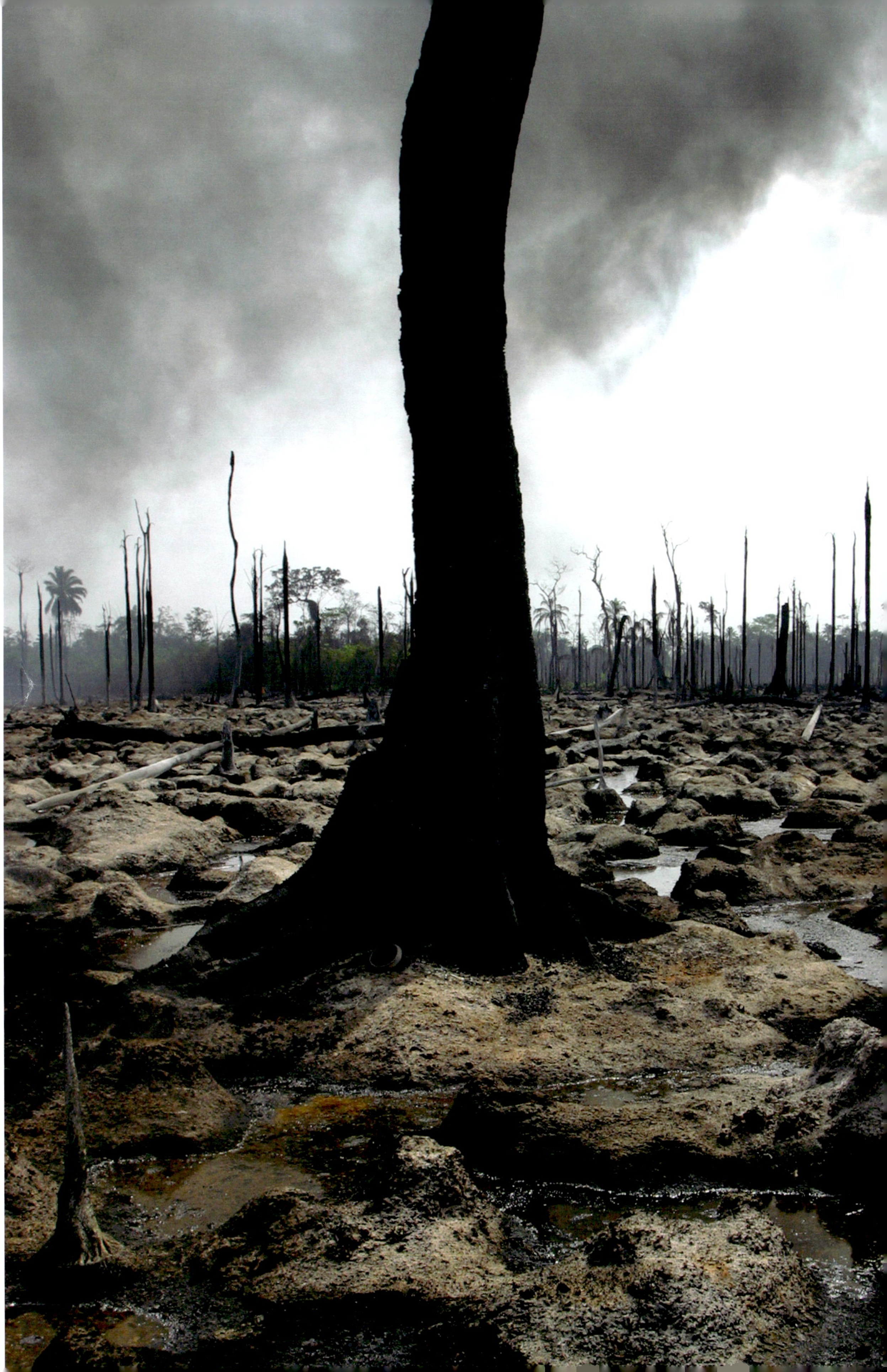

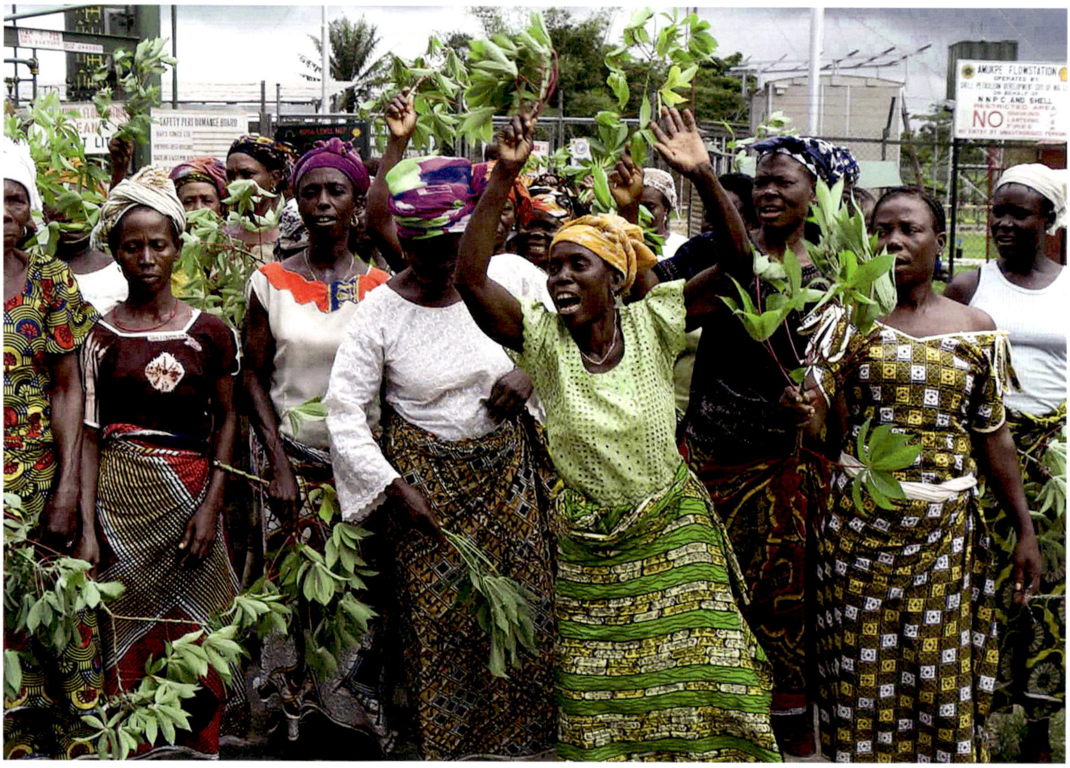

Nigerian women protest at a Shell oil installation they have occupied in the town of Amukpe, 2003. The women's protest forced the company to shut the pumping station,which normally accounts for the production of 40,000 barrels of crude a day. The female occupiers, village women aged between 25 to 60, were demanding the company's Nigeria subsidiary keep its promises of jobs and other benefits for villages in the swampy forested Niger Delta.

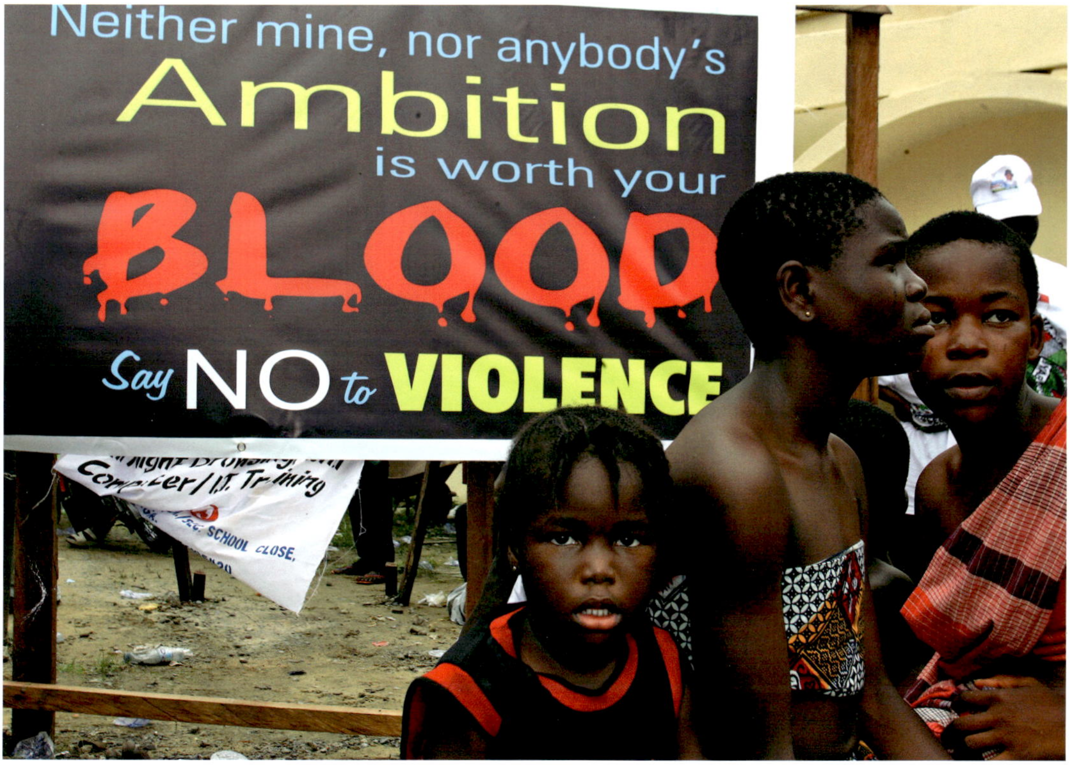

NAKED OPTION

The Ugborodo protest by 600 women for 10 days outside the Escravos tank farm of the Chevron plant, 2002, saw them threaten to go naked.

"WE GO NICKED FOR UNA
WE GO DEY ON NICKED
BECAUSE YOU PEOPLE WANT WE TO SUFFER
AND WATIN UNA SEE FOR INSIDE MAKE UNA TAKE AM LIKE THAT"

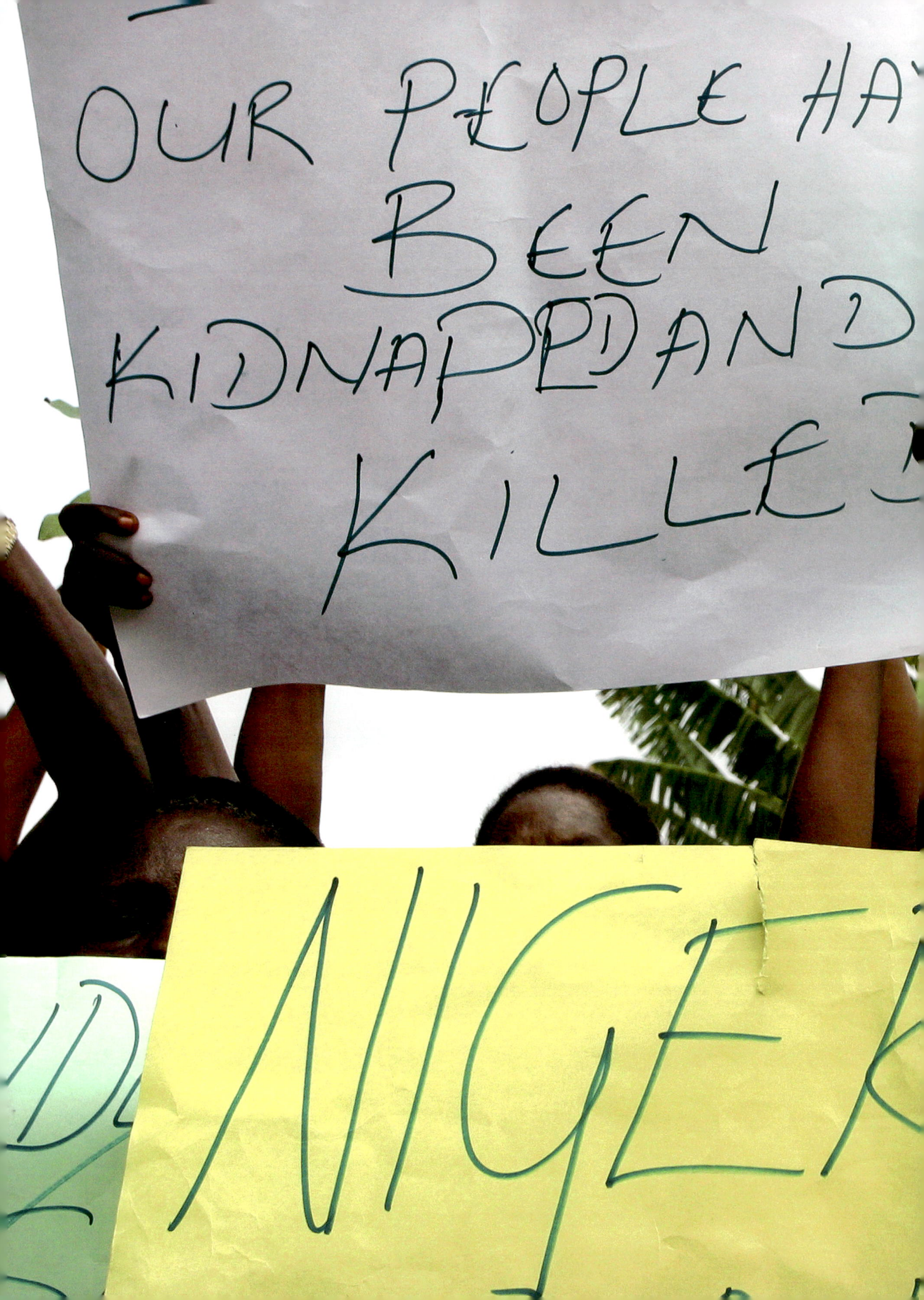
OUR PEOPLE HA
BEEN
KIDNAPED AND
KILLE
NIGE

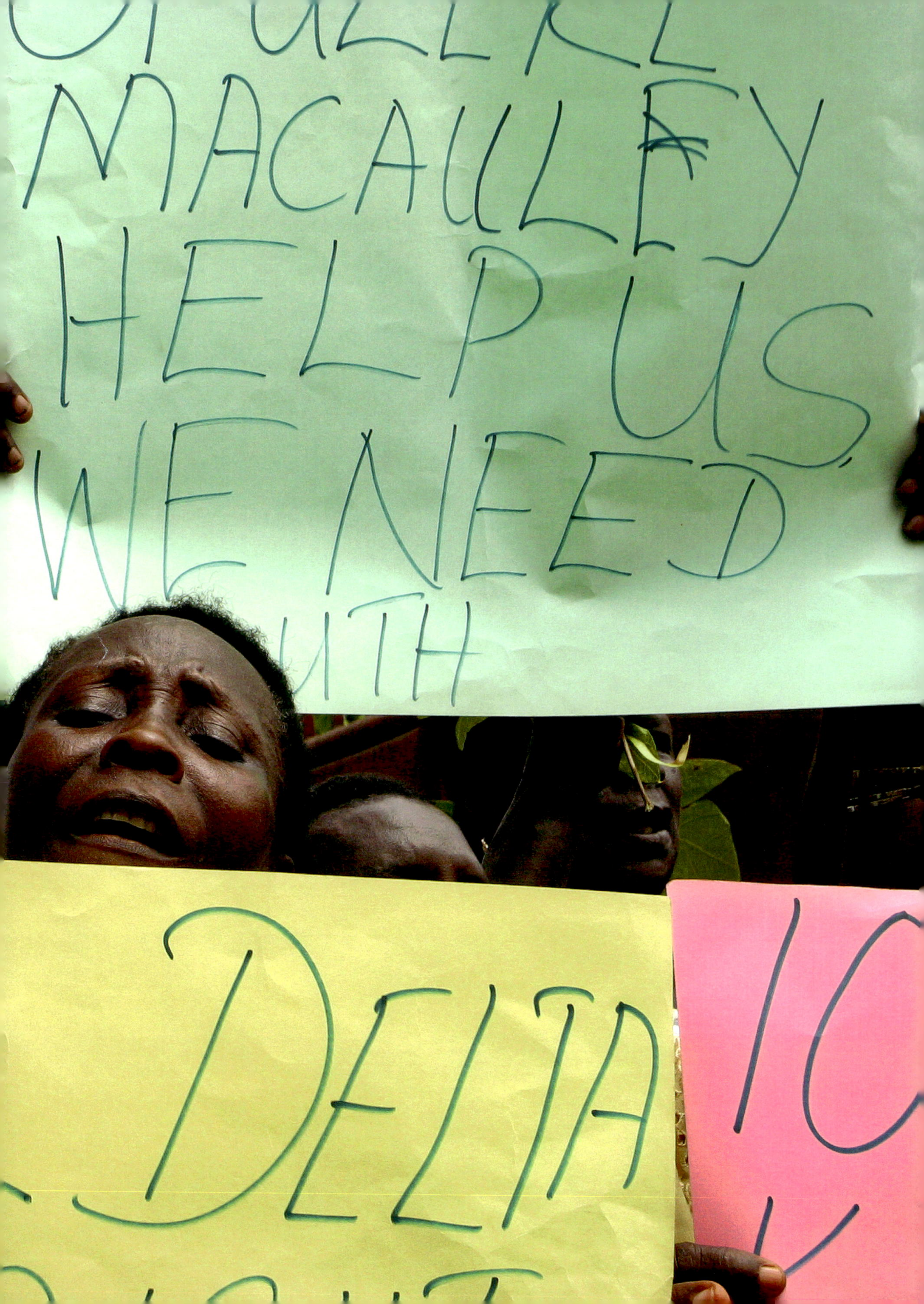
MACAULEY
HELP US
WE NEED
DELTA

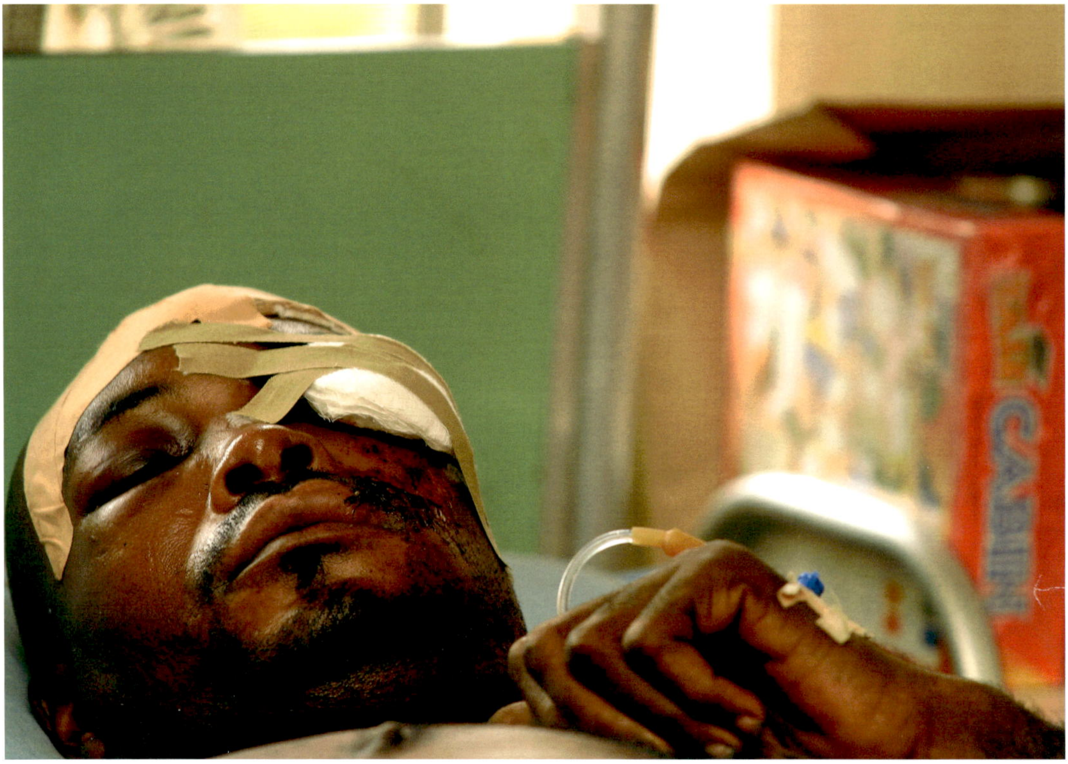

Above: A man receives treatment at the general hospital in Warri after he was injured during a conflict between protesting youths and the Army following the protest against Chevron oil in Escavros, which demanded that the company redeem his pledge on the MOU agreed, entailing job and contract creation for the people of Ugborodo.

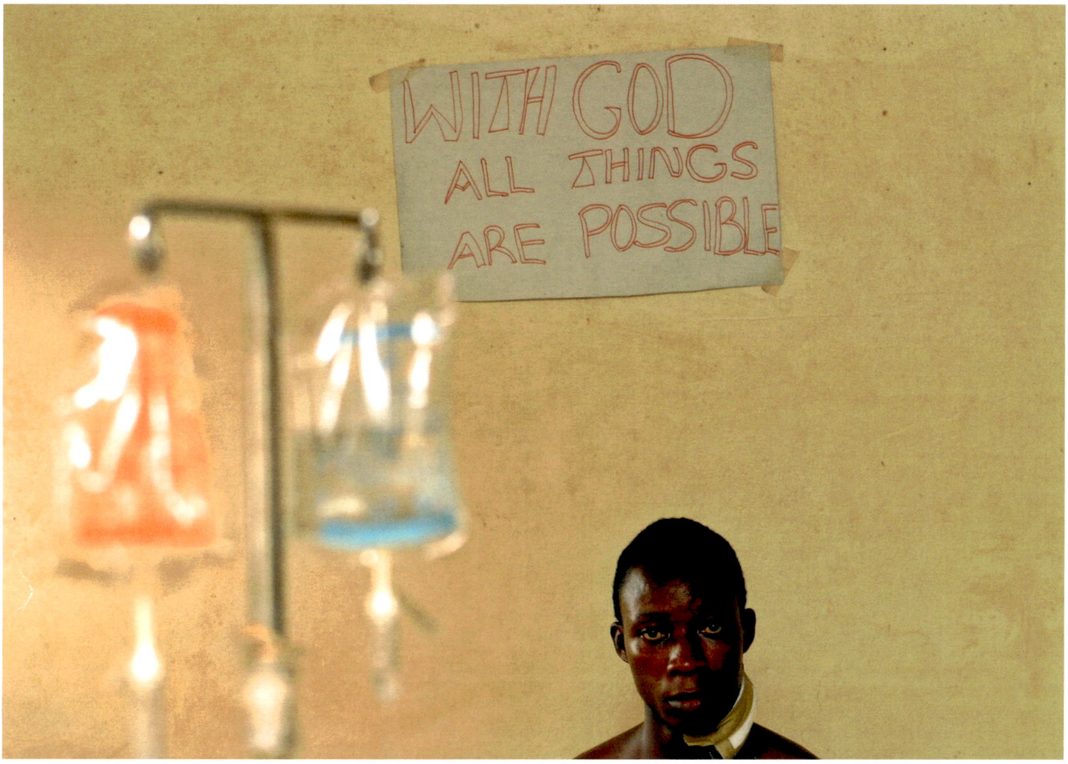

Above: A youth receives treatment at the general hospital in Warri after he sustained a gun shot wound during the Ugborodo protest, 2003.

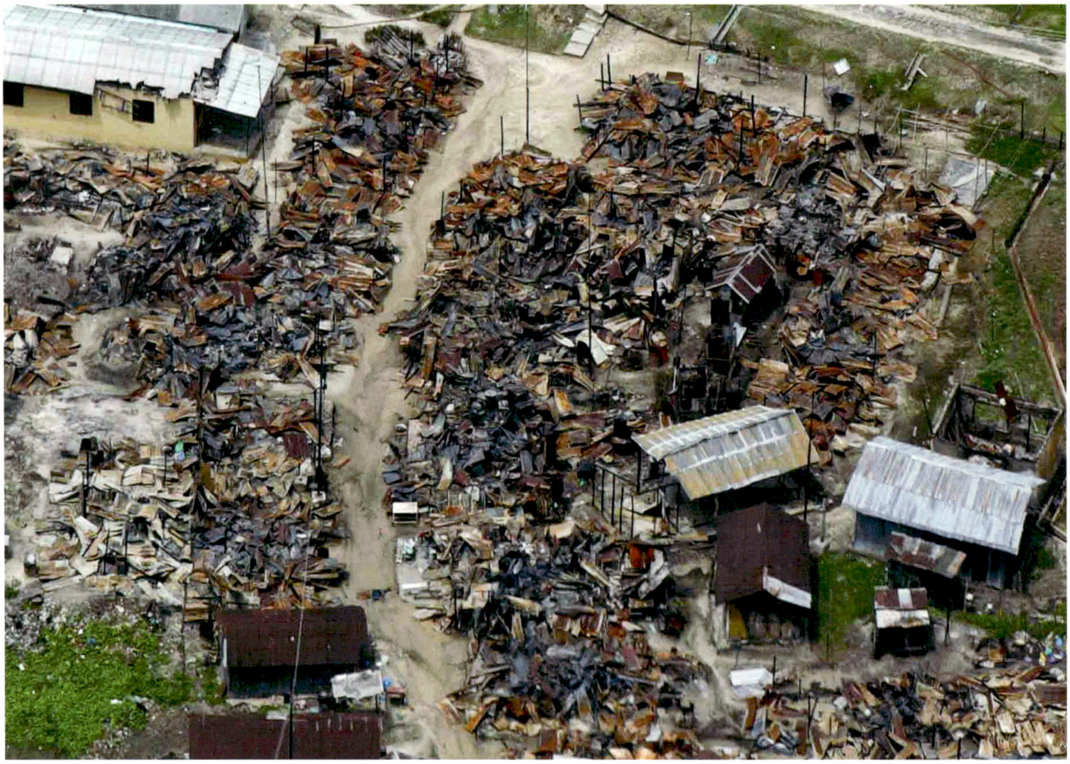

At the heart of the violence, are claims and counterclaims to the ownership of oil-rich land. The individuals and communities who control the land eat up many of the benefits that can be extracted from the oil companies whose wells have been drilled there. These include rent and compensation payments, access to jobs, a quota of which are reserved for local people, and the provision of social amenities such as clean drinking water, schools and health centres. Such manna from heaven is in an environment of widespread poverty and mass unemployment - as a result it is considered well worth fighting over.

Rival local communities frequently fight each other over their claims to the oil-rich land and sometimes they disrupt the operations of the oil companies themselves to back demands for greater benefits.

"Many of these people have been neglected by successive governments, and since the transnationals are in partnership with the government they hold them responsible as the only arm of

Above: On the 6th of May 1999, Madangho an Itsekiri community was razed down allegedly by the Ijaws, during ethnic clashes that left many dead and homeless.

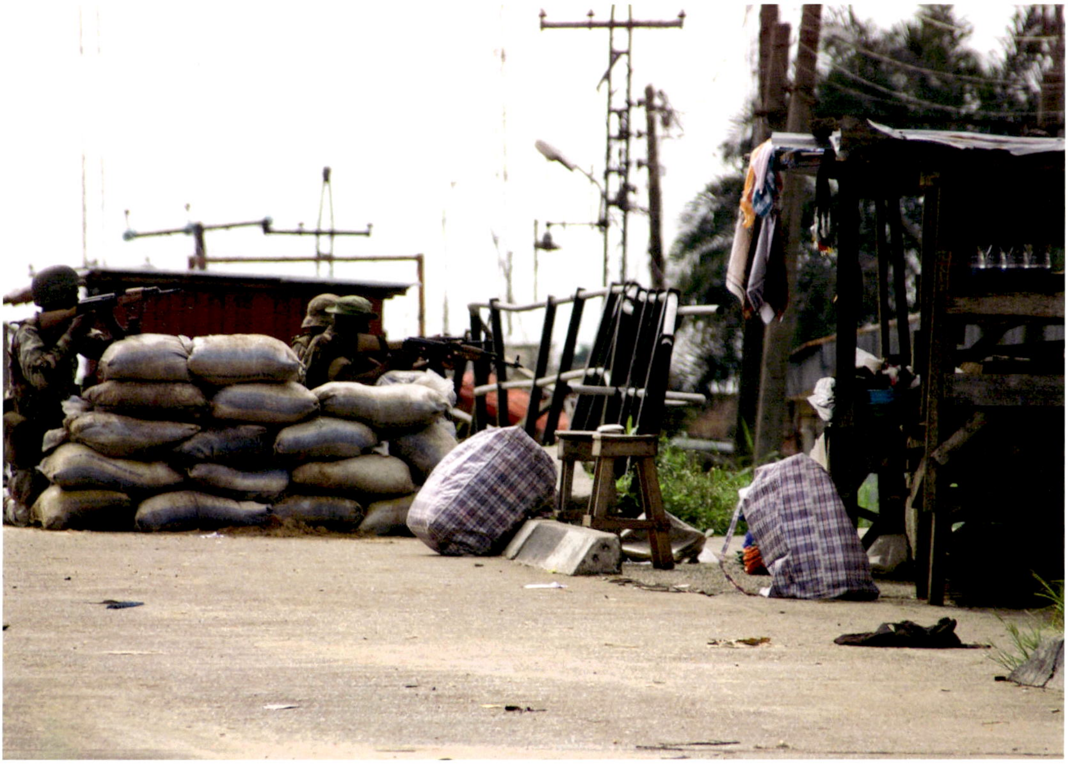

government they see. Tension has floated close to the surface in Warri for the past decade, but disputes over the demarcation of electoral boundaries fueled a new outbreak of fighting in the run-up to general elections in April, 2003. The Ijaw, the largest tribe in the Niger Delta, numbering an estimated five to eight million people, accused the smaller but influential Itsekiri tribe of rigging the boundaries of electoral wards, in connivance with the government of President Olusegun Obasanjo, to win undue political advantage. Fighting first erupted between the Urhobo, a small Delta tribe, and the Itsekiri. This resulted in the deaths of more than two dozen people. Over 5,000 more were made homeless after their houses were burned in the ethnic clashes. After the Ijaws joined forces with the Urhobo against the Itsekiri during March and April, the death toll topped 100, more villages were set alight and thousands more people were forced to abandon their homes.

Above: Military forces engage Ijaw militants, unseen in gun battle near the shores of the NPA sea port in Warri, Niger-Delta, 2003.

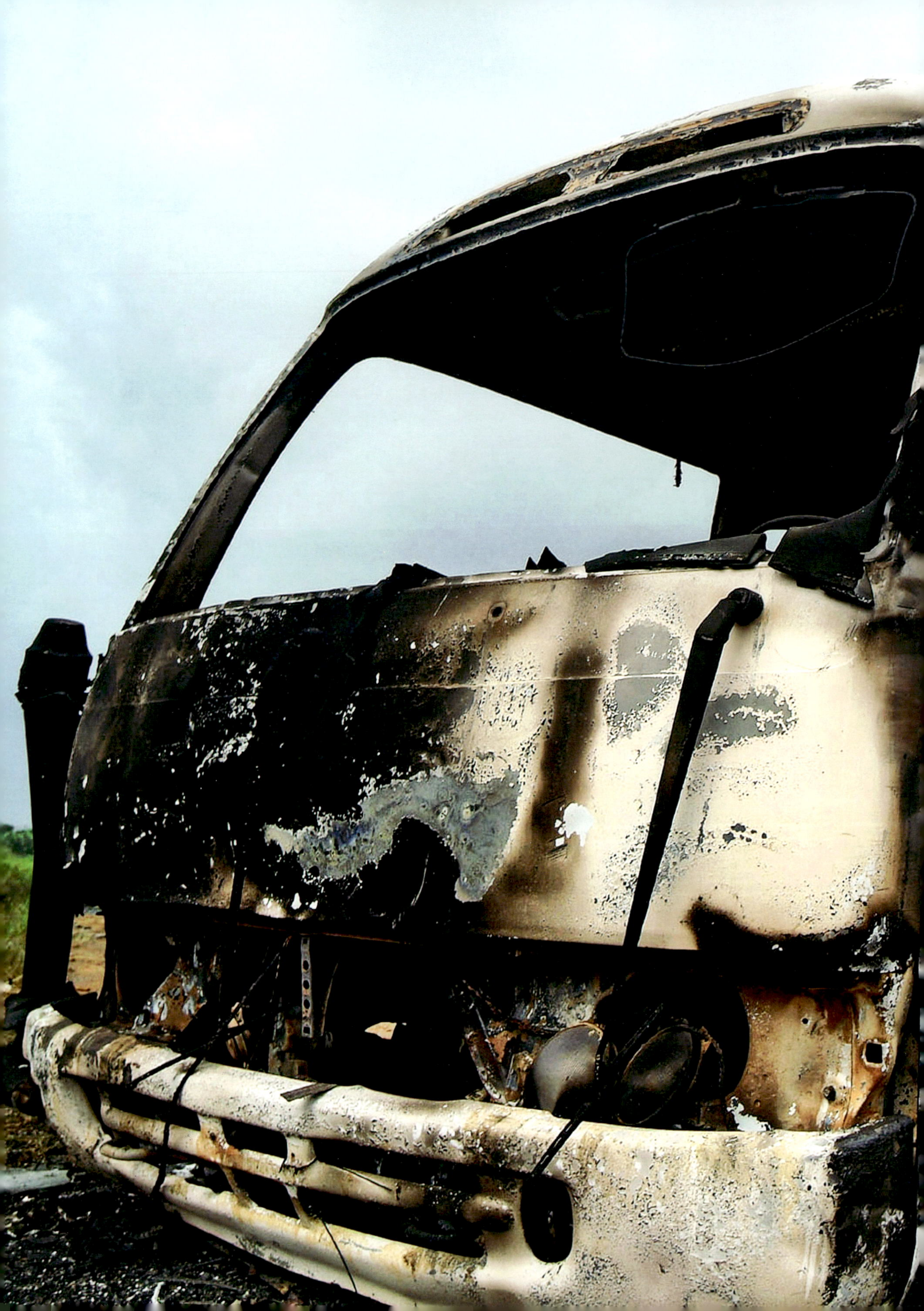

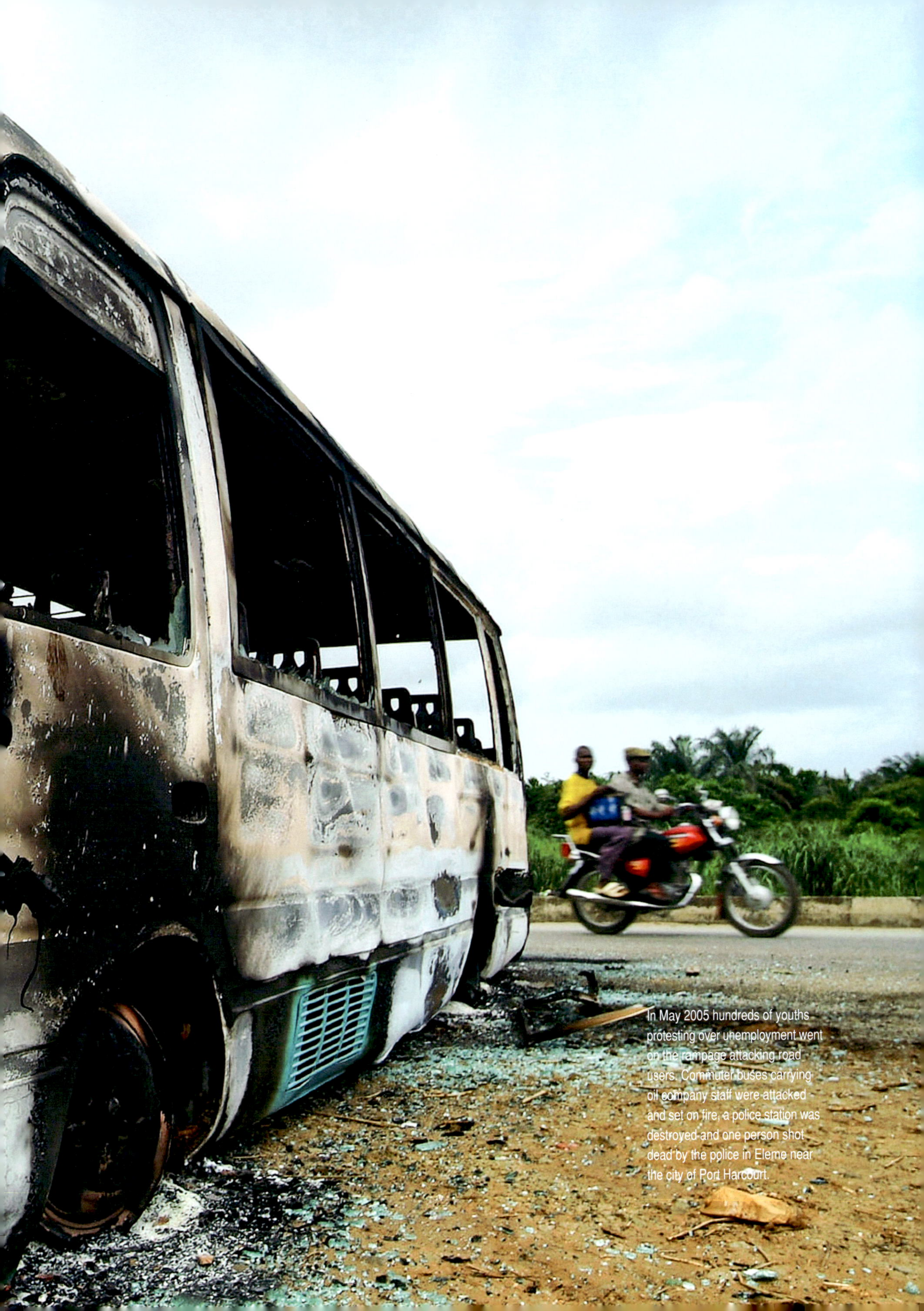

In May 2005 hundreds of youths protesting over unemployment went on the rampage attacking road users. Commuter buses carrying oil company staff were attacked and set on fire, a police station was destroyed and one person shot dead by the police in Eleme near the city of Port Harcourt.

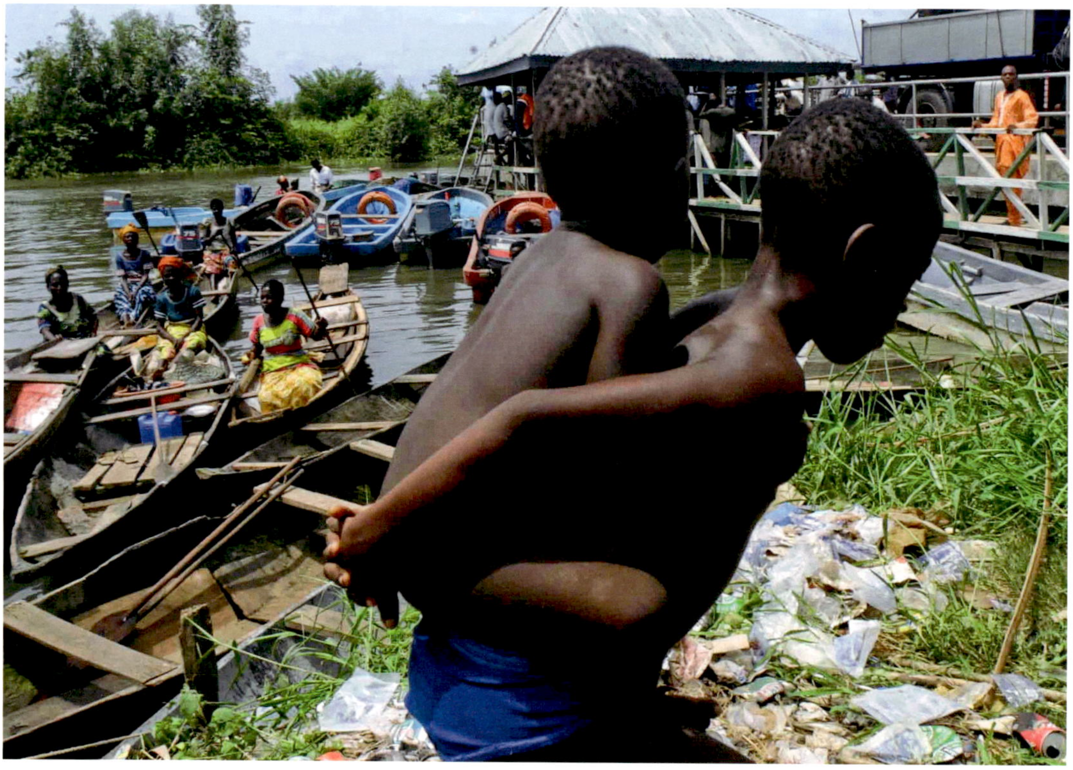

Displaced people arrive in Ogbe-Ijoh village from the conflict zone following renewed clashes between the Ijaw youths, security forces and the Itsekiri youths in the riverine area of the Delta, 2003. Scores of people lost their lives in the ethnic clashes.

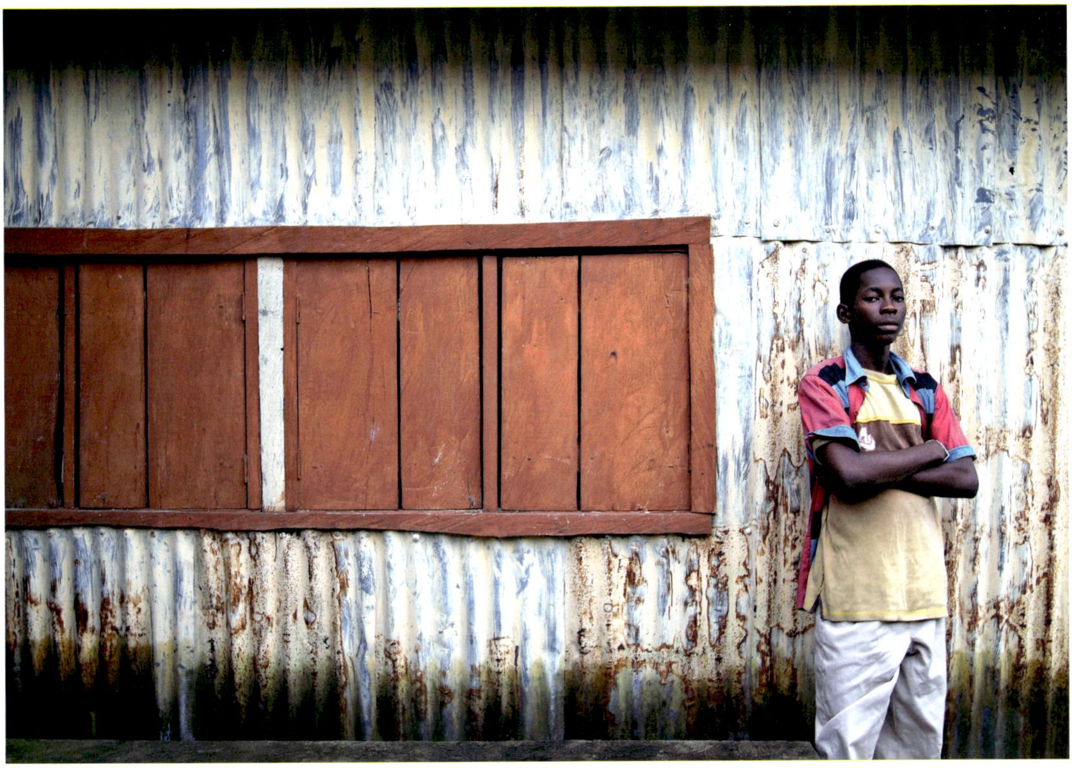

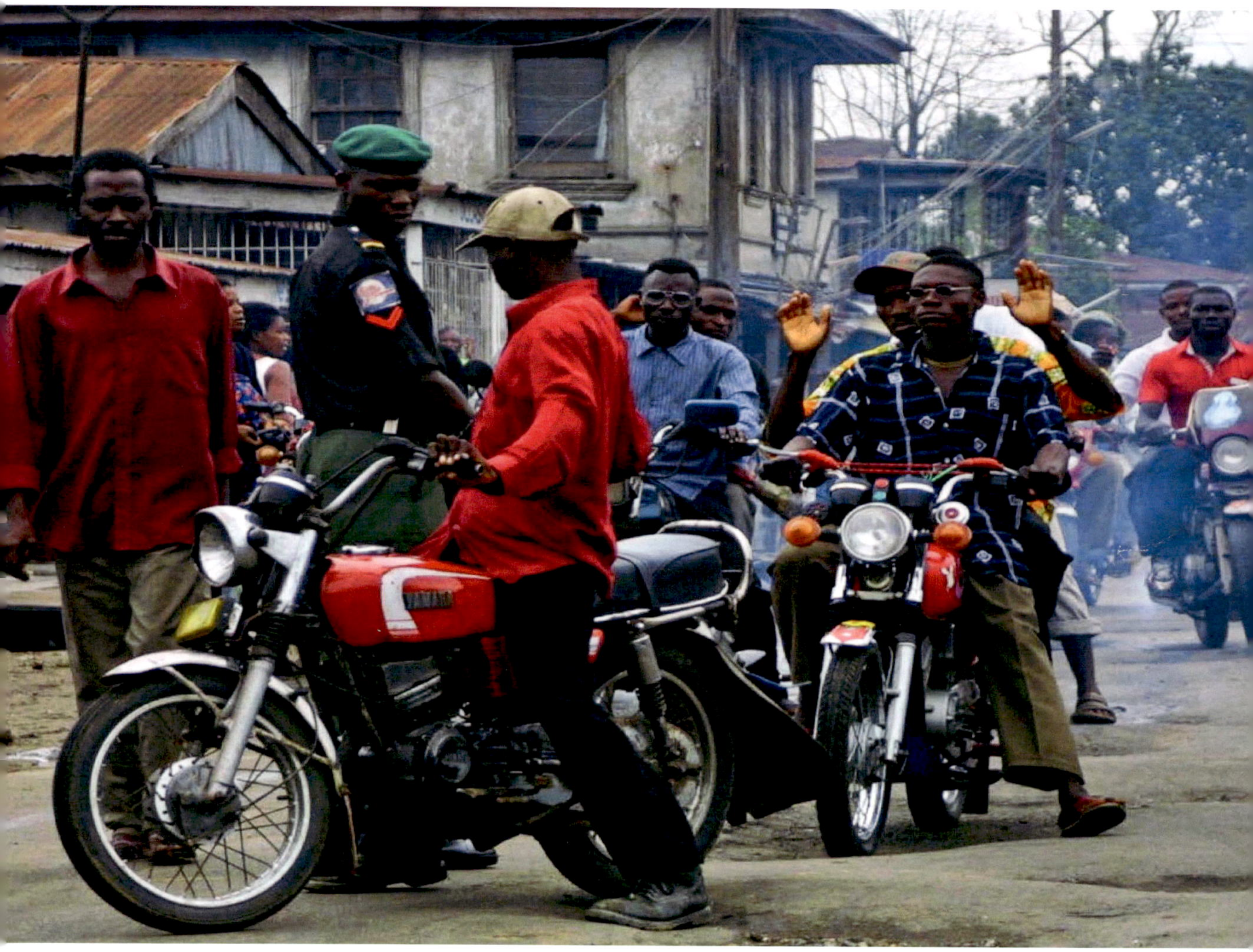

Above and right: Police are deployed to restore order following renewed clashes between the Ijaw and Itsekiri ethnic youths during the April 2003 legislative and general elections, Jakpa junction, Warri, Delta State.

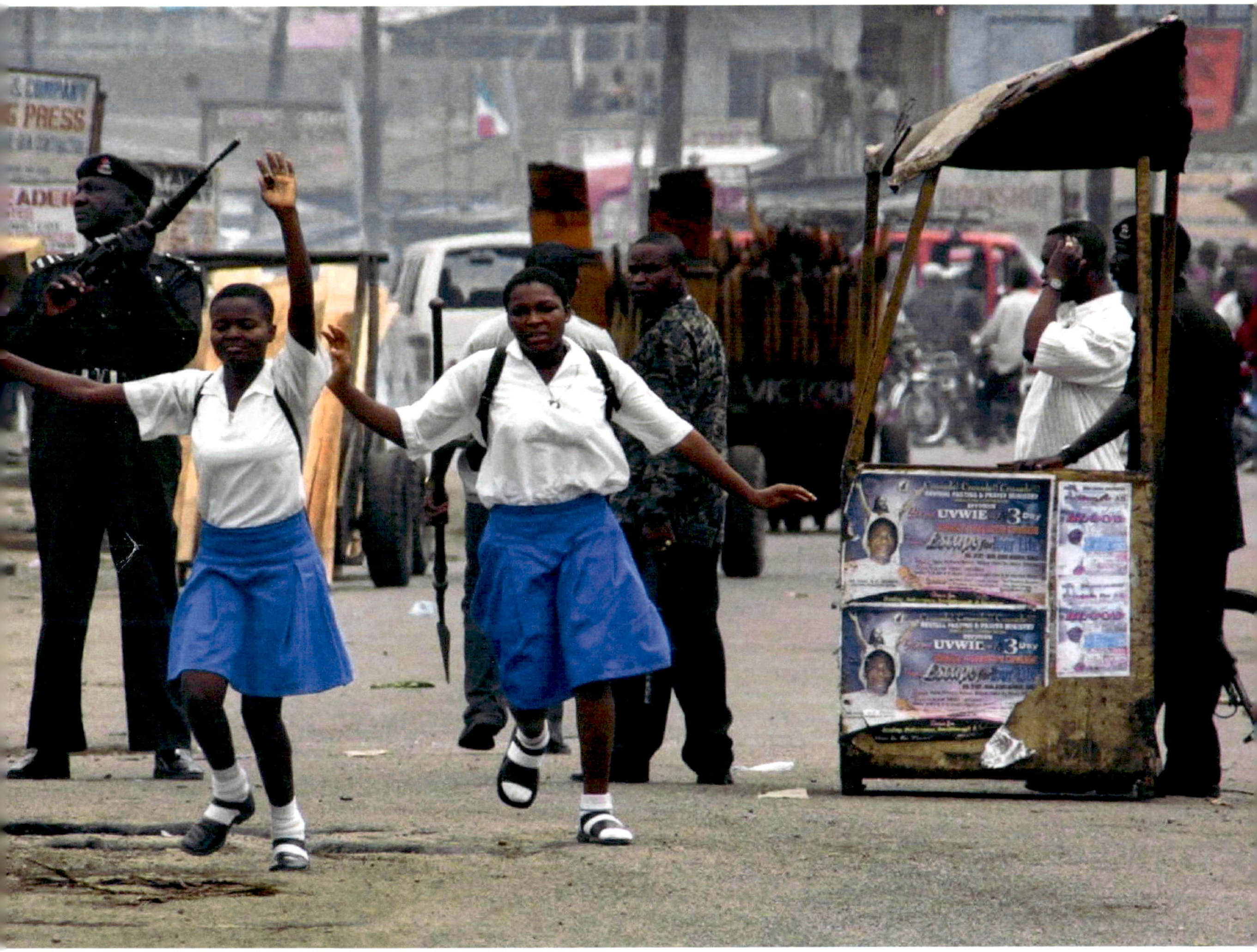
& COMPANY
G PRESS
UVWIE
UVWIE

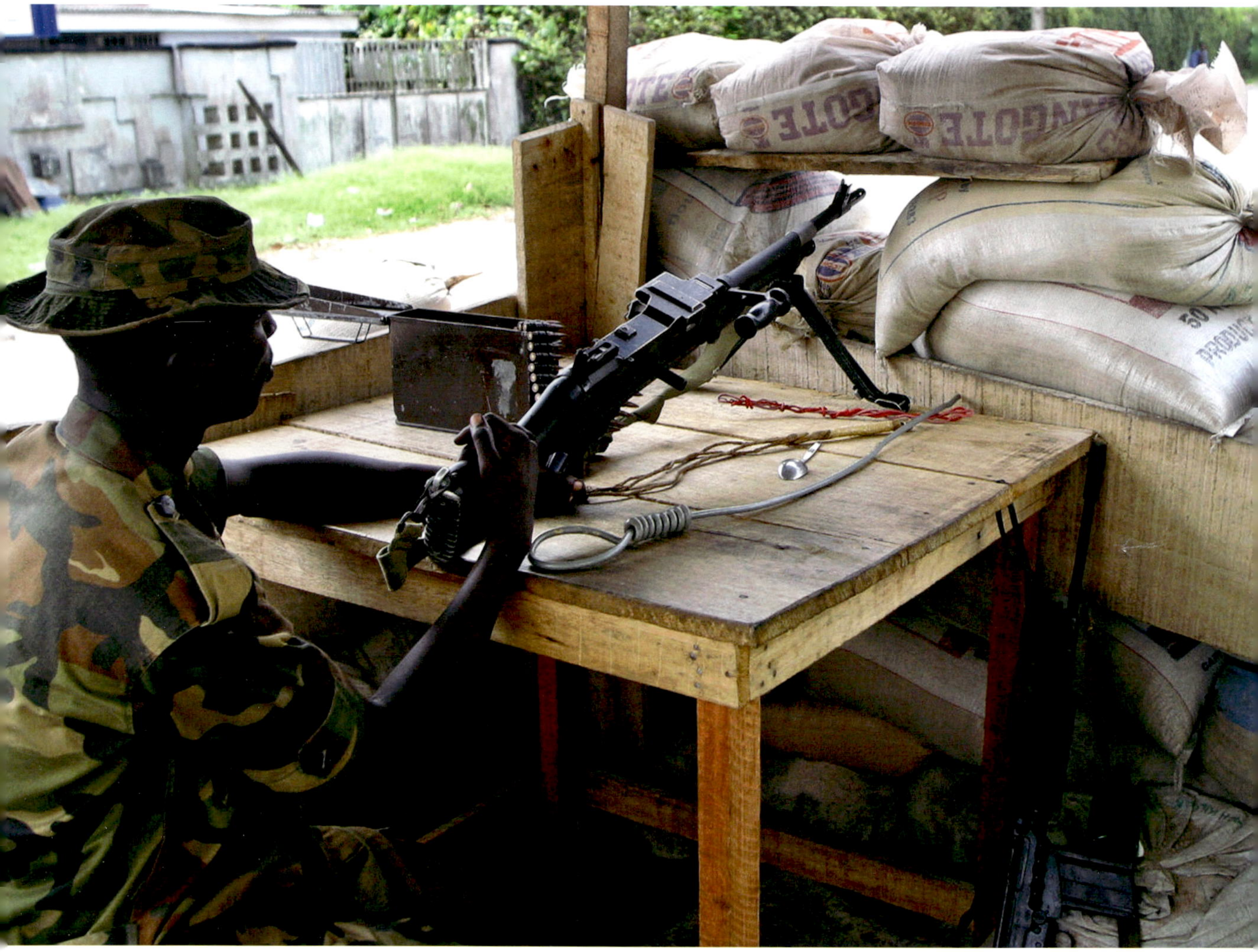

Nigerian military during unrest,
Warri, 2003.

STATE GOVERNMENT SOLUTIONS TO THE NIGER DELTA TROUBLES: KEY CHALLENGES, NEEDS AND STRATEGIES DR. EMMANUEL UDUAGHAN, EXECUTIVE GOVERNOR, DELTA STATE.

Chatham House, London, 26 February, 2009.

The crisis in Delta State arose from inter-ethnic problems between three ethnic groups. In 1997, these ethnic problems were resolved but the situation allowed for other issues to come in.

Without peace and security, there can be no infrastructure. This is why we have three interwoven strategies for dealing with crime. When I refer to 'insecurity', I mean those problems which have to do with the people there on the ground and the events that actually occur. 'Instability' refers to the conditions which lead up to these events. The pattern of oil in the Niger Delta is the same with crude oil as it was for palm oil in 1898. Agriculture was originally the mainstay of the economy, but it has been driven to the background.

Since oil companies have come into Delta State, ethnic communities are now fighting over ownership of the land where oil is drilled. Criminals are also active because of the oil. Oil company workers looking to use their technical knowledge for a greater profit sometimes also partake in criminal activity. Of the players in crime, the Godfathers are active both within and outside the state structure.

The co-operative framework which we will put into practice in Delta State has development on one side and law and order on the other. At the federal level, law and order is maintained by an armed forces Joint Task Force and at the state level, by the police and local security measures.

My concern about the communities that host oil companies is that workers do not live with the community, they are flown in from Lagos or Abuja, and live within the company compound. This is why the hostility of the host community grows. If more people from the local community could be engaged in unskilled work within oil companies, the companies would have less trouble from their host communities. This will mean added security for the oil companies. Most of the small arms used in criminality in Delta State are not produced in Nigeria.

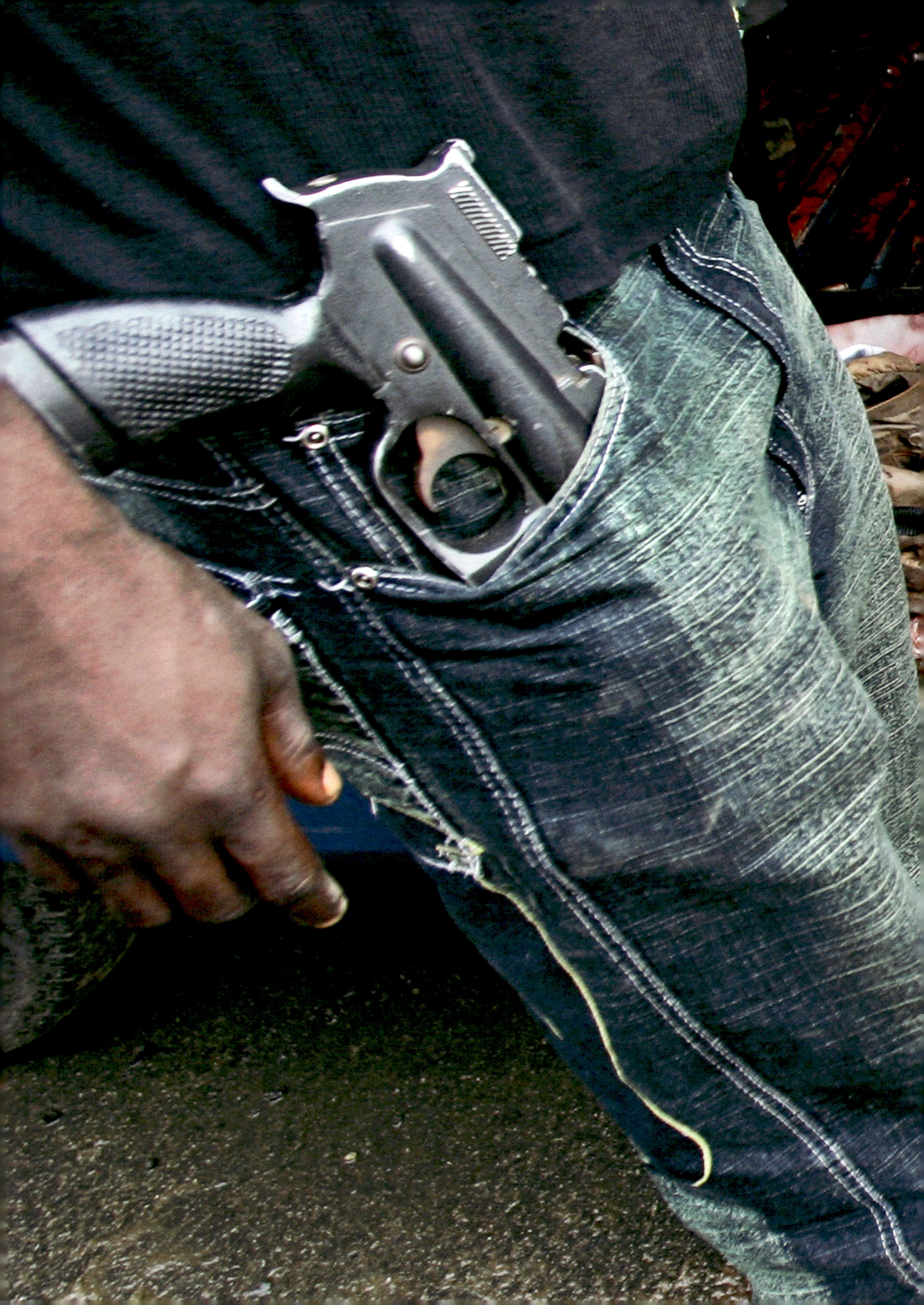

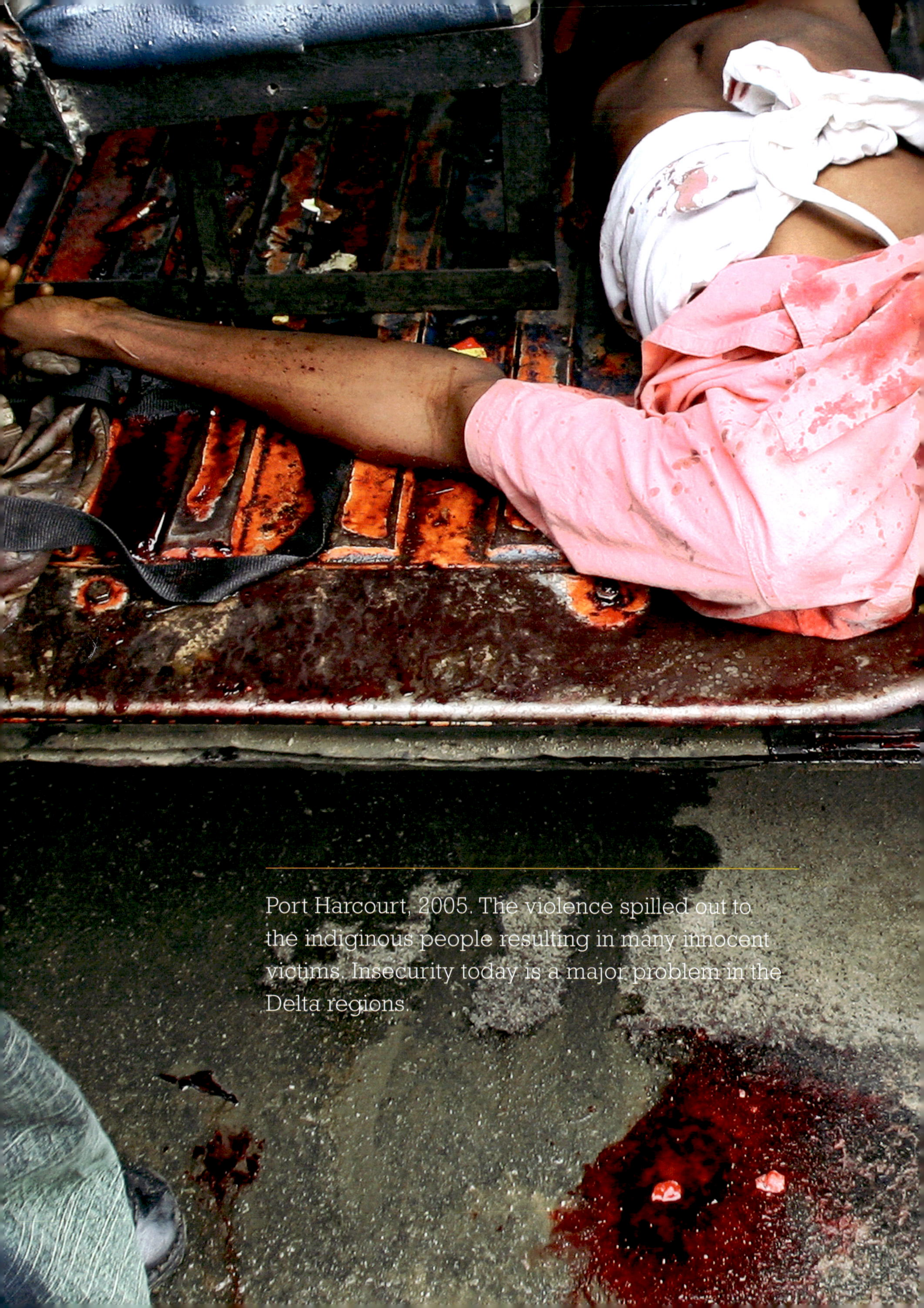

Port Harcourt, 2005. The violence spilled out to the indiginous people resulting in many innocent victims. Insecurity today is a major problem in the Delta regions.

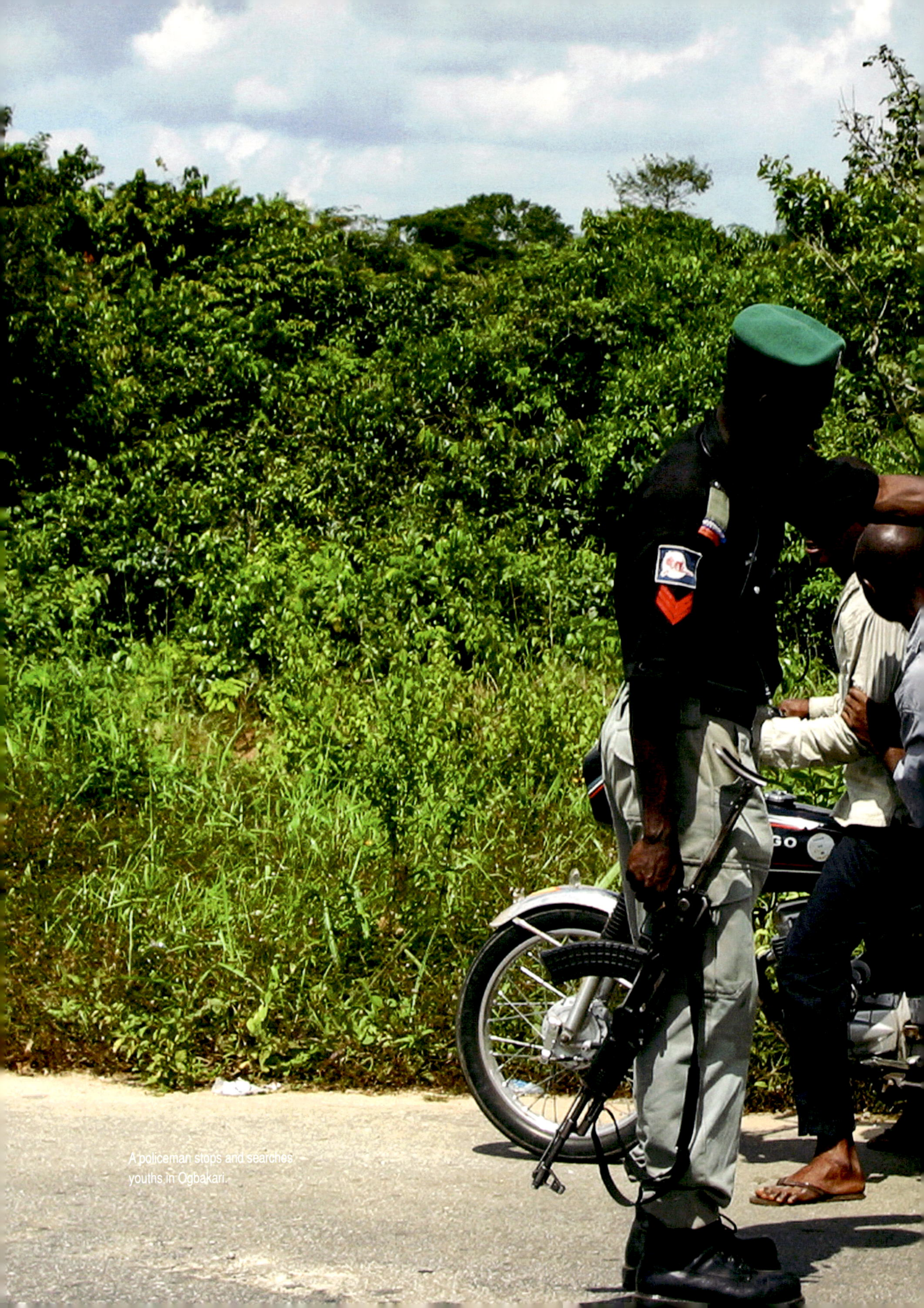

A policeman stops and searches youths in Ogbakari.

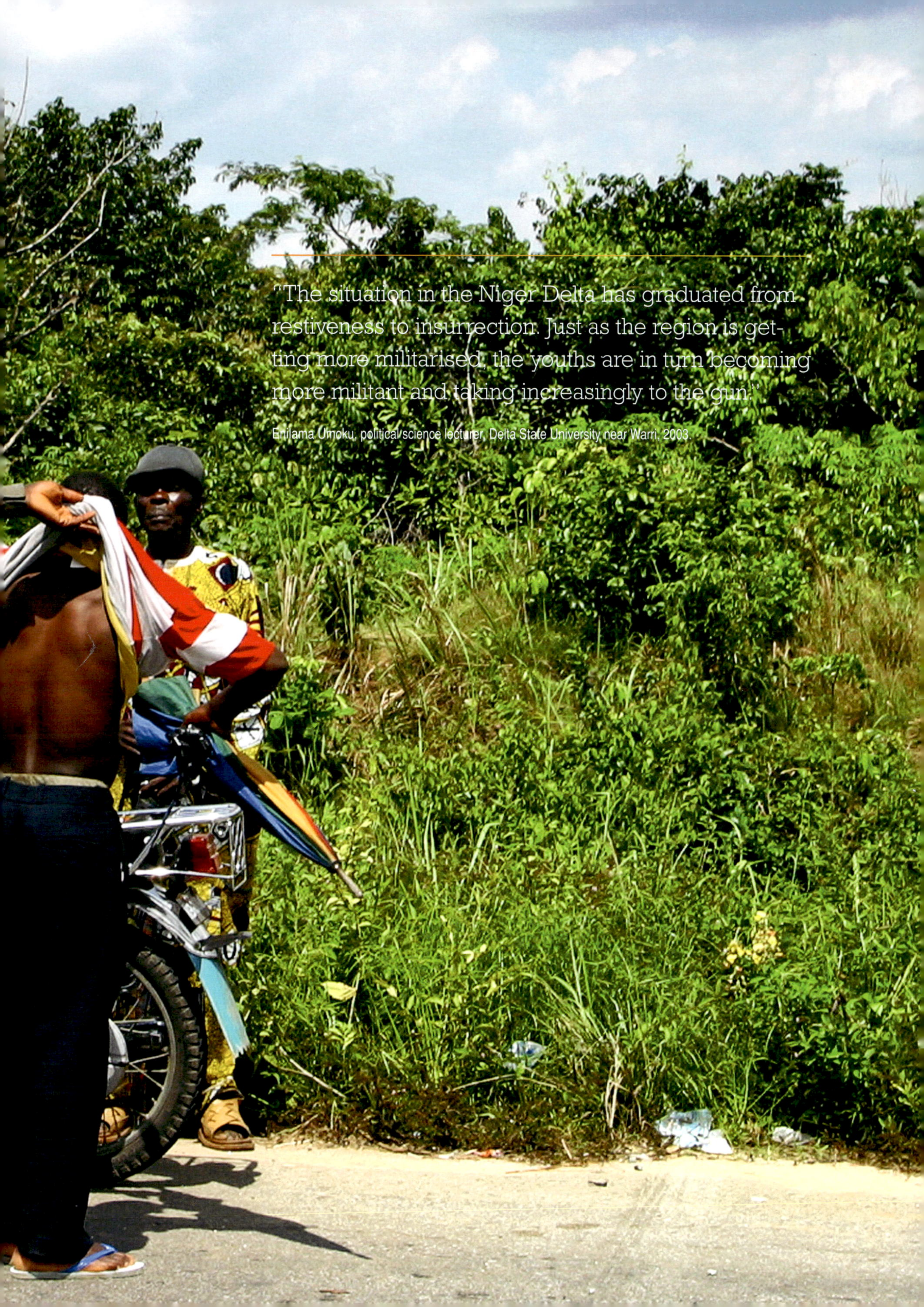

"The situation in the Niger Delta has graduated from restiveness to insurrection. Just as the region is getting more militarised, the youths are in turn becoming more militant and taking increasingly to the gun."

Enilama Umoku, political science lecturer, Delta State University near Warri, 2003.

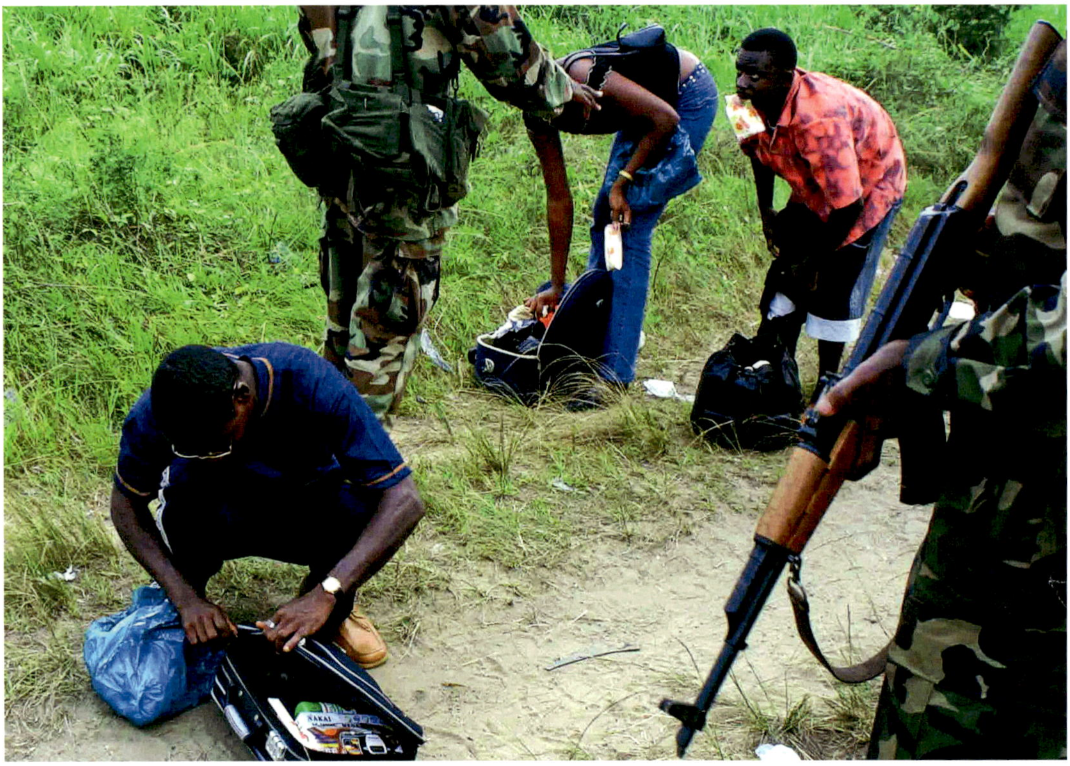

Above: Soldiers deployed to keep peace stop and search people, Warri, 2003.

Right: Naval officials stop and whip some suspected youths in Warri, 2003.

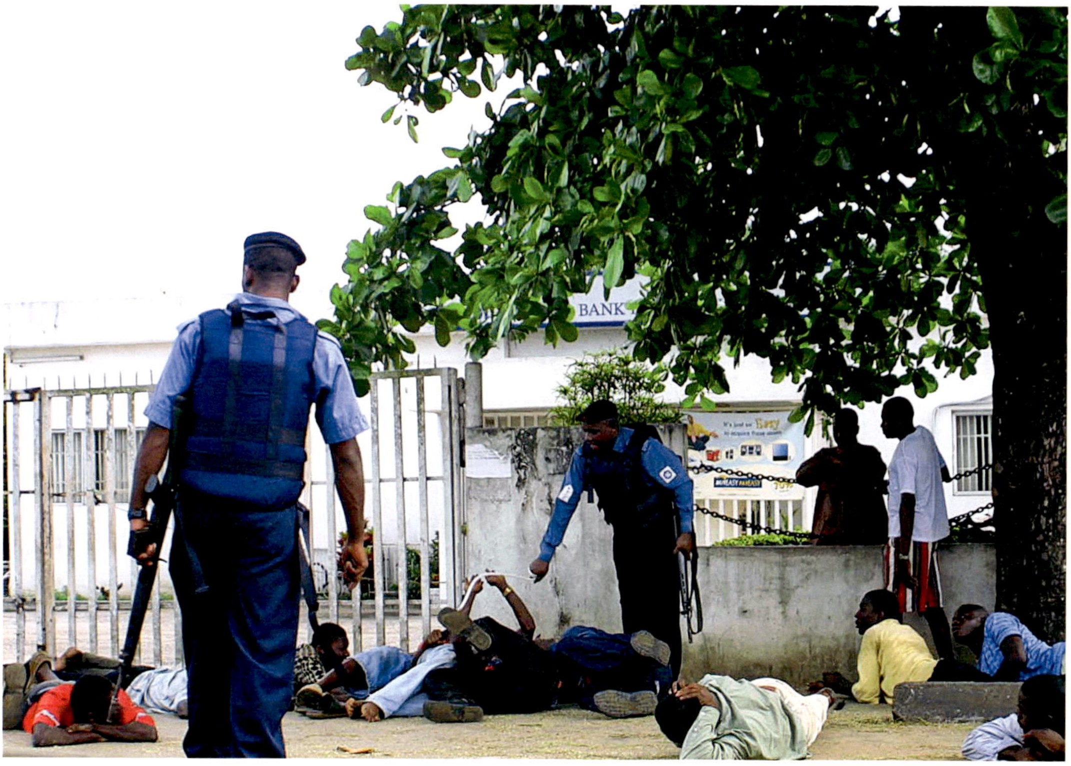

Oil theft and kidnappings for ransom have become lucrative trades in the Niger Delta, where the majority of residents live on less than $2 a day. Hostage taking of local civilians has increased rapidly in recent years, anyone seen to have money is now a target, and it is not restricted only to foreign nationals working for the oil companies.

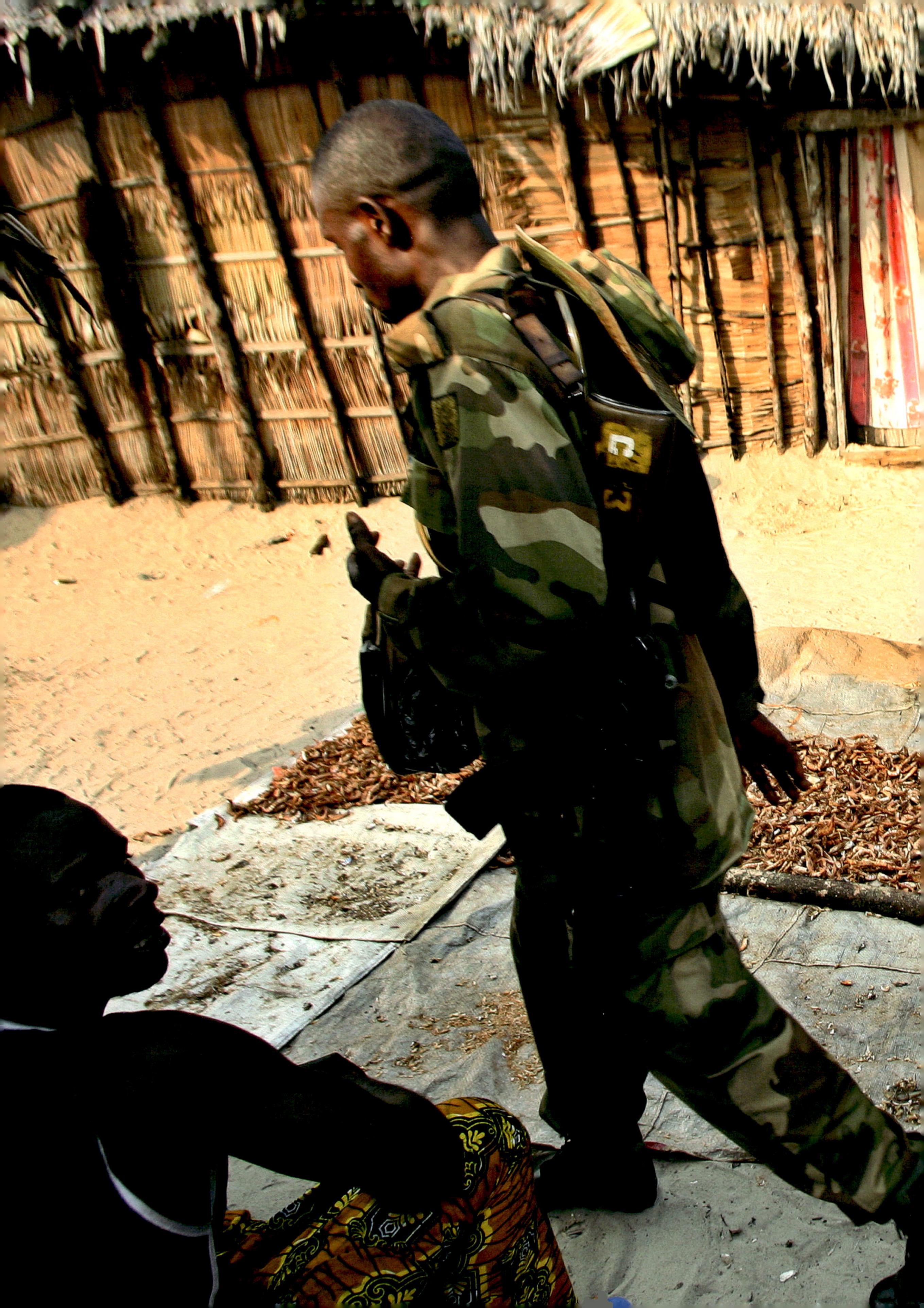

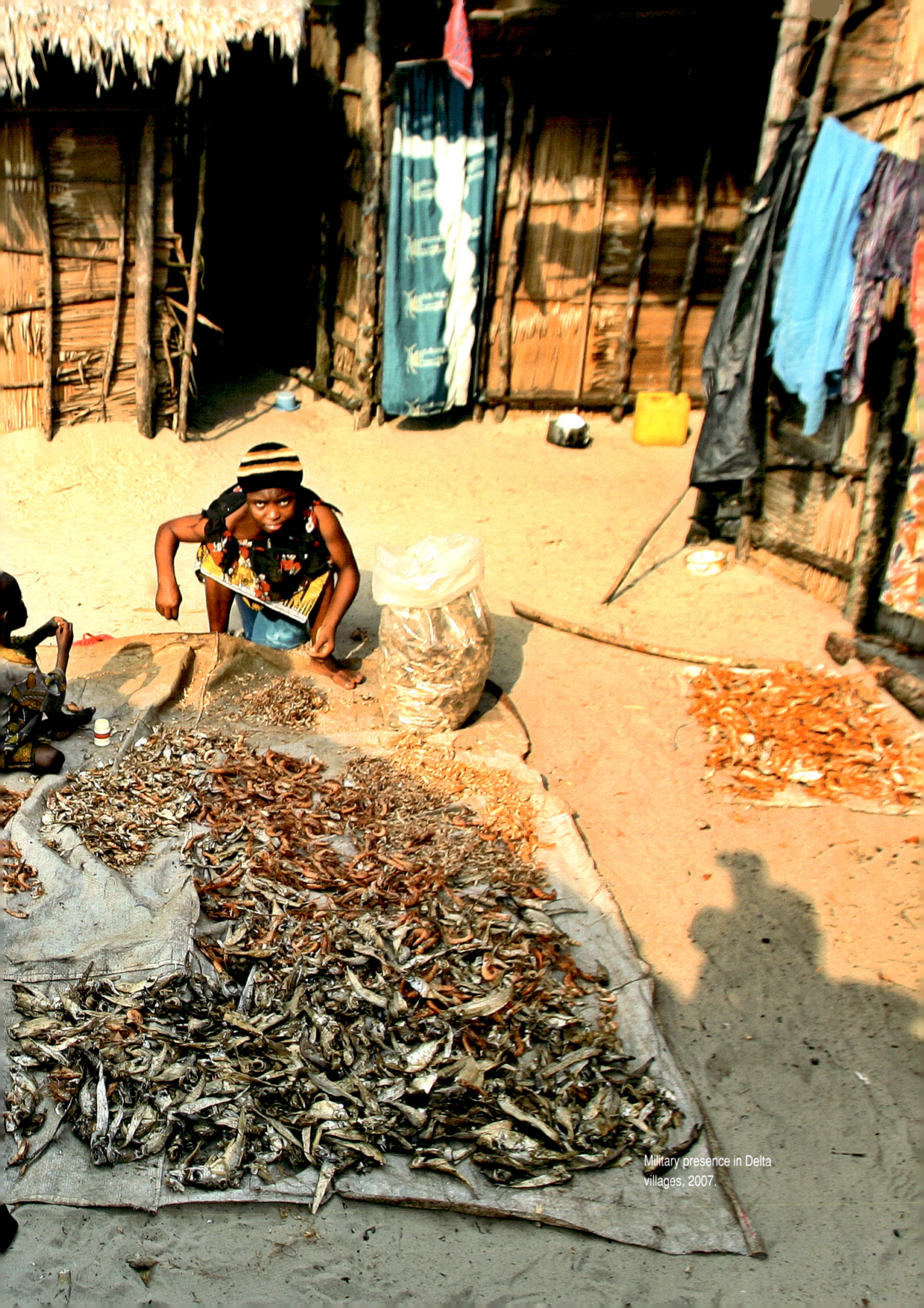

Military presence in Delta villages, 2007.

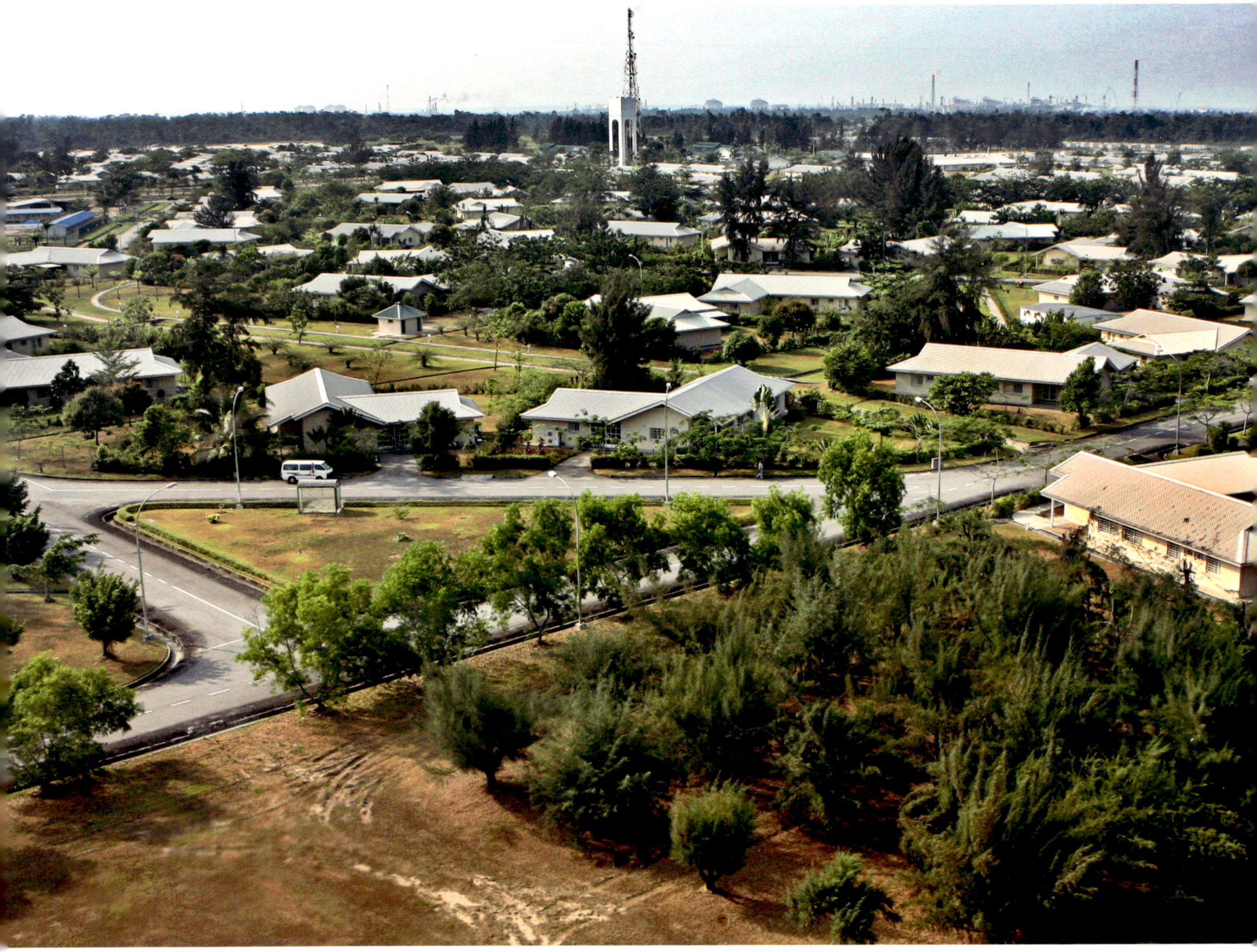

Above: Oil workers apartments, Finima, Bonny Island. Homes are heavily guarded and surrounded by barbed wire fences against militants and kidnappings - in sharp contrast to the living conditions of the local communities.

Right: An aerial view shows oil storage tanks at Chevron Texaco's Escravos export terminal in 2003, the scene of a massive protest the preceding year.

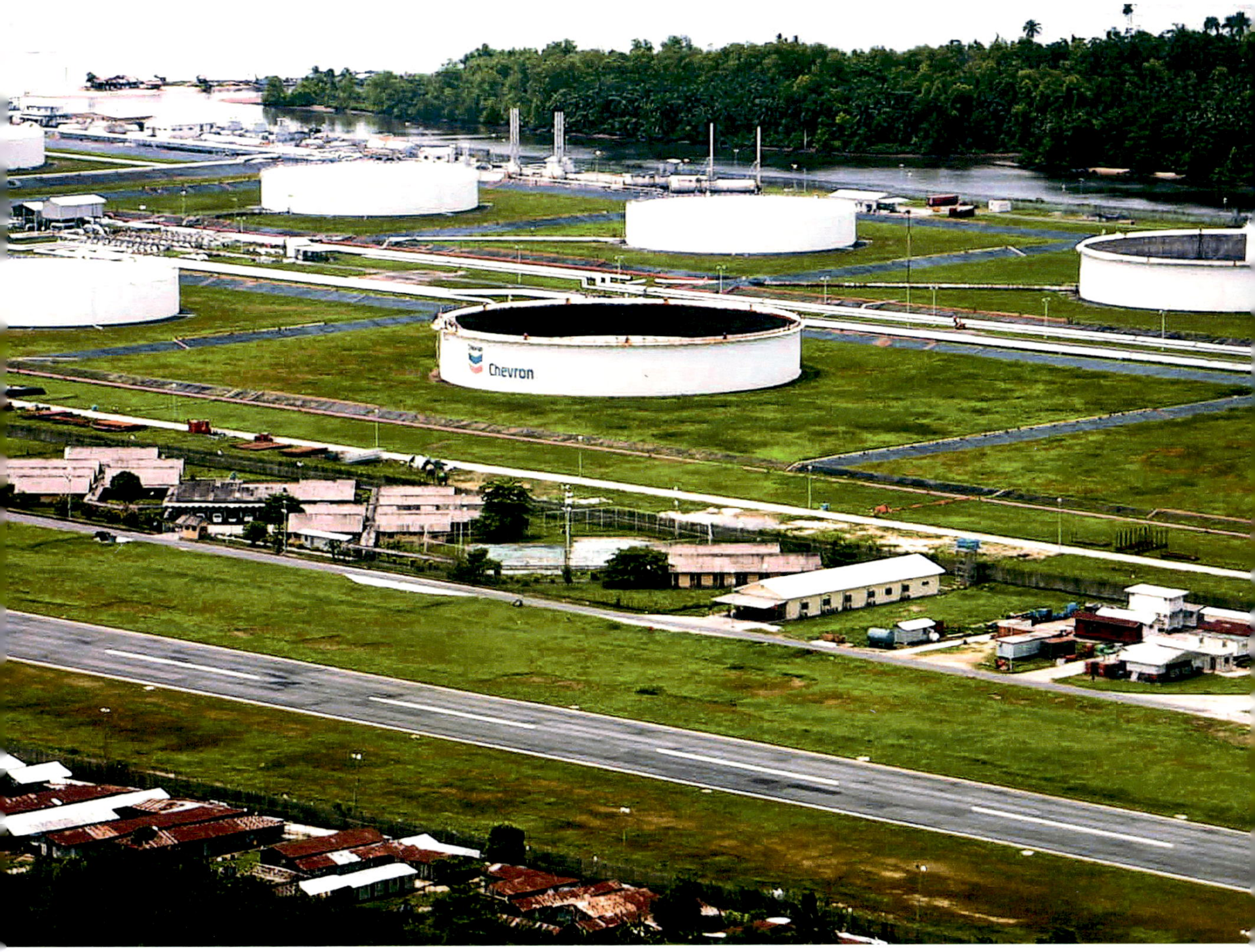

In 2002, hundreds of protesting youths, women and children of Ugborodo embarked on a protest to Chevron Nigeria's tank farm near their homes. The protesters were allegedly shot by Joint Task Force government security personnel who opened gunfire at them resulting in injuries from stray bullets.

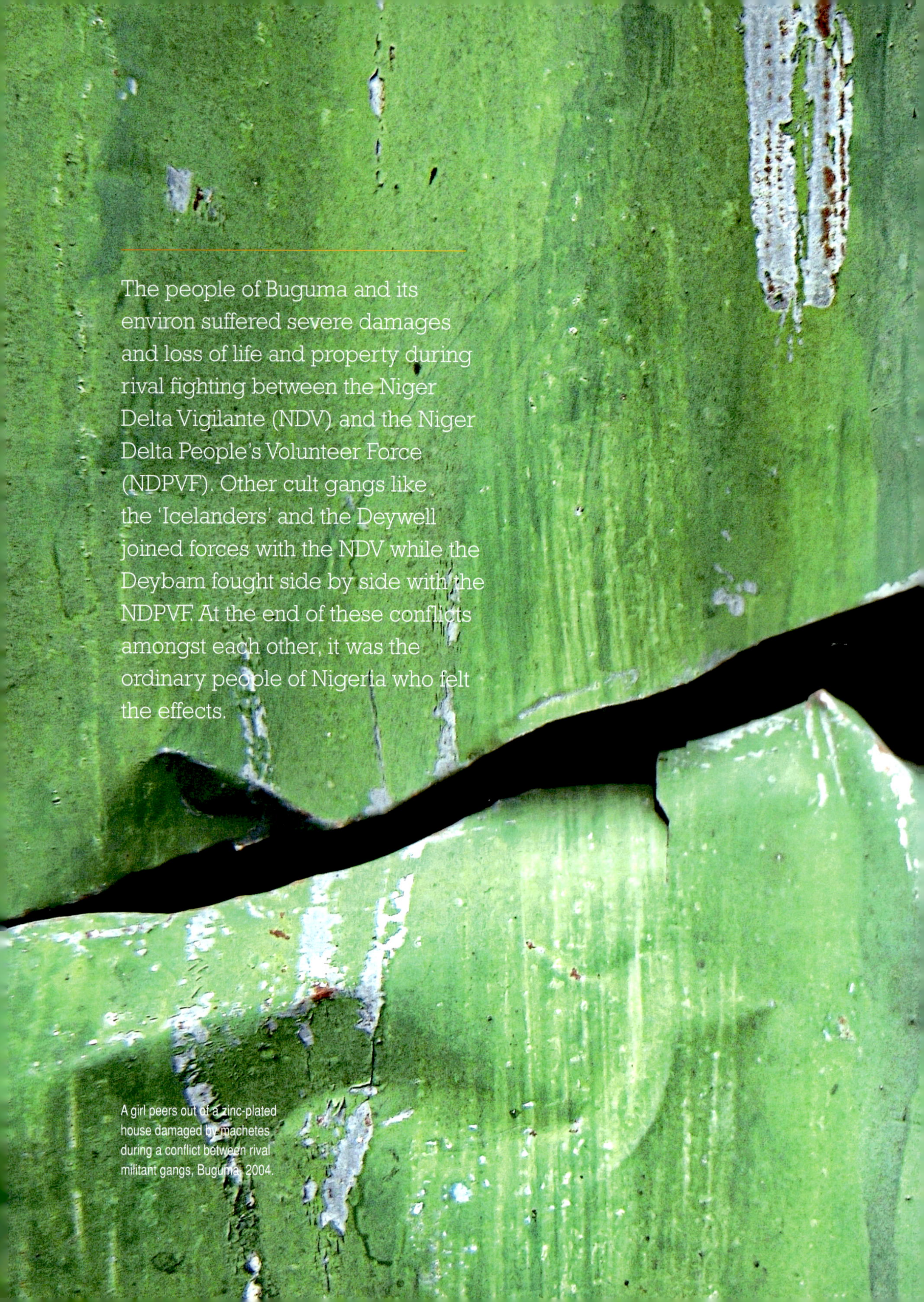

The people of Buguma and its environ suffered severe damages and loss of life and property during rival fighting between the Niger Delta Vigilante (NDV) and the Niger Delta People's Volunteer Force (NDPVF). Other cult gangs like the 'Icelanders' and the Deywell joined forces with the NDV while the Deybam fought side by side with the NDPVF. At the end of these conflicts amongst each other, it was the ordinary people of Nigeria who felt the effects.

A girl peers out of a zinc-plated house damaged by machetes during a conflict between rival militant gangs, Buguma, 2004.

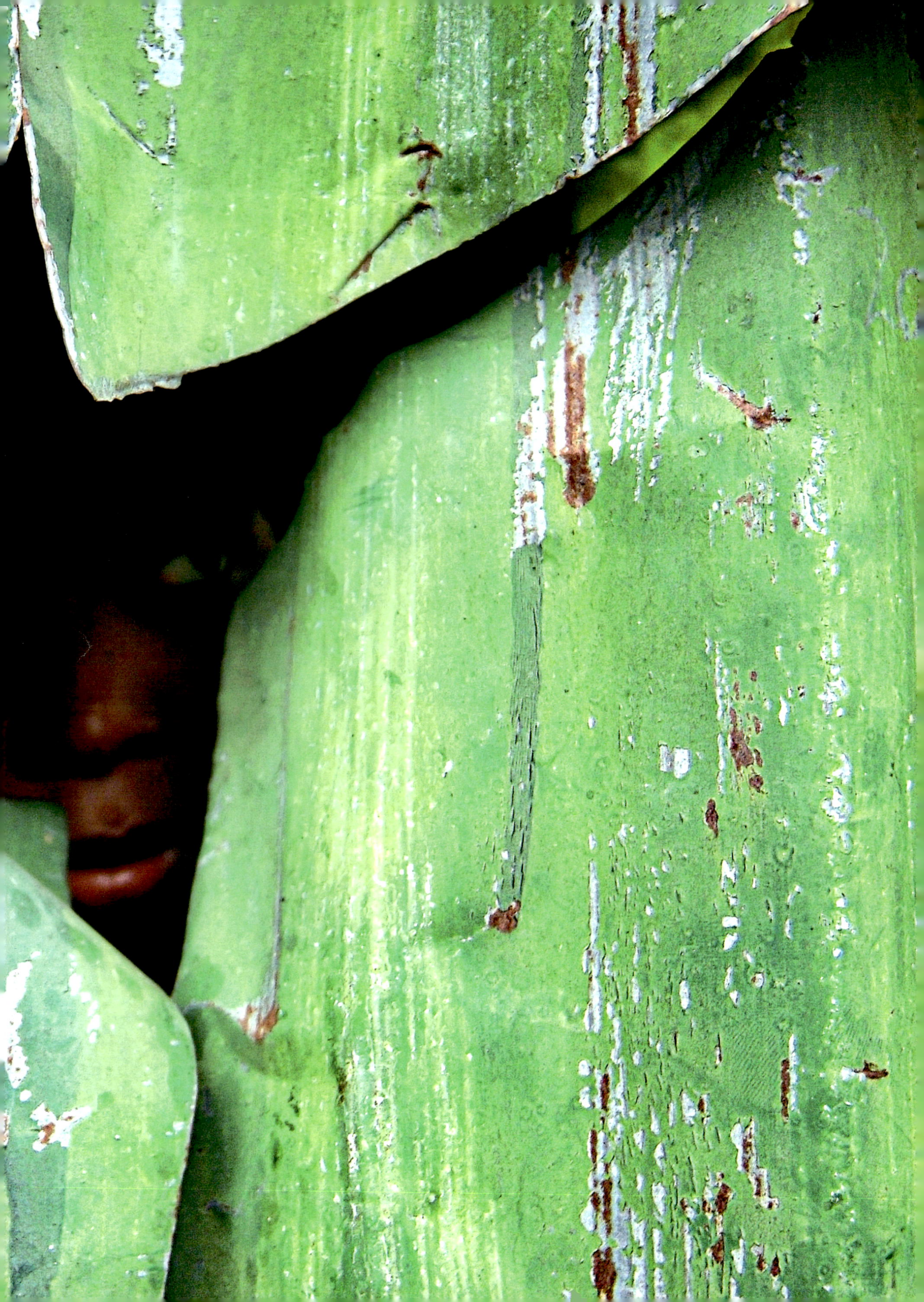

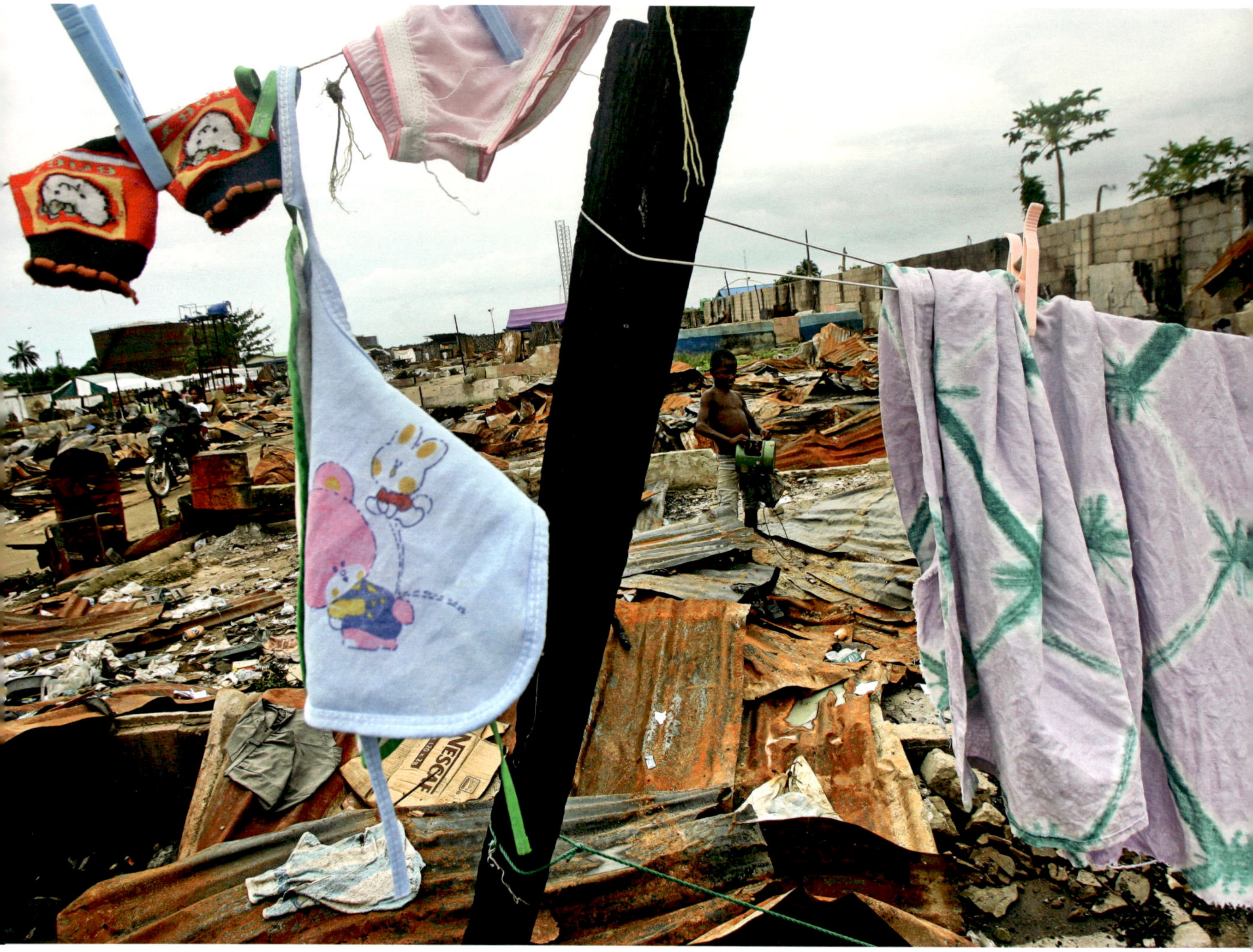

Above: A boy picks up burnt household material in the Aker Base slum area, Port Harcourt, 2006.

Right: Okotie Atigan's house was one among many houses destroyed in Koko during the Ijaw and Itsekiri crisis, caused by fighting over oil territories in 2003.

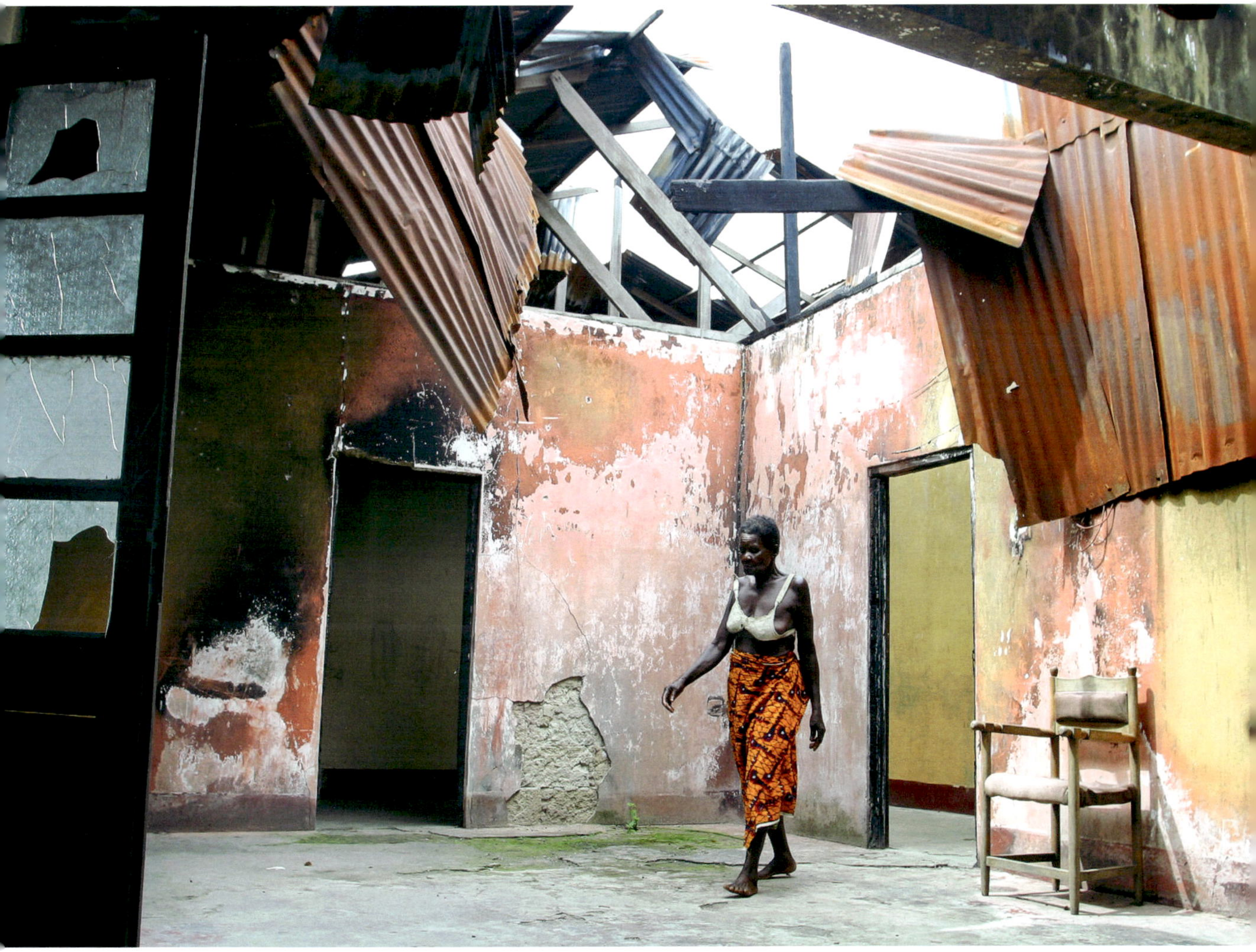

Two expatriates, including an Italian oil worker working for Saipem, were kidnapped from Aker Base, Port Harcourt, on August 24, 2006. The Italian was kidnapped from a bar in the slum area and his bodyguard was shot dead by a band of militants. This lead to the Nigerian army destroying the entire slum, leaving scores of people dead and many injured.

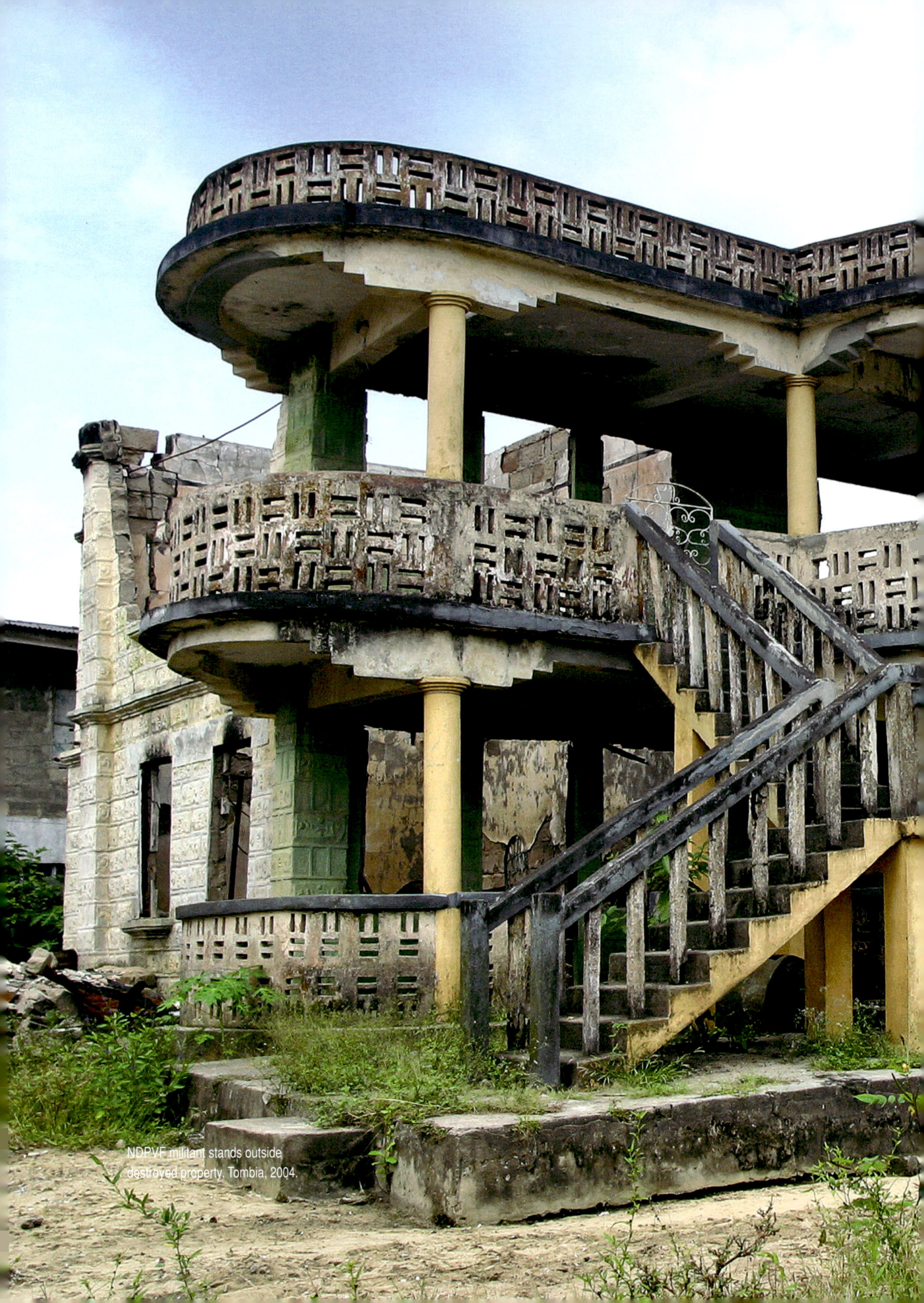

NDPVF militant stands outside destroyed property. Tombia, 2004.

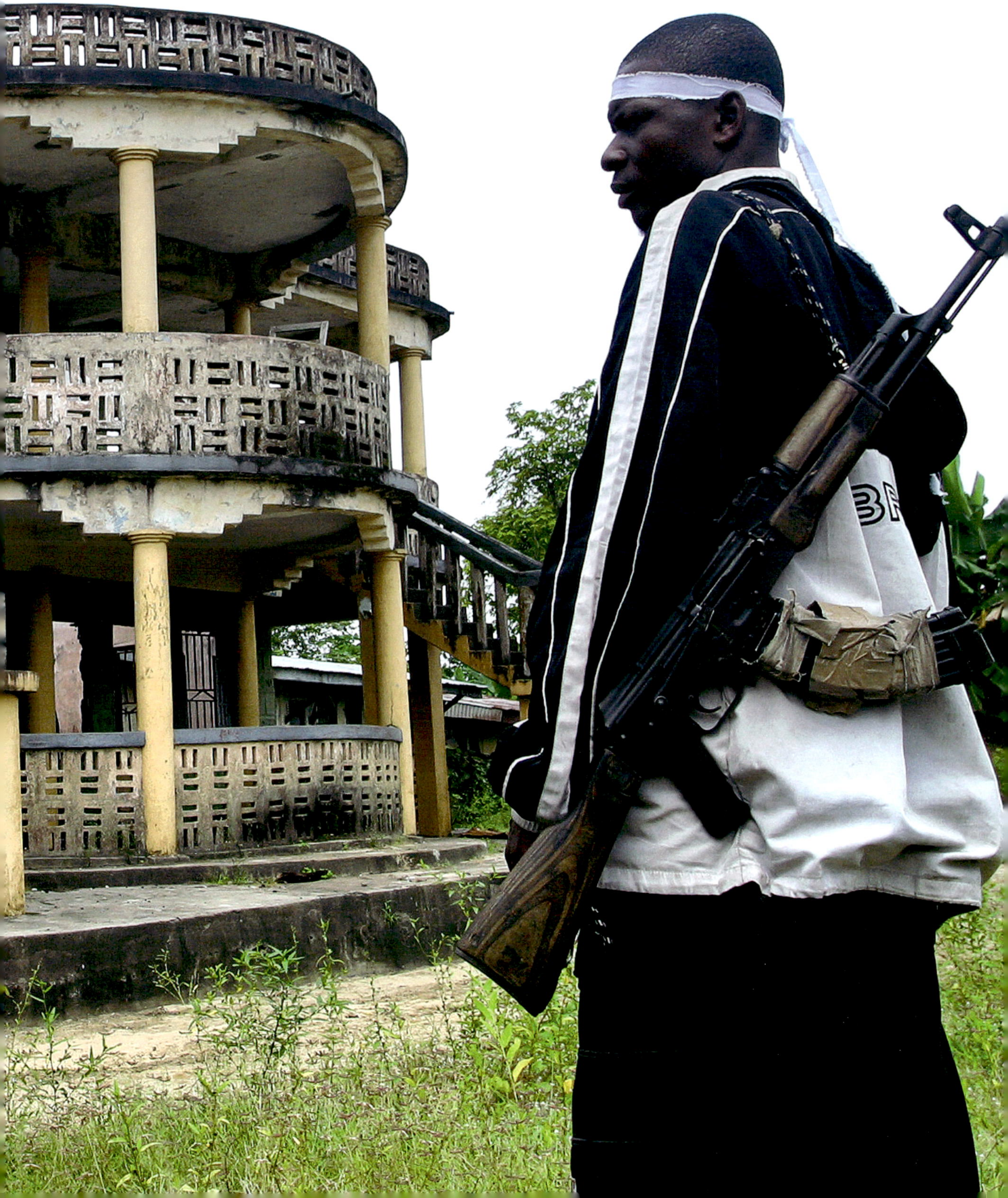

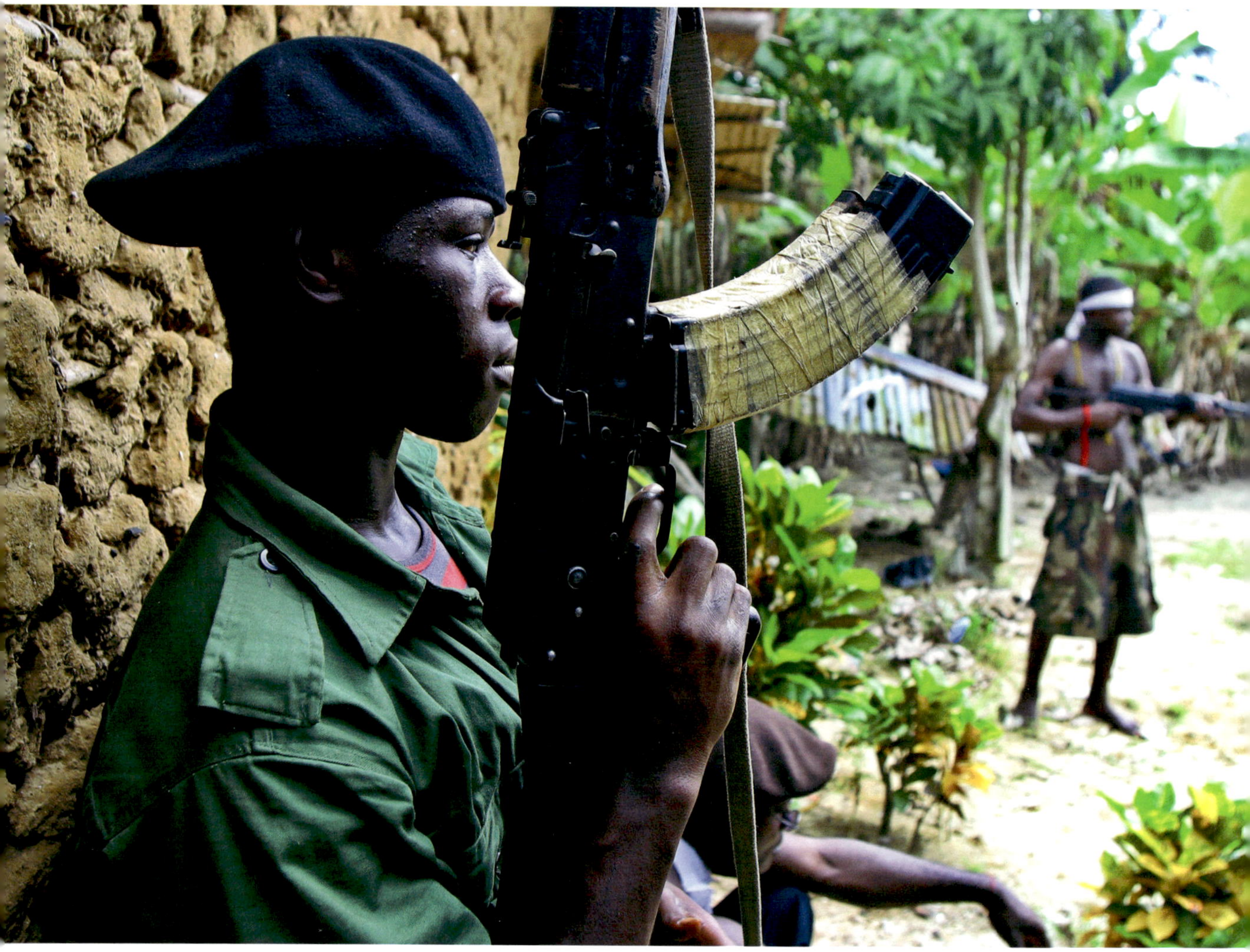

Above: NDPVF militants stand guard, Ogbakiri, 2004.

Right: A young NDPVF militant, Tombia, 2004.

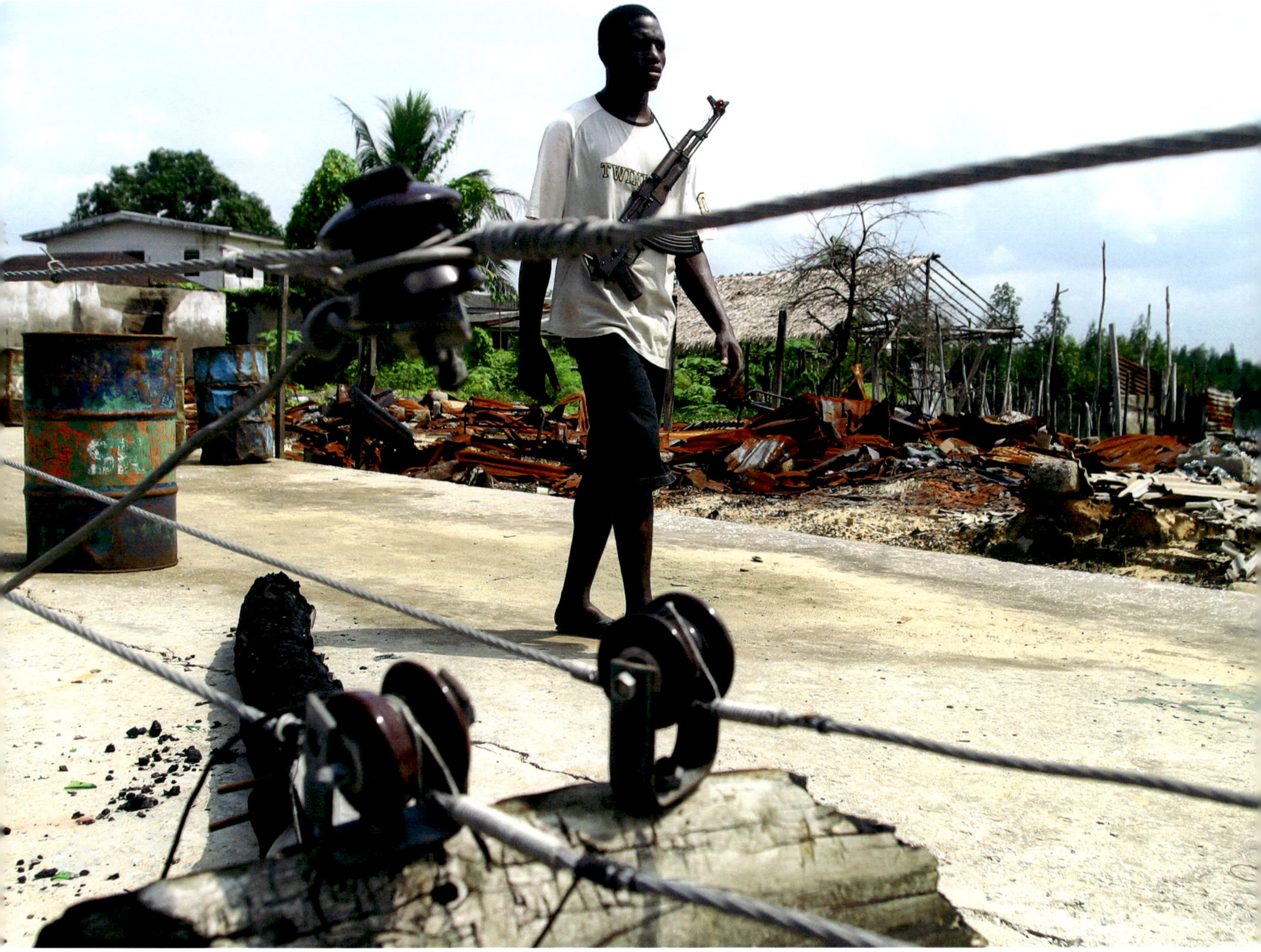

In 2004 Tombia was a scene of conflict between the two main rival gangs - NDPVF and NDV - the first major operation between them.

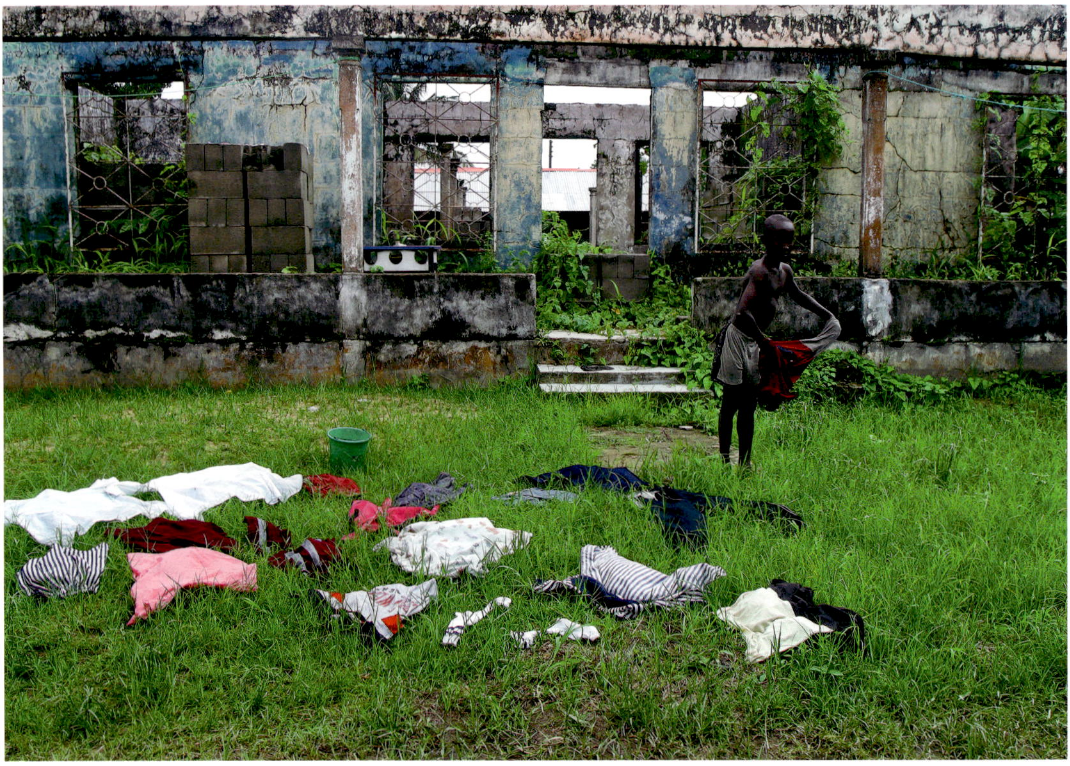

OPERATION HAKURI II

In November 1999, the Nigerian army troops stormed Odi, a sleepy village right in the heart of the troubled Niger Delta. They went on a mission to trail suspects who had allegedly kidnapped and killed 12 mobile policemen serving in the area. Frustrated by their failure to bring in the suspects, the troops sacked the entire village and razed it to the ground apparently as a punitive measure, amidst accusations of rape and other violations of human rights.

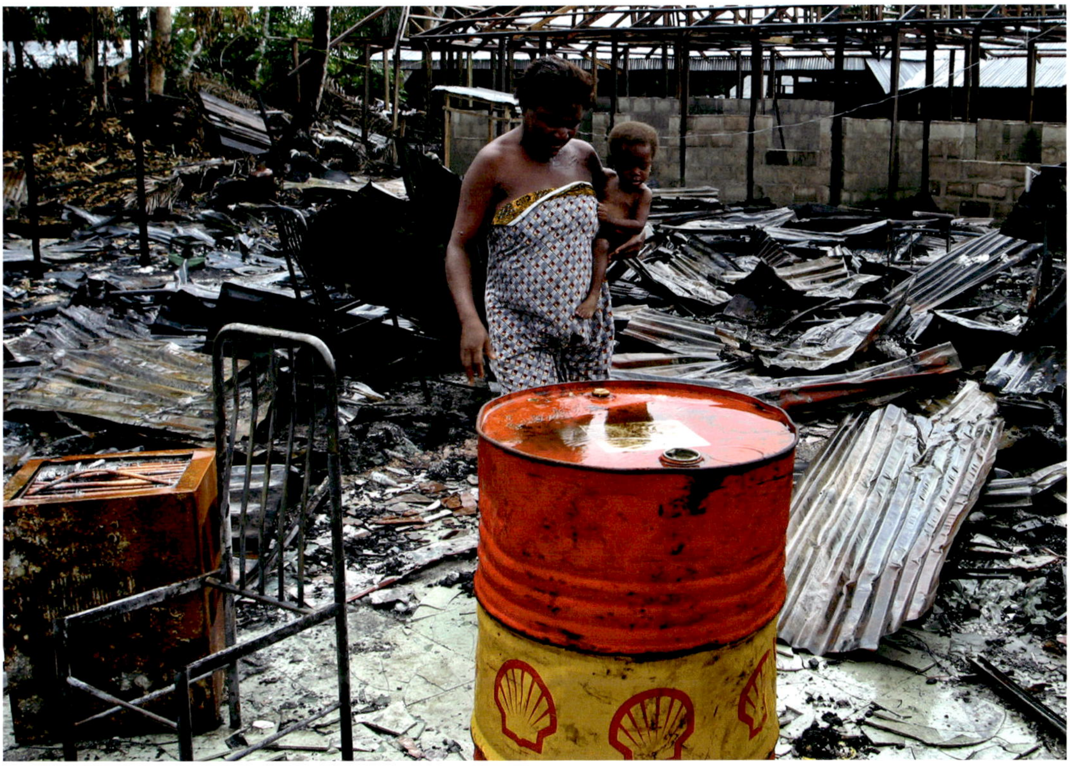

Left: A young boy spreads clothes outside his house, destroyed during the Army attack in Odi, 2006.

Above: One of the camps destroyed by a military raid in the heat of the crisis in search of Asari Dokubo, leader of the NDPVF, Ogbakiri, 2004.

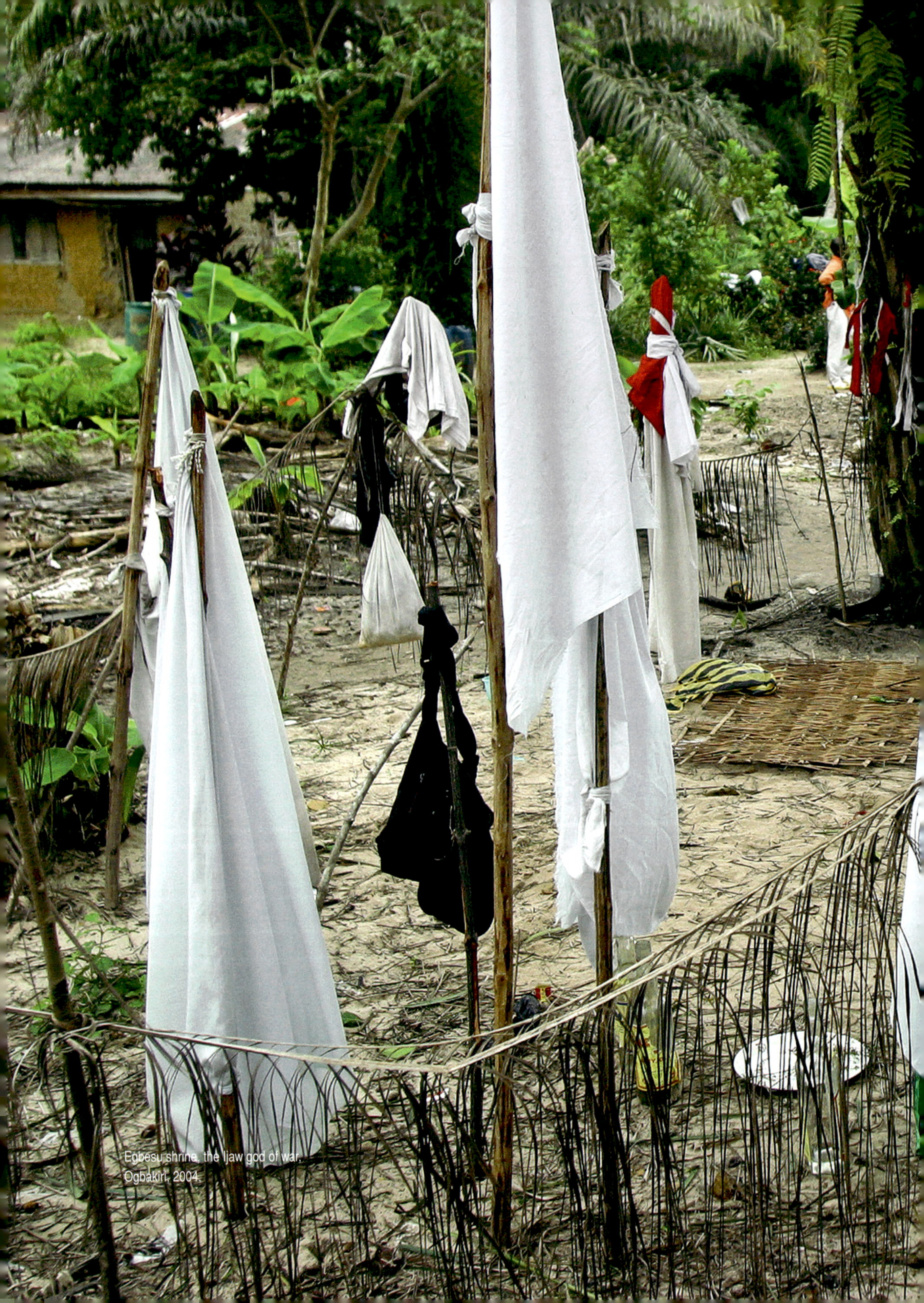

Egbesu shrine, the Ijaw god of war,
Ogbakiri, 2004.

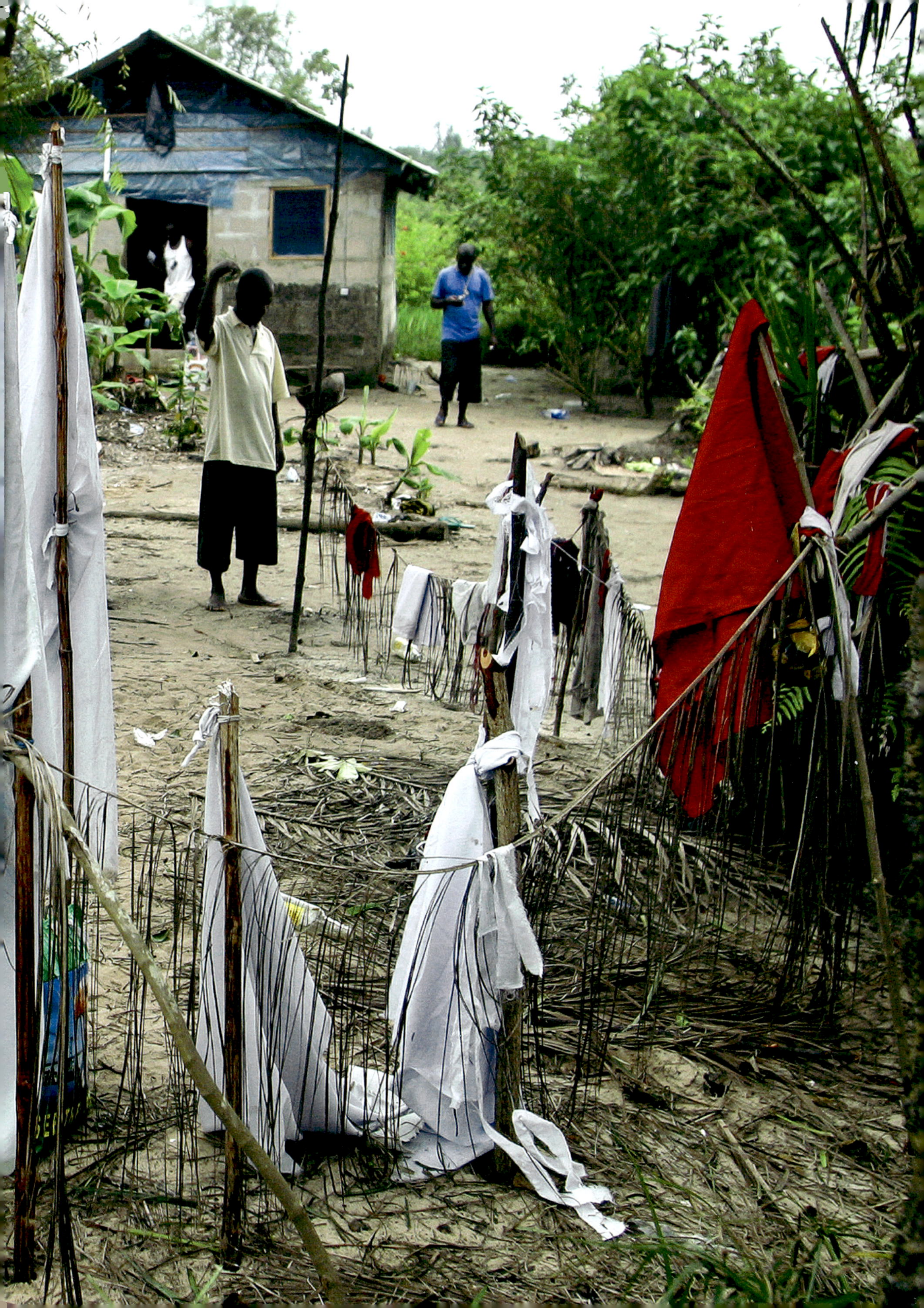

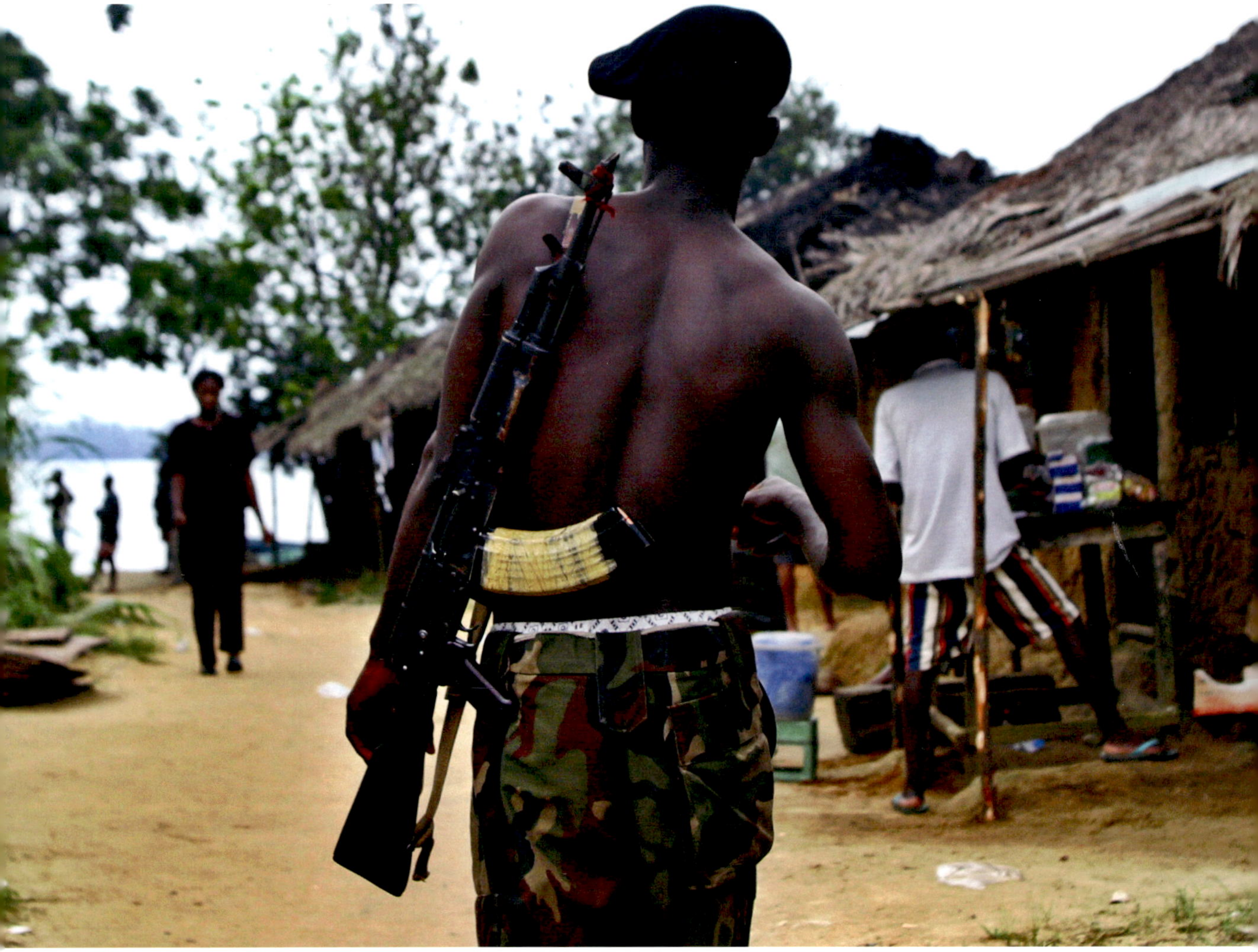

Above: Young militants patrol the Ogbakiri community carrying AK47s, Rivers State, 2004.

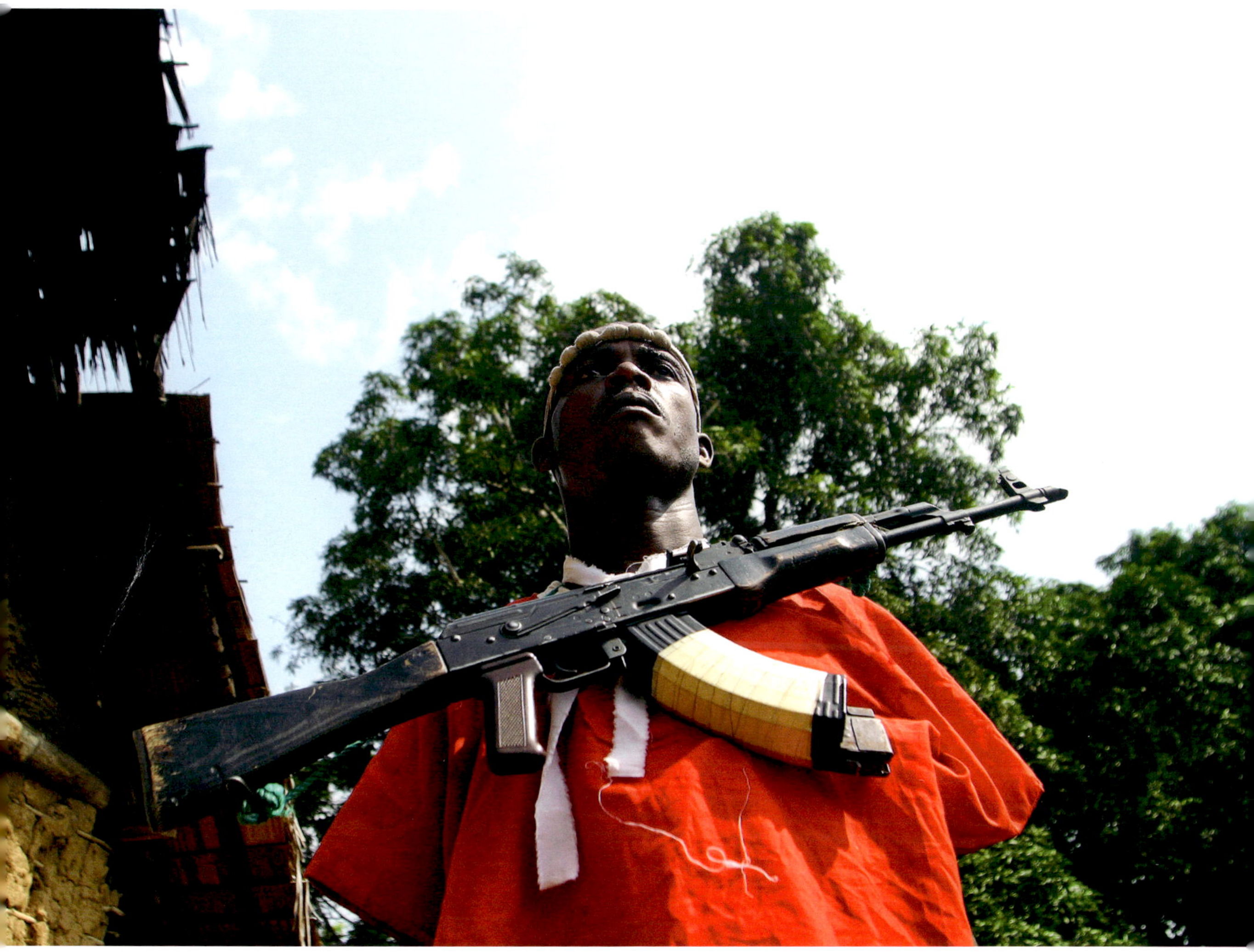

Above: A miltant stands guard in a NDPVF camp, Ogbakiri, 2004.

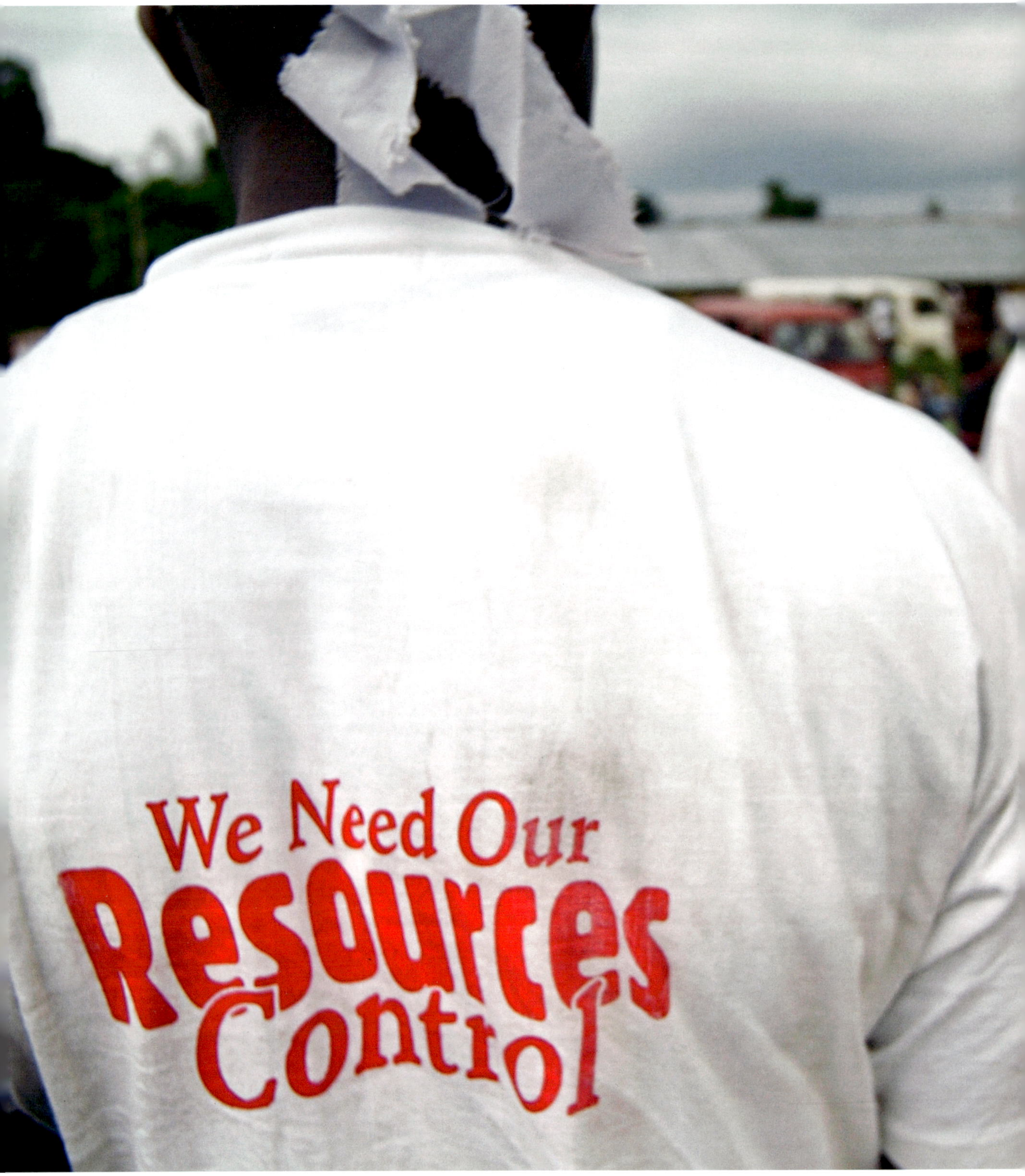
We Need Our
Resources
Control

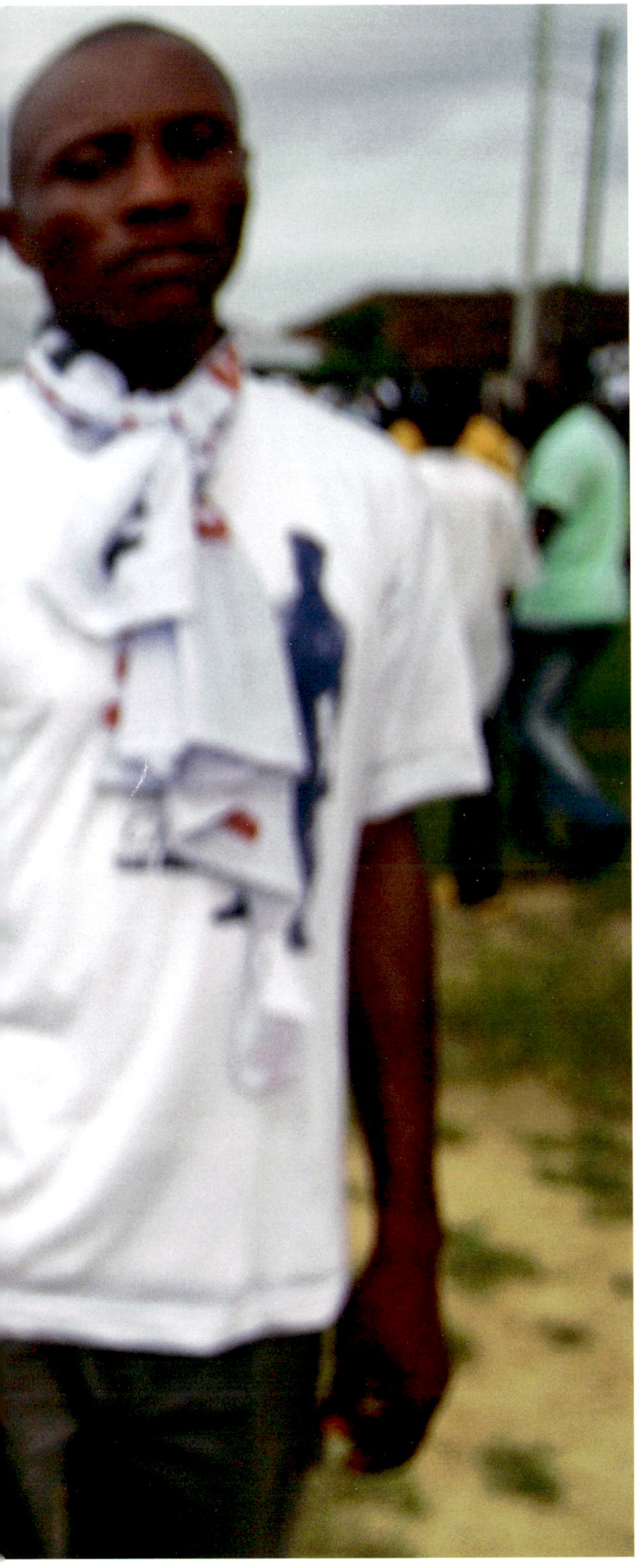

Youths protest resource control,
Harry's Town, 2005.

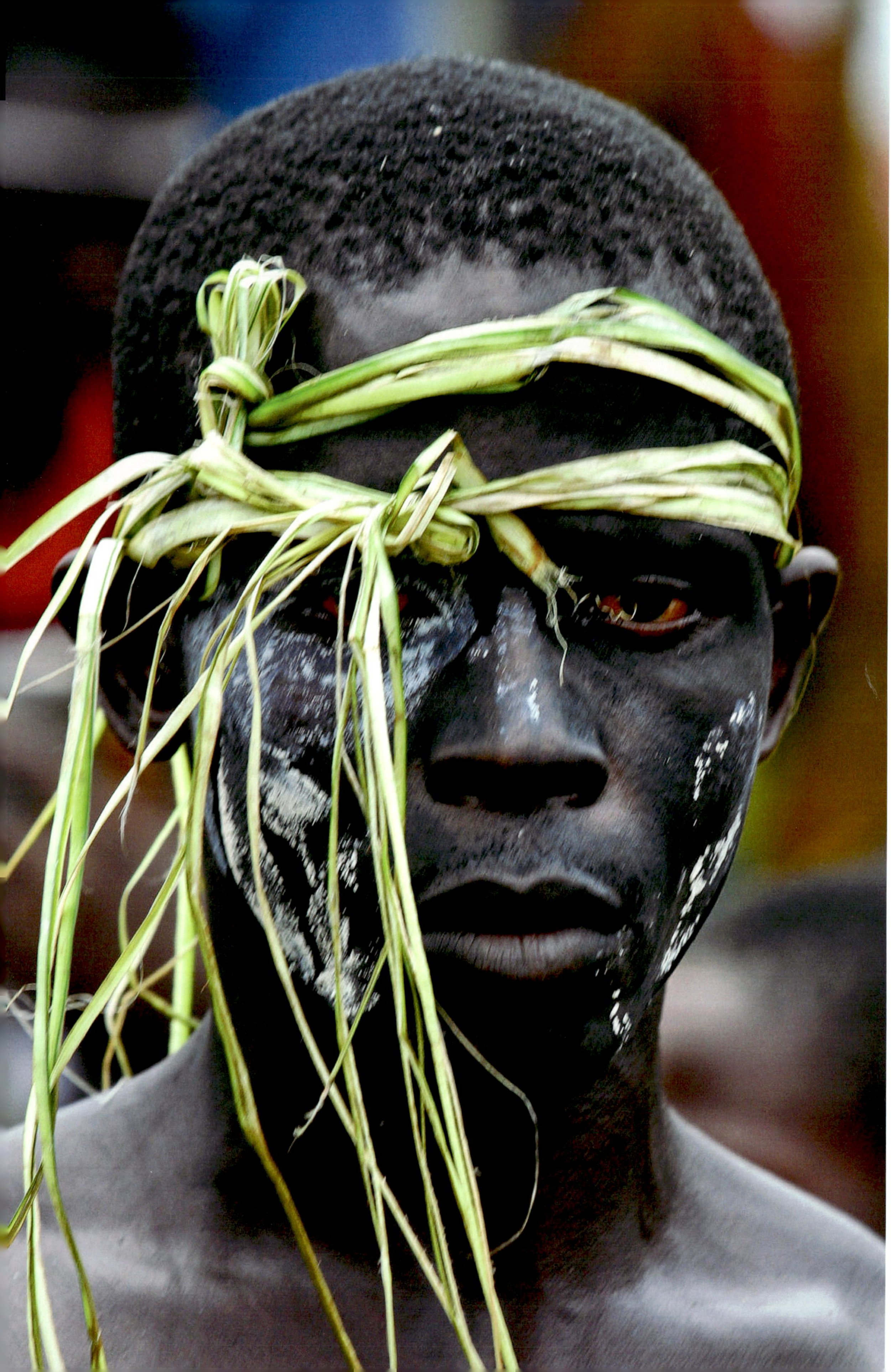

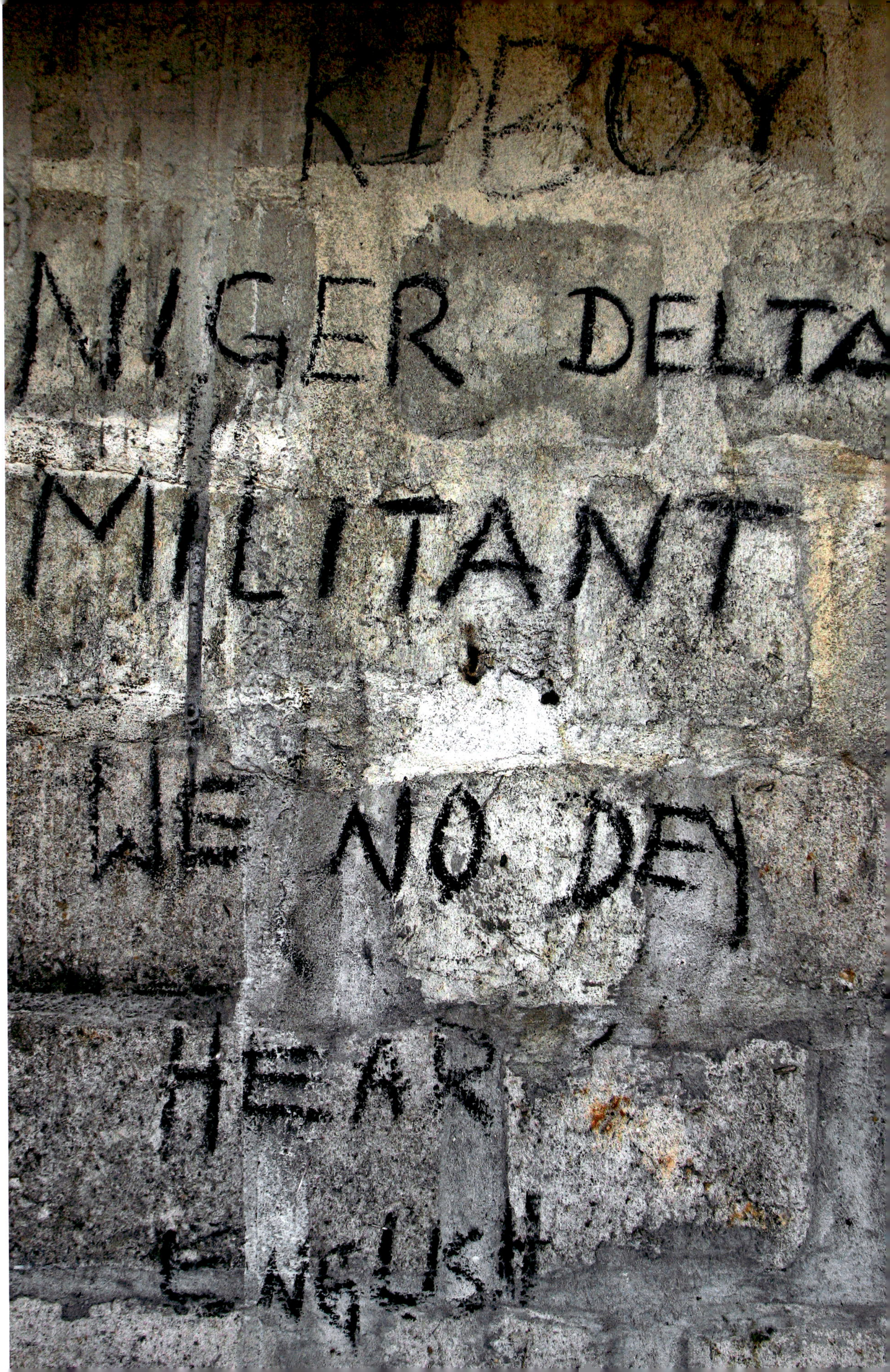
NIGER DELTA
MILITANT
WE NO DEY
HEAR
ENGLISH

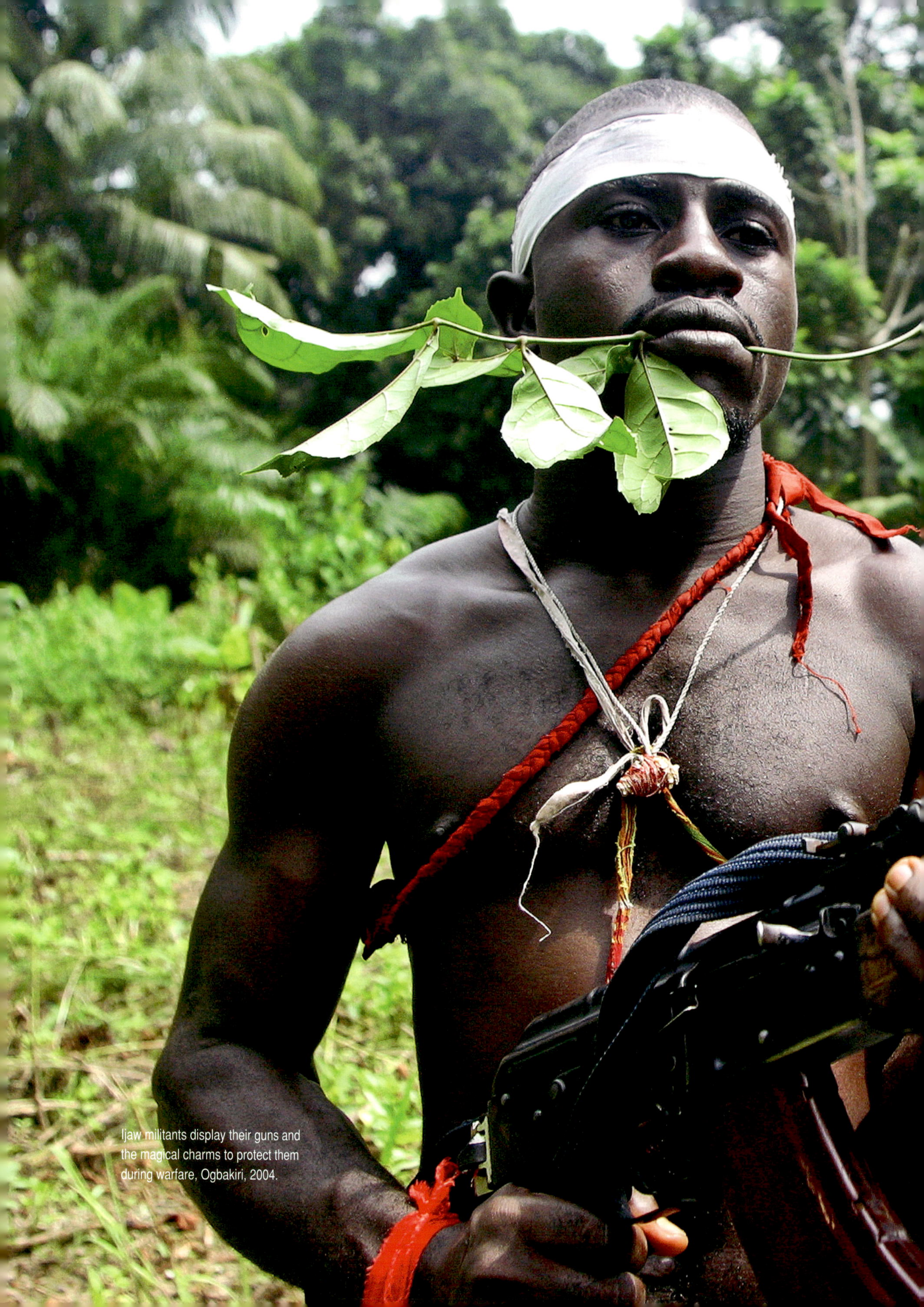

Ijaw militants display their guns and the magical charms to protect them during warfare, Ogbakiri, 2004.

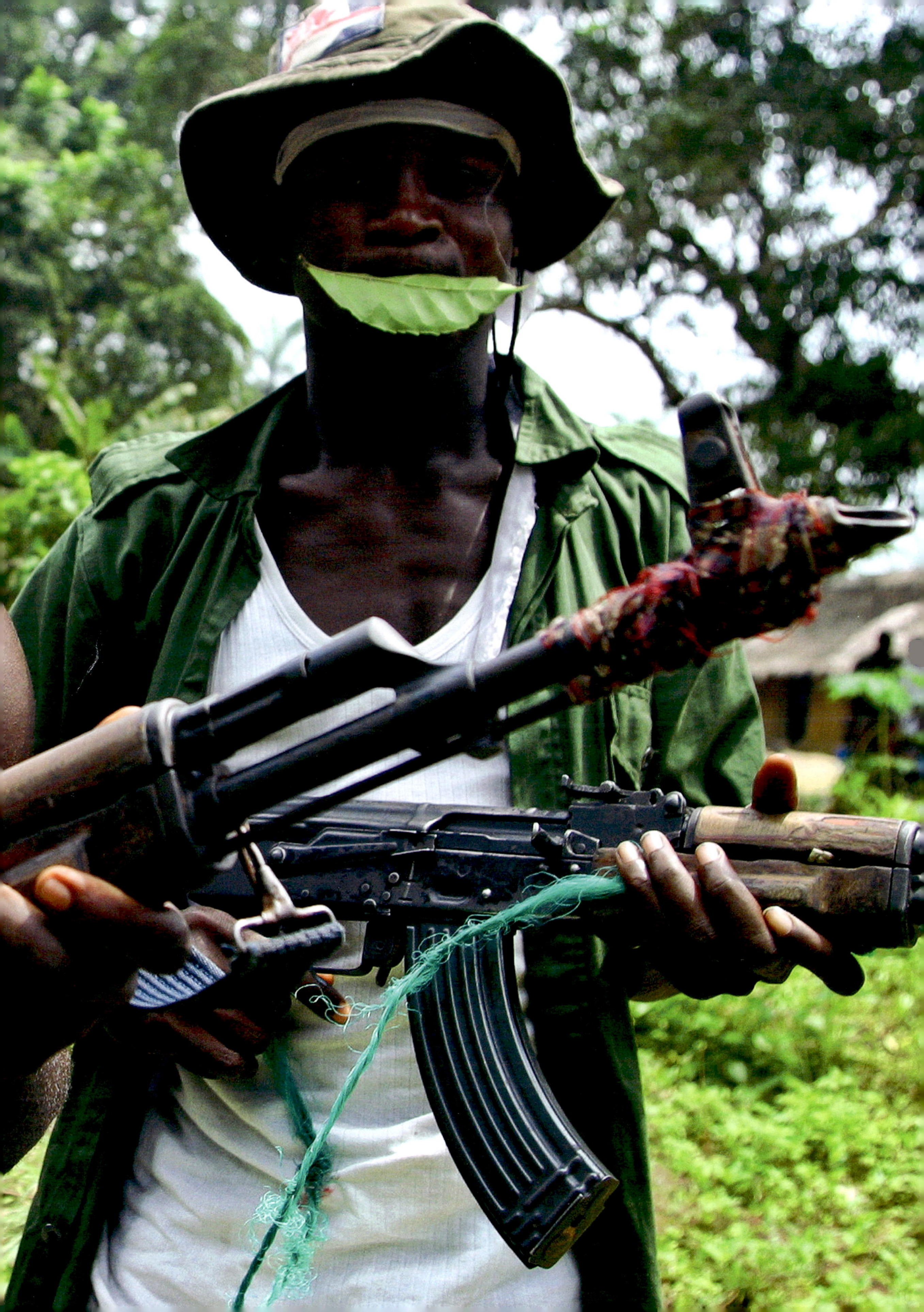

MEND (Movement for the Emancipation of the Niger Delta) is at war with its government and the victim is often the oil companies operating in the country. The group says it will not quit until all foreign companies leave the Delta and no longer deal with the Nigerian government.

MEND militant, undisclosed location, 2006.

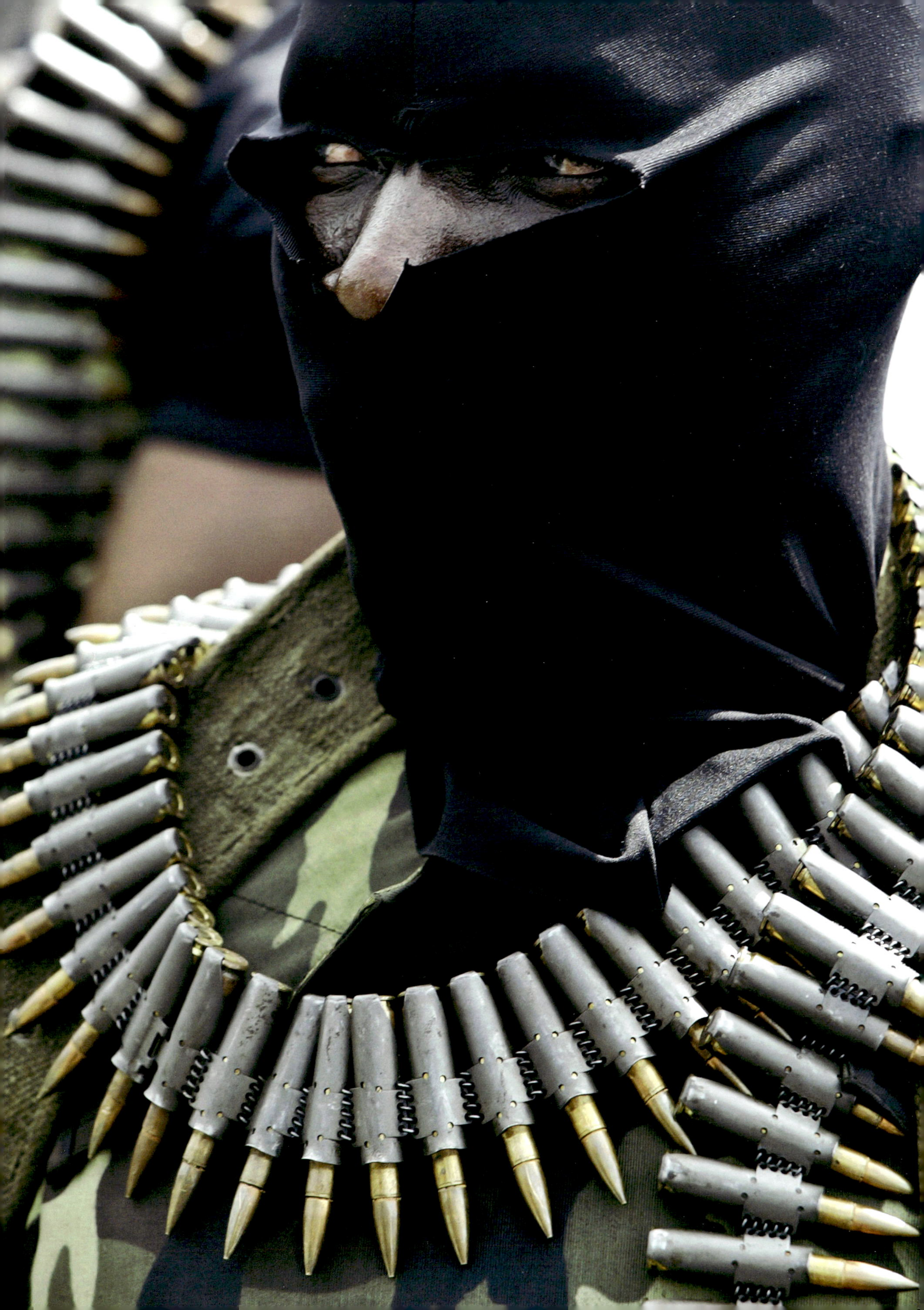

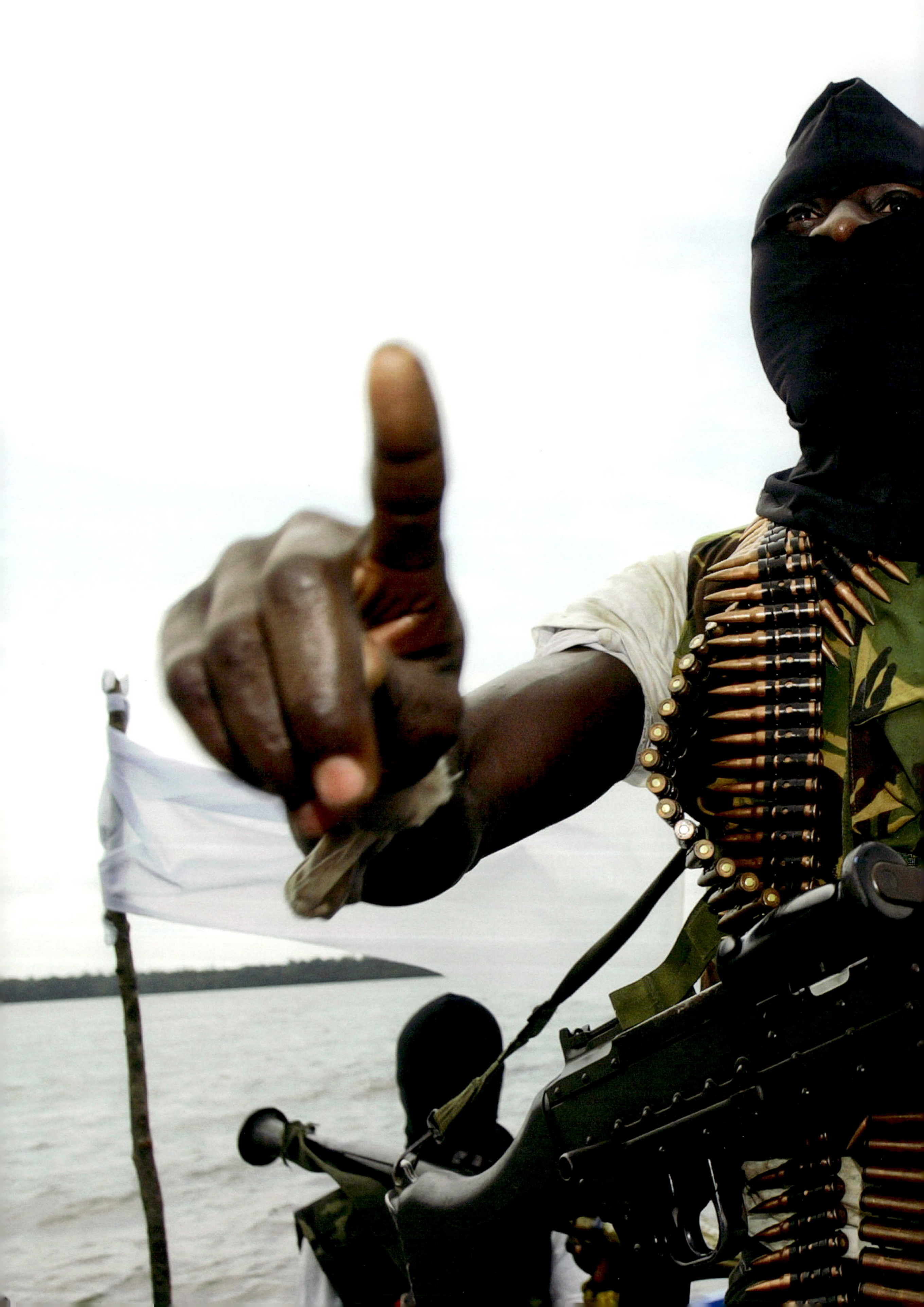

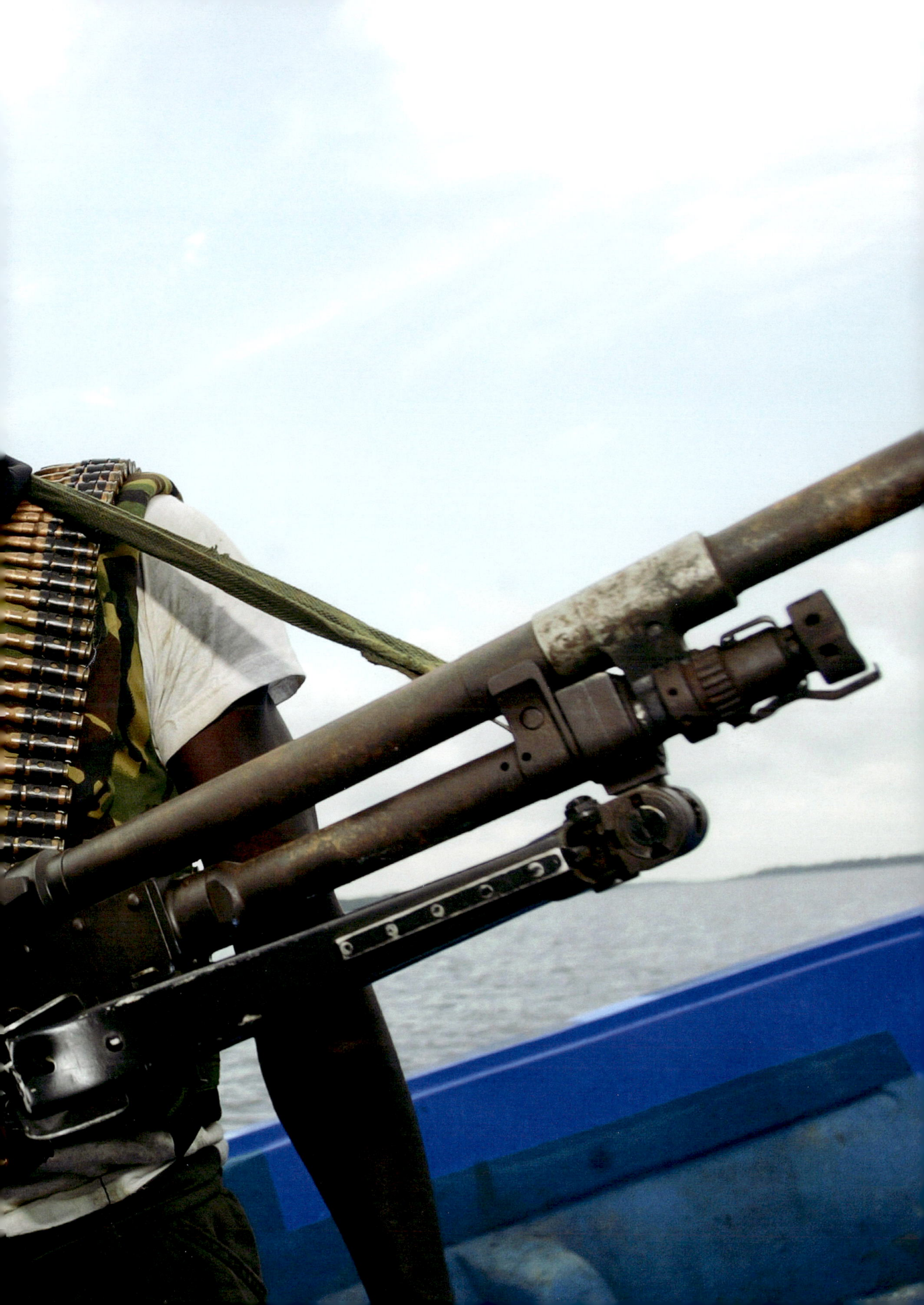

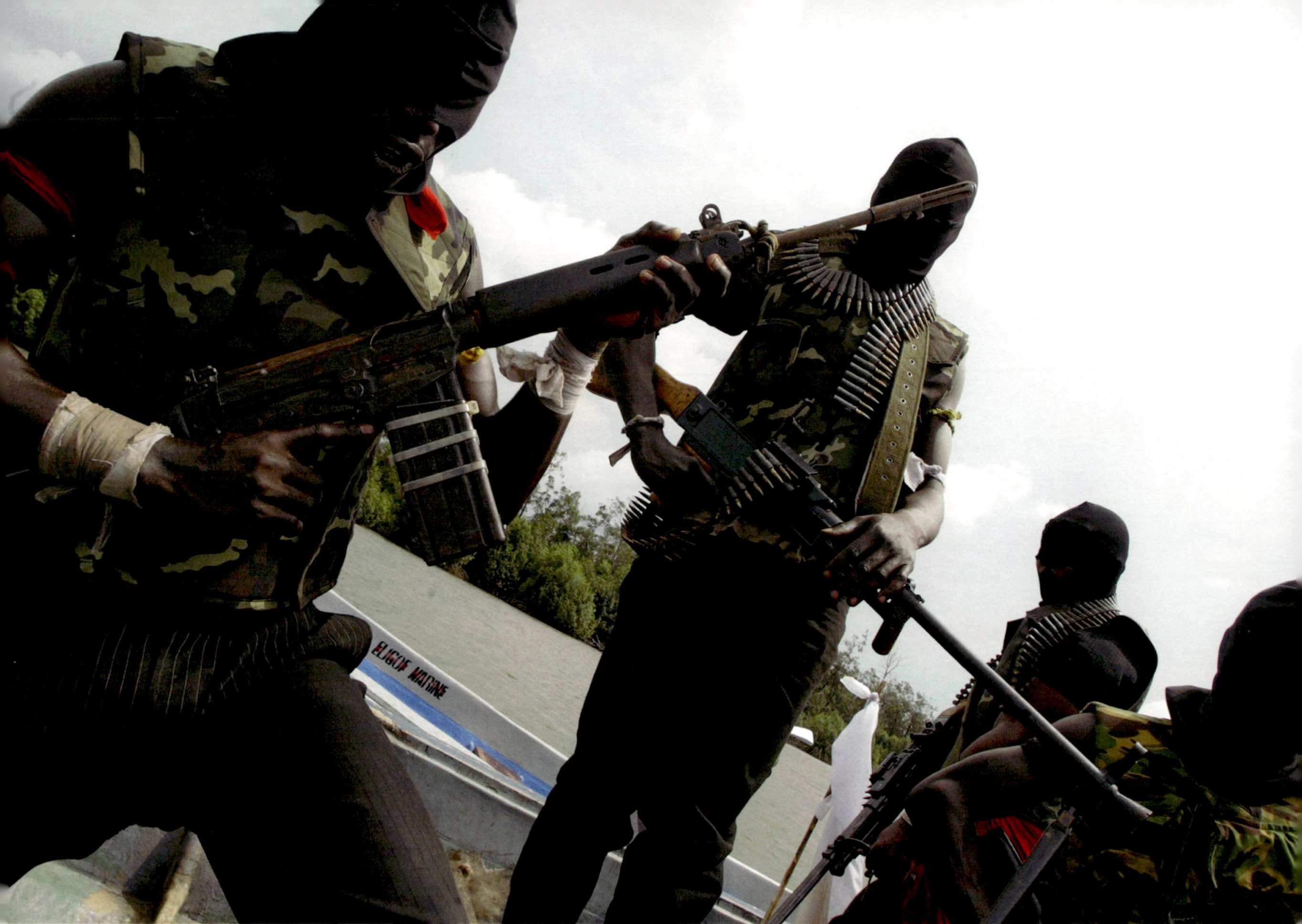

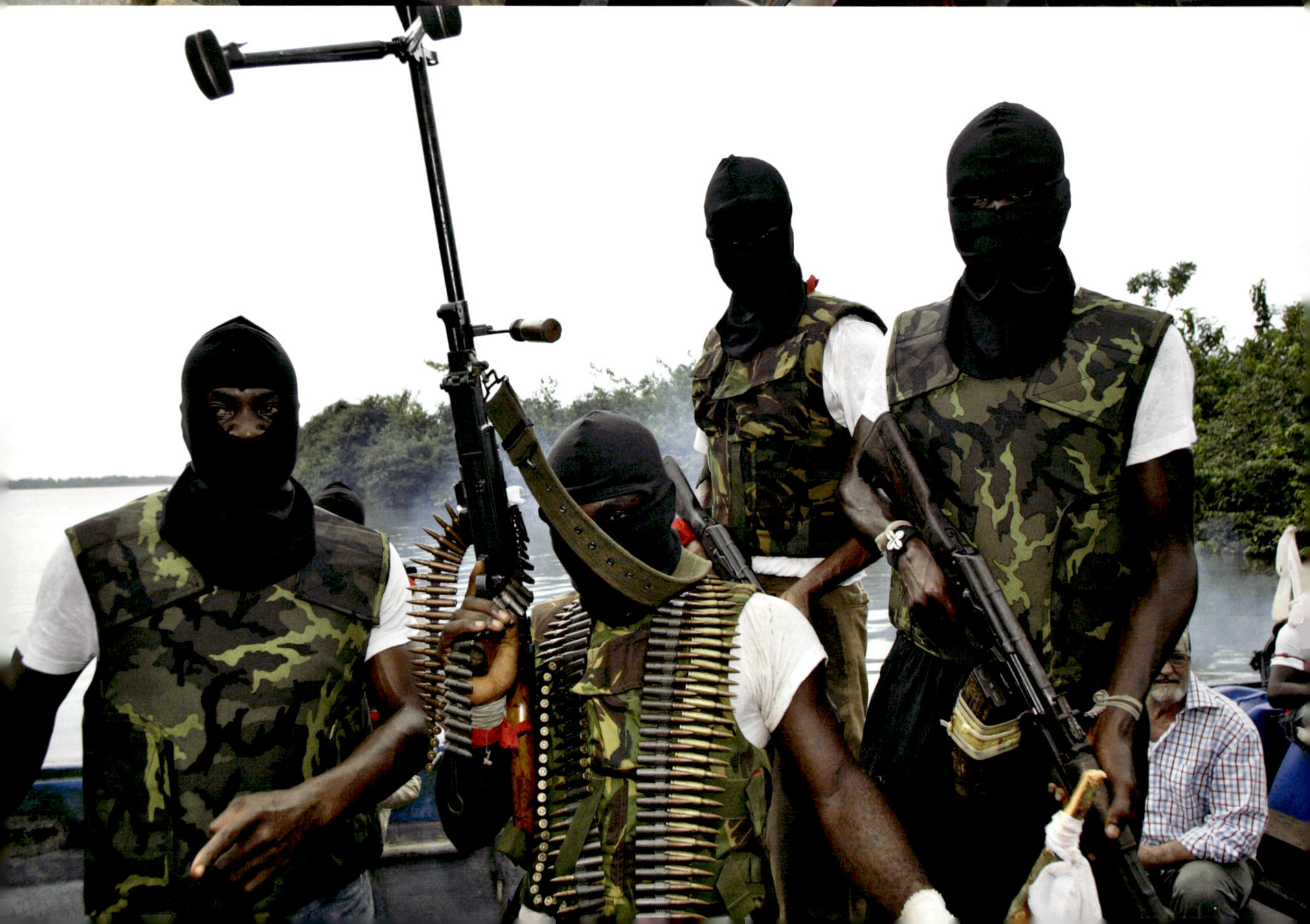

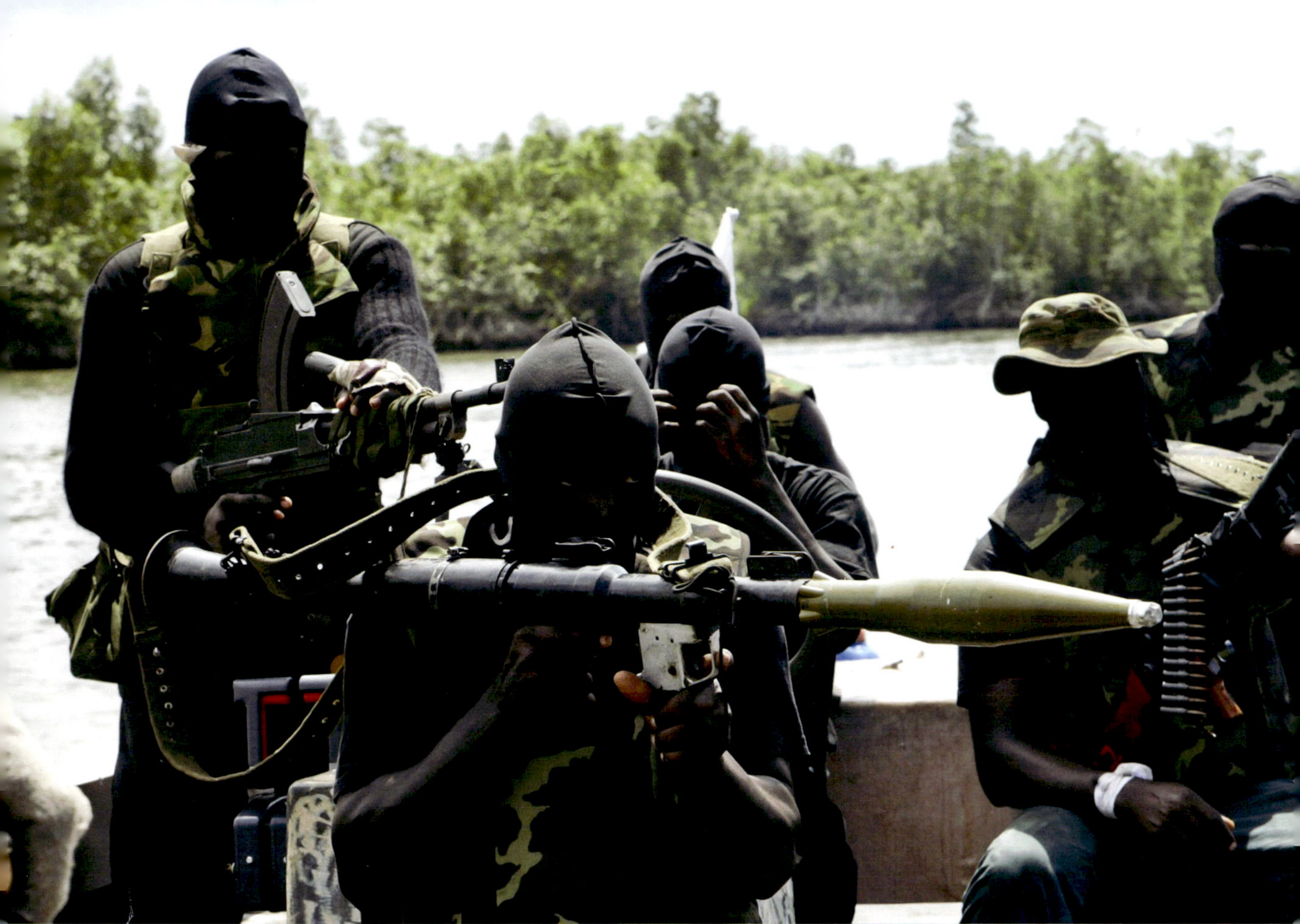

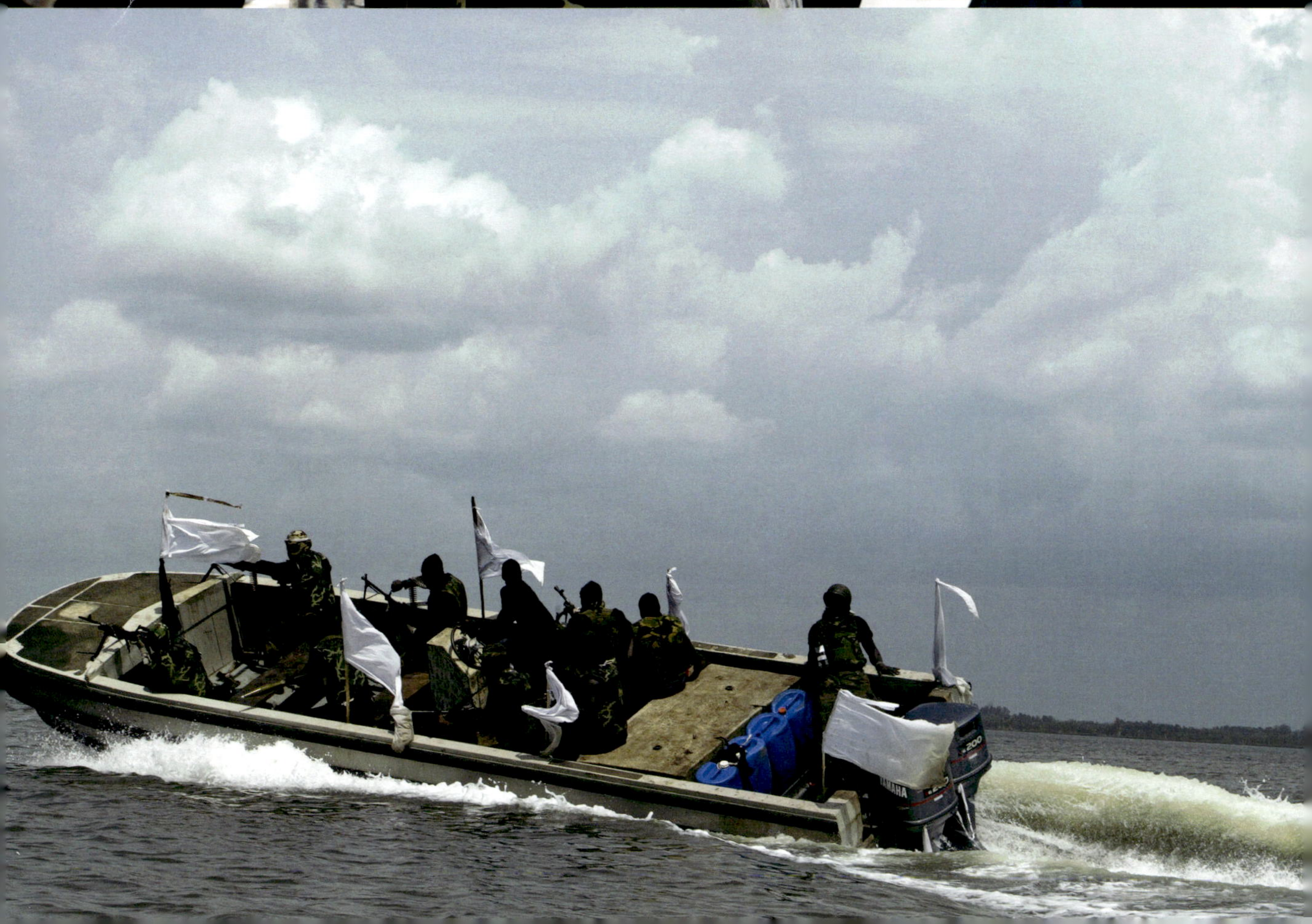

"They wanted us to look at these little villages. They all have dirt floors, and there were no schools. It was a painful life they lived. I didn't like being a captive but then looking back on it I think those people did what they thought they had to do to try to get something out of all that billions of dollars. I can't hold it against them. They want a fair share and they'll tell you right quick what they want. And it's not unreasonable."

"They get nothing out of the oil, and they produce all of the oil. They're tired of it, so they're going to fight, and they're going to fight until death."

Macon Hawkins, released US oil worker taken hostage by MEND in 2006.

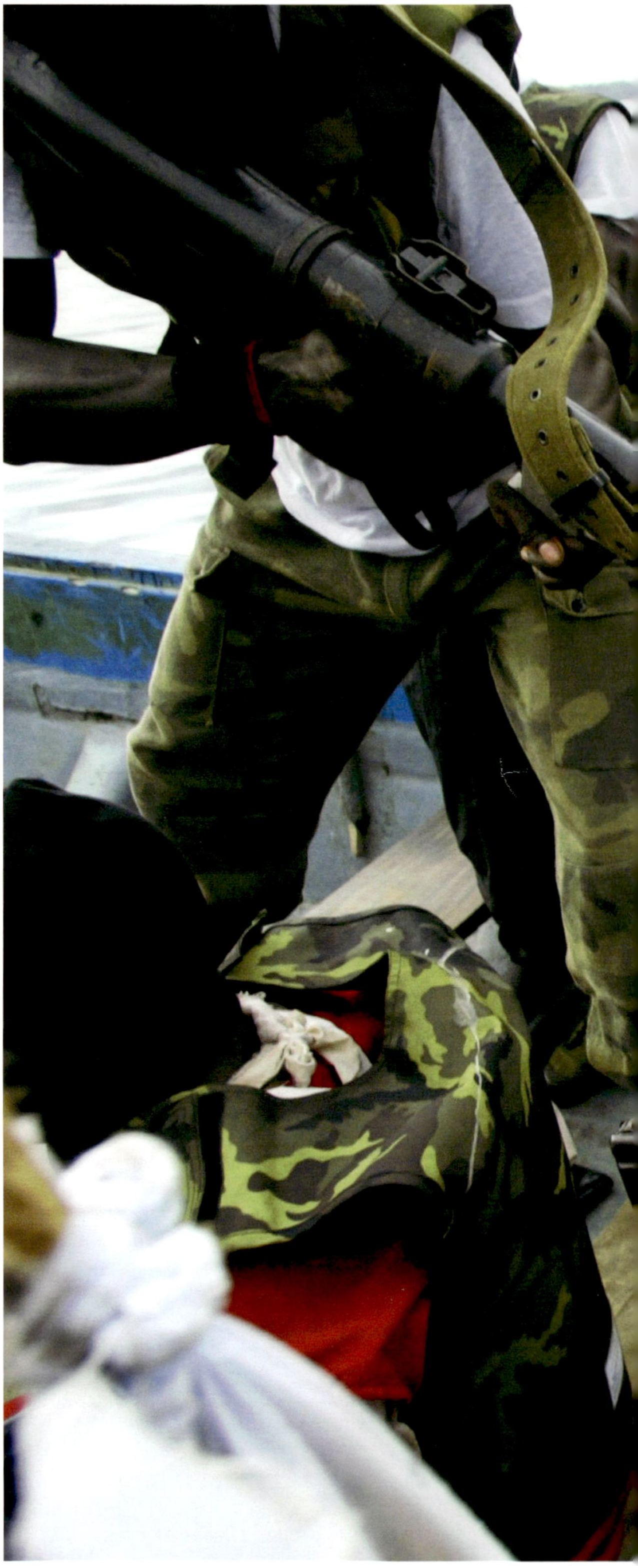

Members of the militant group MEND prepare to release American oil worker Macon Hawkins, who along with eight others, was being held hostage in the Delta region, 2006.

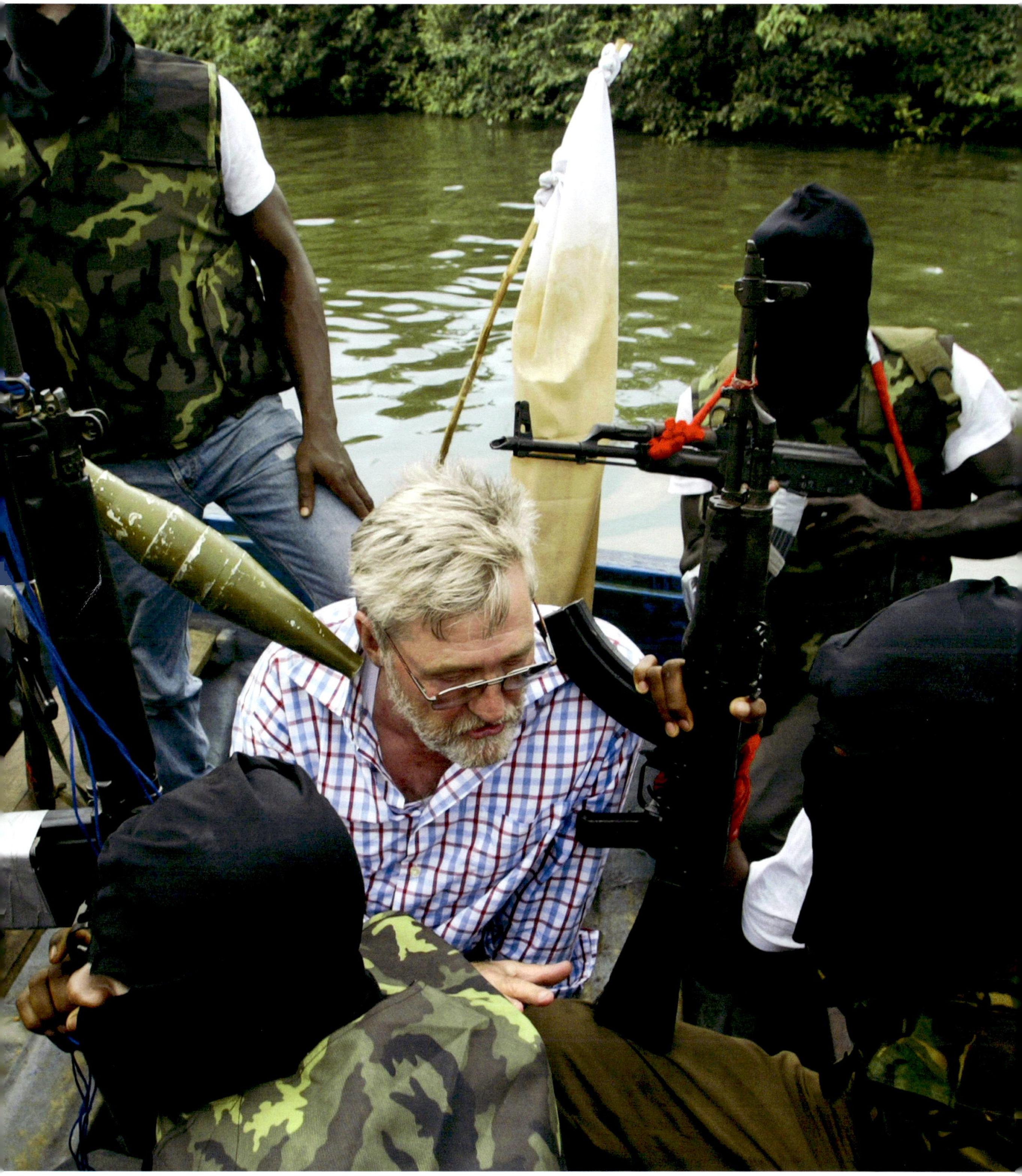

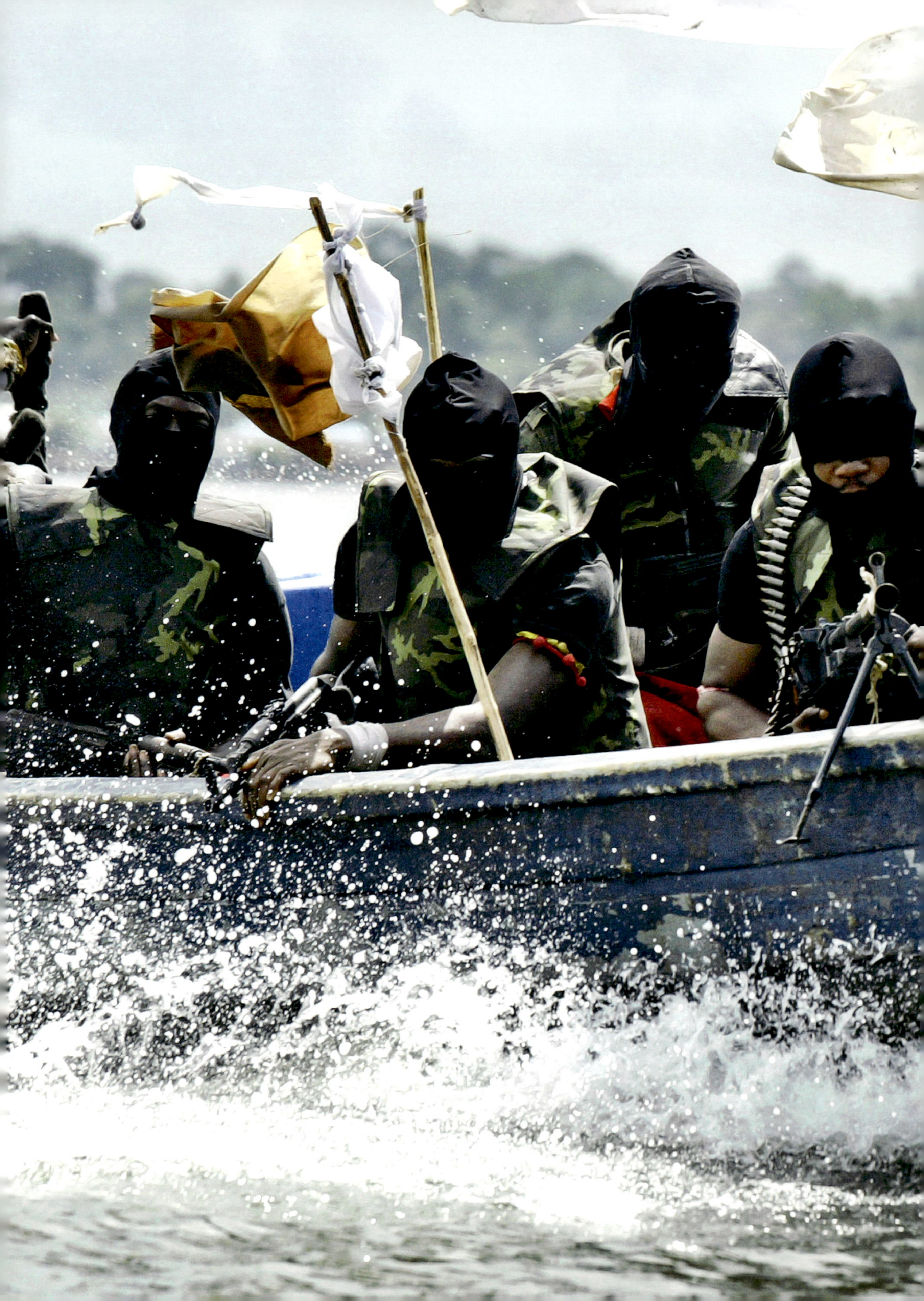

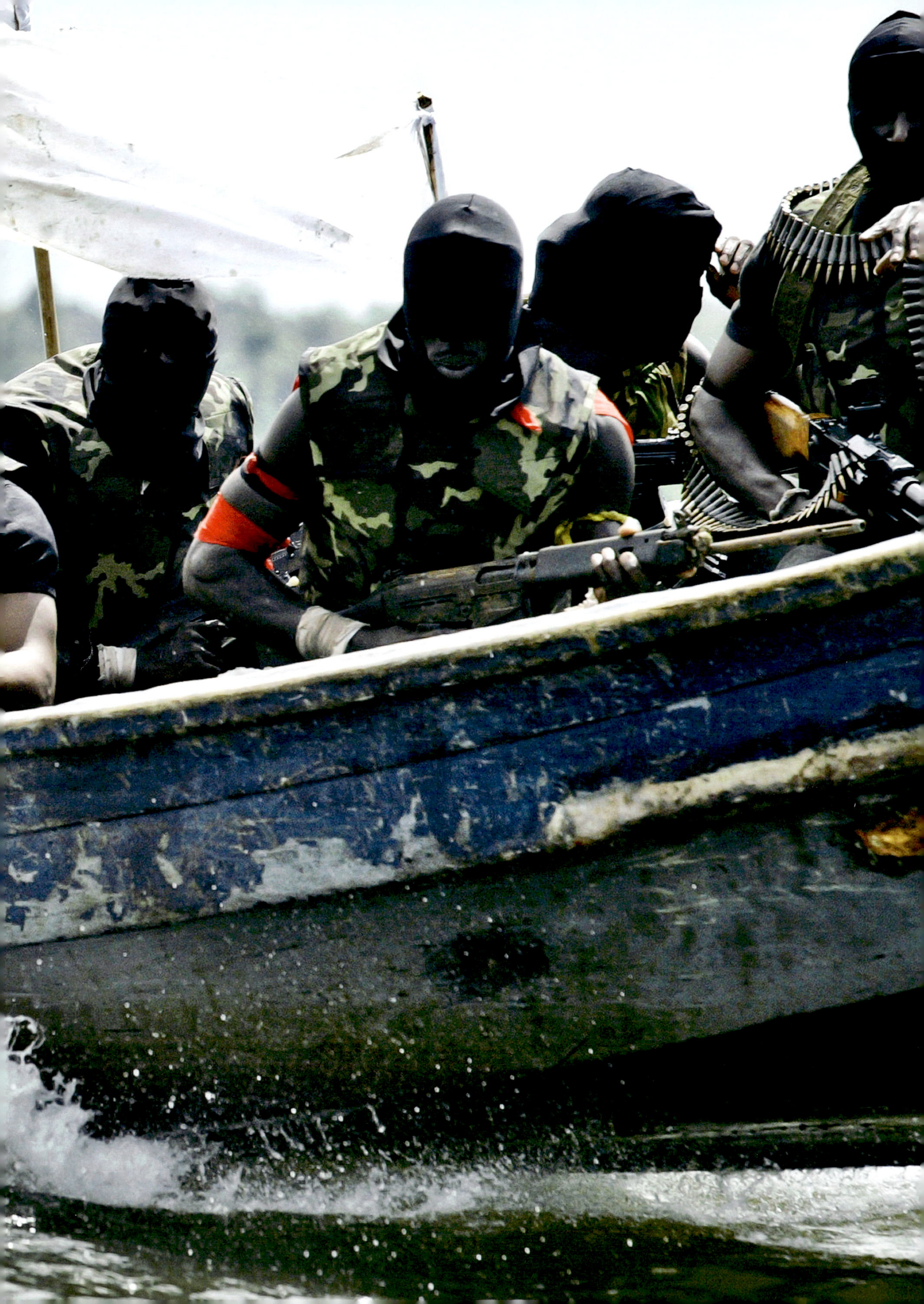

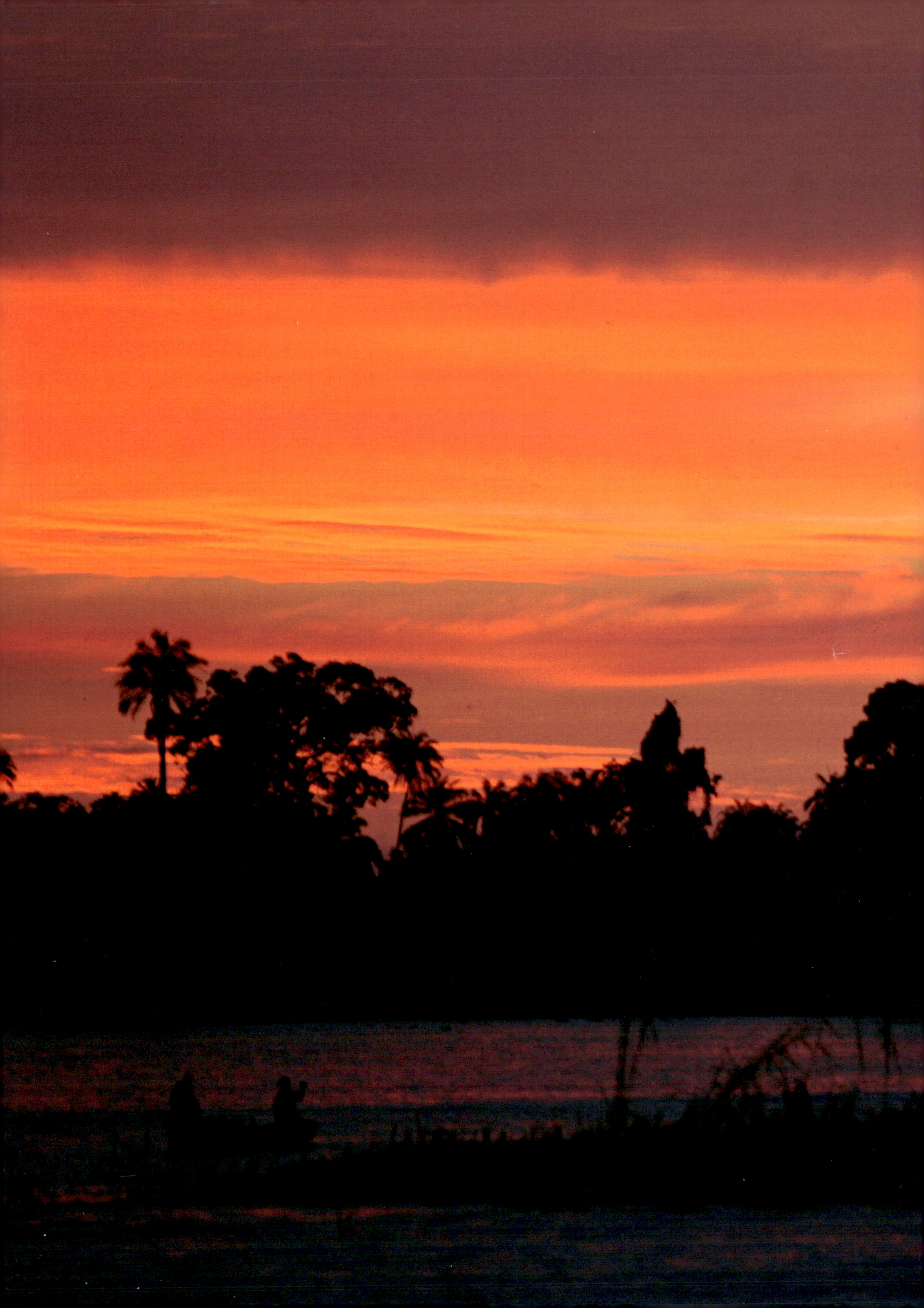

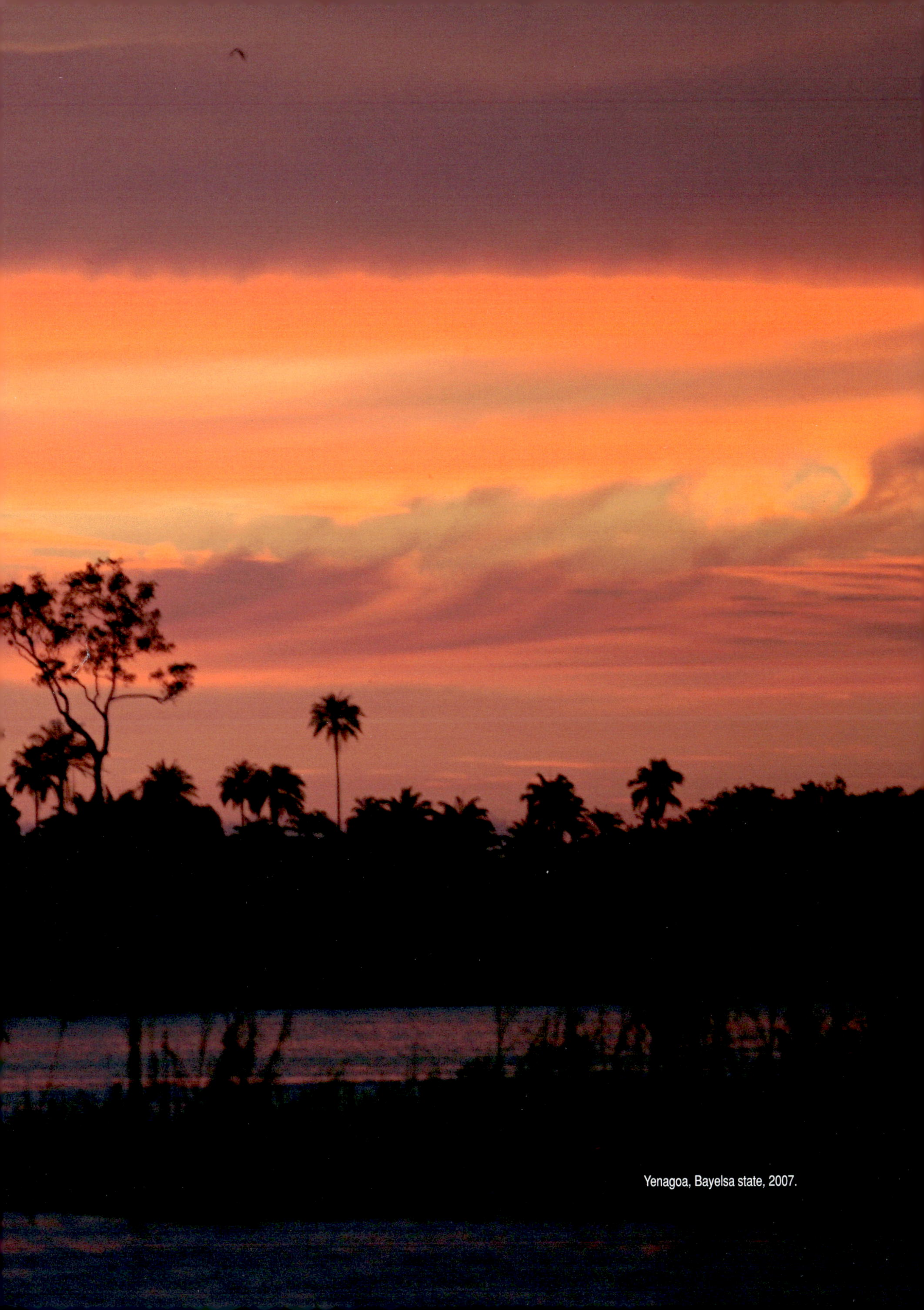

Yenagoa, Bayelsa state, 2007.

INSECURITY IN THE DELTA 2009 AMNESTY

In 2009 President Umaru Yar'Adua granted an unconditional amnesty to all of the Niger Delta militants, with most of the major militant groups accepting to lay down their weapons, with just a few groups refusing to join in the programme. Although peace has returned relatively with the increase in oil production, there has also been an increase in the level of insecurity in the region, with various unknown groups and other criminal elements. Kidnapping even local indiginous people for a ransom has become a lucrative trade. However, recently the special Joint Task Force (JTF) issued an order for increased surveillance near oil installations in the oil-rich south, giving militants a deadline for surrender.
"The JTF has issued a deadline of September 18, 2011 to all militants still bearing arms in the Niger Delta to surrender such weapons at any of the JTF units. Anyone found with arms after the deadline will be arrested and treated as a criminal," said JTF spokesman, Timothy Antigha.

Right: Calendar showing satirical cartoon of the money situation from oil in Nigeria.

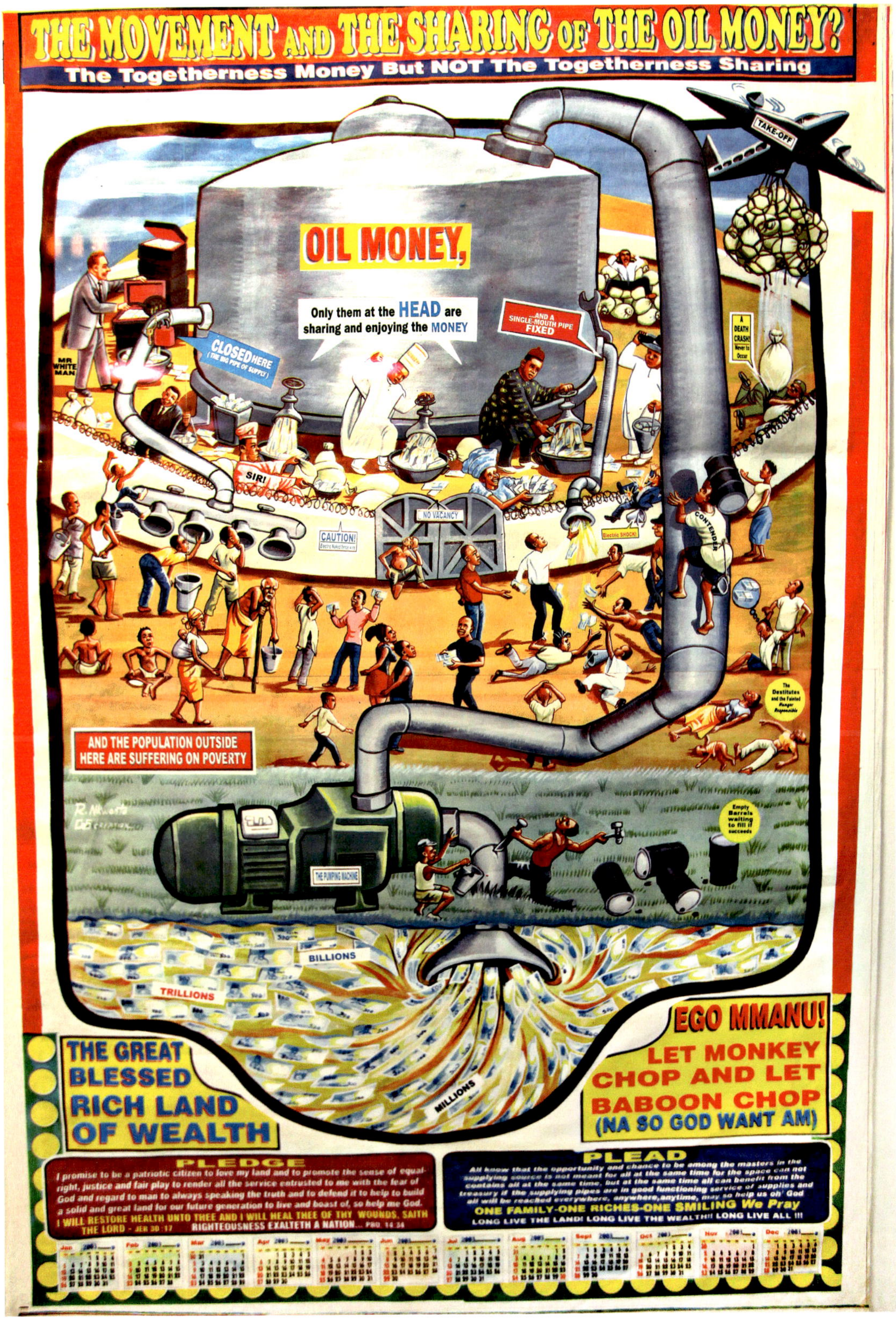
THE MOVEMENT AND THE SHARING OF THE OIL MONEY?
The Togetherness Money But NOT The Togetherness Sharing
TAKE-OFF
OIL MONEY,
Only them at the HEAD are sharing and enjoying the MONEY
...AND A SINGLE-MOUTH PIPE FIXED
CLOSED HERE (THE BIG PIPE OF SUPPLY)
MR WHITE MAN
A DEATH CRASH! Never to Occur
SIR!
NO VACANCY
CAUTION!
Electric SHOCK!
CONTENDER
The Destitutes and the Fainted Hunger Responsible
AND THE POPULATION OUTSIDE HERE ARE SUFFERING ON POVERTY
Empty Barrels waiting to fill if succeeds
THE PUMPING MACHINE
BILLIONS
TRILLIONS
MILLIONS
THE GREAT BLESSED RICH LAND OF WEALTH
EGO MMANU!
LET MONKEY CHOP AND LET BABOON CHOP (NA SO GOD WANT AM)
PLEDGE
I promise to be a patriotic citizen to love my land and to promote the sense of equal-right, justice and fair play to render all the service entrusted to me with the fear of God and regard to man to always speaking the truth and to defend it to help to build a solid and great land for our future generation to live and boast of, so help me God.
I WILL RESTORE HEALTH UNTO THEE AND I WILL HEAL THEE OF THY WOUNDS, SAITH THE LORD - JER 30:17
RIGHTEOUSNESS EXALTETH A NATION... PRO. 14:34
PLEAD
All know that the opportunity and chance to be among the masters in the supplying source is not meant for all at the same time for the space can not contains all at the same time, but at the same time all can benefit from the treasury if the supplying pipes are in good functioning service of supplies and all will be reached everywhere, anywhere,anytime, may so help us oh' God
ONE FAMILY-ONE RICHES-ONE SMILING We Pray
LONG LIVE THE LAND! LONG LIVE THE WEALTH!! LONG LIVE ALL !!!

ACKNOWLEDGEMENTS

Shooting in the Niger Delta was never easy, especially when carrying a camera.

Special thanks to my wife Maureen Osodi for her understanding in allowing my frequent travels.

Thanks to Damka Pueba,Chris Newsom, Ledum Metee, Oronto Douglas, Dulue Mbachu, Messio and Famous the driver, Glenn Mckenzie, Ed Harris, Mike Kamber, Asari Dokubo and Ateke Tom, Koyo Kouoh, Roger Buergel and Ruth Noack, Bisi Silva, Professor. David Aradeon, Professor Ebiegberi J. Alagoa, Ziggi Golding, Martin Bell, Hannah and Gigi from Trolley.

And my profound gratitude goes out most to the people of the Delta region, for giving me access to their lives.

BIOGRAPHIES

Dr. Francis Ugiomoh is Senior Lecturer in Art History/Head of Fine Art at the University of Port Harcourt, Nigeria and author of a series of pivotal critical texts on African visual arts, including: 'The Crises of Modernity: Art and the Definition of Cultures in Africa'.

Bisi Silva is the founder and artistic director of the Centre for Contemporary Art, Lagos. She was co-organiser of the Dak'Art Biennale 2006 in Dakar. She has written for among others Art Monthly, Untitled, and Third Text, and curated two exhibitions as part of the Biennial of African Photography in Bamako in 2007.

Tam Fiofori is a photographer and writer born in Port Harcourt, Nigeria. He has since lived and worked as a filmmaker/photographer/writer in England, U.S.A., France, Scandinavia & Africa. His work has been published in magazines and newspapers in Britain, U.S.A., France, Germany, Scandinavia, Japan, Kenya, Ghana & Nigeria.

Chinwe Azubuike is a contemporary African Poet, born in Lagos, Nigeria. Much of her work explores the relationship between traditional beliefs and modernity, with her poetry highlighting the complicated issues and beauty of the people of Africa. She has collaborated with artists in installations and exhibitions at a number of venues including the Victoria and Albert Museum, London.

Nneka Egbuna is a Nigerian-German hip hop/soul singer and songwriter. She sings in both English and Igbo and has won several international awards including Best African Act at the 2009 MOBO Awards.

GEORGE OSODI (B. LAGOS)

'Oil Rich Niger Delta' is a body of work that Nigerian photographer George Osodi developed over the course of many years of research in the Niger Delta region. The work has toured extensively internationally including the prestigious DOCUMENTA 12 in 2007 in Kassel. He began the project in 2003, and over the following six years compiled an expansive body of work. The result is a gripping visual testimony to the situation faced by the people of the Niger Delta.

Osodi's compassionate but realistic eye catches not only the crudeness of a life enmeshed in a regional tragedy; his pictures also succeed in capturing the general beauty of 'being' and tell the story of human strength and resilience. His lens is thus not one of pity, but one of utmost respect. Although a sense of right and wrong is present, Osodi's aesthetic sensitivity and formalistic expertise clearly places this work within an artistic discourse, as much as it renders a visual documentation of a humanitarian, political and environmental scandal of fifty years of exploitation and mal-governance.

George Osodi began working as a photojournalist for the now defunct Comet Newspaper in Lagos from 1999-2002. He joined the Associated Press News Agency in Lagos from 2002-2008. He has covered many assignments for both local and international media, including the New York Times, Time, the Guardian, the International Herald Tribune and CNN. Osodi has won several awards including Fuji African Photojournalist Of The Year, 2004 and was nominated for the Prix Pictet award in 2008.

DELTA NIGERIA EXHIBITIONS

2011

Rencontres de Bamako
Tang Museum, Skidmore College, New York
Anderson Gallery, Richmond, USA
Raw Material Company, Dakar, Senegal

2010

Museumslandschaft Hessen Kassel
SUBTE, Centro Cultural Solar Barão, Montevedeo, Uruguay
Museu da Gravura Cidades de Curitiba, Brazil
Centre d'art le LAIT, Albi, France
Akademie der Künste, Berlin
Nottingham Contemporary, Nottingham
Lazarides Gallery, Old Vic Tunnels, London
Fondazione Fotografia, Modena, Italy

2009

National Museum of Contemporary Art, Athens
National Museum of Contemporary Art, Oslo, and since travelling to museums in Norway.
Gallery BMB Mumbai, India
Forum Stadtpark, Graz, Austria
OCA, Sao Paulo, Brazil
Biennal Del Fin del Mundo, Argentina
Centre de Cultura Contemporània de Barcelona

2008

Kulturkiken Skare, Haugesund, Norway
Northern Gallery Of Contemporary Art, UK
Centre for Contemporary Art, Lagos, Nigeria
Martin Margulies Warehouse, Miami

2007

Documenta 12, Kassel
Lagos Uncelebrated, Goethe Institut Lagos

2003

Nimbus Art Gallery, Lagos, Nigeria

Published in Great Britain in 2011
by Trolley Ltd

www.trolleybooks.com

ISBN 978-1-904563-85-3

A catalogue record for this book is available from the British Library.

Printed in Italy 2011 by Grafiche Antiga

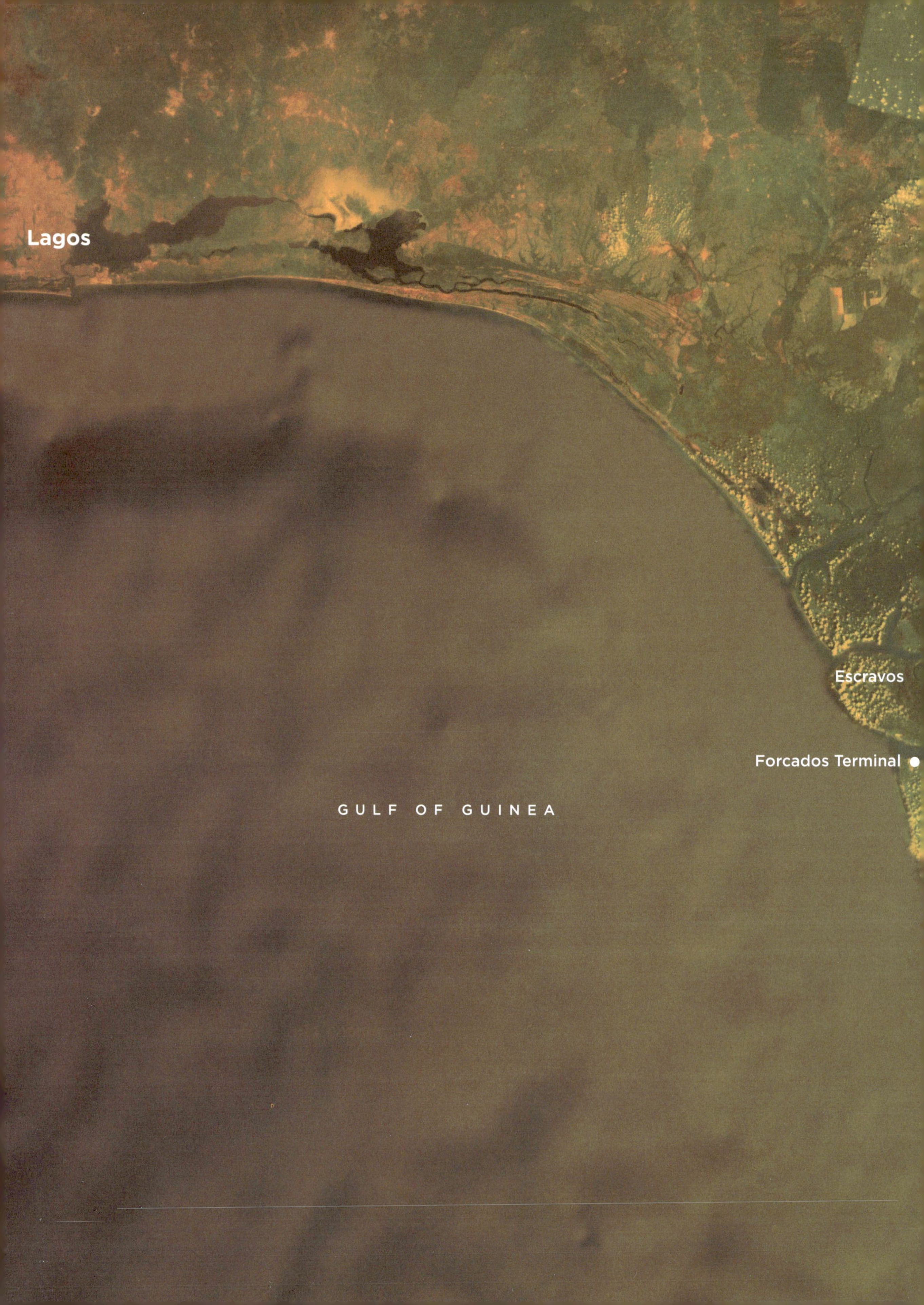
Lagos
Escravos
Forcados Terminal
GULF OF GUINEA